EXECUTIVE FUNCTION

Executive Function: Development Across the Life Span presents perspectives from leading researchers and theorists on the development of executive function from infancy to late adulthood and the factors that shape its growth and decline. Executive function is the set of higher-order cognitive processes involved in regulating attention, thoughts, and actions. Relative to other cognitive domains, its development is slow and decline begins early in late adulthood. As such, it is particularly sensitive to variations in environments and experiences, and there is growing evidence that it is susceptible to intervention – important because of its link to a wide range of important life outcomes.

The volume is made up of four sections. It begins with an overview of executive function's typical development across the life span, providing a foundation for the remainder of the volume. The second section presents insights into mechanisms of executive function, as provided by a variety of methodological approaches. The third and fourth sections review the current research evidence on specific factors that shape executive function's development, focusing on normative (e.g., bilingualism, physical activity, cognitive training) and clinically relevant (e.g., substance use, neurodegenerative disease) developmental pathways.

Sandra A. Wiebe is a developmental cognitive scientist who studies executive function in childhood. She is an Associate Professor in the Department of Psychology and the Neuroscience and Mental Health Institute at the University of Alberta, Canada.

Julia Karbach is a developmental psychologist who studies neurocognitive development and plasticity across the life span. She is a Professor at the Department of Psychology at the University of Koblenz-Landau, Germany.

Frontiers of Developmental Science
Series Editors
Martha Ann Bell and Kirby Deater-Deckard
www.routledge.com/Frontiers-of-Developmental-Science/book-series/
FRONDEVSCI

Frontiers of Developmental Science is a series of edited volumes that aims to deliver inclusive developmental perspectives on substantive areas in psychology. Interdisciplinary and life-span oriented in its objectives and coverage, the series underscores the dynamic and exciting status of contemporary developmental science.

Published

Social Cognition
Jessica A. Sommerville and Jean Decety

Executive Function
Sandra A. Wiebe and Julia Karbach

Forthcoming

Emotion Regulation
Pamela M. Cole and Tom Hollenstein

Reach-to-Grasp Behavior
Daniela Corbetta and Marco Santello

Genetics and Epigenetics
Stephen A. Petrill and Christopher W. Bartlett

EXECUTIVE FUNCTION

Development Across the
Life Span

*Edited by Sandra A. Wiebe and
Julia Karbach*

NEW YORK AND LONDON

First published 2018
by Routledge
711 Third Avenue, New York, NY 10017

and by Routledge
2 Park Square, Milton Park, Abingdon, Oxon, OX14 4RN

Routledge is an imprint of the Taylor & Francis Group, an informa business

© 2018 Taylor & Francis

The right of Sandra A. Wiebe and Julia Karbach to be identified as the authors of the editorial material, and of the authors for their individual chapters, has been asserted in accordance with sections 77 and 78 of the Copyright, Designs and Patents Act 1988.

All rights reserved. No part of this book may be reprinted or reproduced or utilized in any form or by any electronic, mechanical, or other means, now known or hereafter invented, including photocopying and recording, or in any information storage or retrieval system, without permission in writing from the publishers.

Trademark notice: Product or corporate names may be trademarks or registered trademarks, and are used only for identification and explanation without intent to infringe.

Library of Congress Cataloging in Publication Data
A catalog record for this book has been requested

ISBN: 978-1-138-65554-6 (hbk)
ISBN: 978-1-138-65555-3 (pbk)
ISBN: 978-1-315-16071-9 (ebk)

Typeset in Bembo
by Wearset Ltd, Boldon, Tyne and Wear

CONTENTS

List of Contributors   xi

Introduction: Development and Plasticity of Executive Function Across the Life Span   1
Sandra A. Wiebe & Julia Karbach

PART I
Characterizing Executive Function Development Across the Life Span   9

1 Emergence of Executive Function in Infancy   11
 Kimberly Cuevas, Vinaya Rajan, & Lauren J. Bryant

2 Executive Function in Early and Middle Childhood   29
 Nicolas Chevalier & Caron A. C. Clark

3 Executive Function Development in Adolescence   44
 Eveline A. Crone, Sabine Peters, & Nikolaus Steinbeis

4 Executive Function Development in Aging   59
 Karen Z. H. Li, Kiran K. Vadaga, Halina Bruce, & Laurence Lai

PART II
Understanding Mechanisms of Executive Function Development and Plasticity 73

5 Neural Mechanisms of Executive Function Development during Early Childhood 75
   Yusuke Moriguchi

6 Aging and the Neural Correlates of Executive Function 91
   Robert West

7 Genetic Influences on Executive Functions Across the Life Span 106
   James J. Li & Delanie K. Roberts

8 Computational Models of Executive Function Development 124
   Aaron T. Buss

PART III
Environmental, Cultural, and Lifestyle Factors That Shape Executive Function Development Across the Life Span 145

9 Adversity and Stress: Implications for the Development of Executive Functions 147
   Jenna E. Finch & Jelena Obradović

10 Parental Influences on Children's Executive Function: A Differentiated Approach 160
   Claire Hughes & Rory T. Devine

11 Bilingualism and the Development of Executive Function in Children: The Interplay of Languages and Cognition 172
   Gregory J. Poarch & Janet G. van Hell

12 Physical Activity, Exercise, and Executive Functions 188
   Nicolas Berryman, Kristell Pothier, & Louis Bherer

13 Cognitive Training to Promote Executive Functions 200
   Matthias Kliegel, Alexandra Hering, Andreas Ihle, & Sascha Zuber

PART IV
Atypical Patterns of Executive Function Development Across the Life Span   **215**

14 Executive Dysfunction in Very Preterm Children and Associated Brain Pathology   217
Elisha K. Josev & Peter J. Anderson

15 Executive Functions and Developmental Psychopathology: Neurobiology of Emotion Regulation in Adolescent Depression and Anxiety   233
Kristina L. Gelardi, Veronika Vilgis, & Amanda E. Guyer

16 Executive Function and Substance Misuse: Neurodevelopmental Vulnerabilities and Consequences of Use   247
Monica Luciana & Emily Ewan

17 Trajectories and Modifiers of Executive Function: Normal Aging to Neurodegenerative Disease   263
G. Peggy McFall, Shraddha Sapkota, Sherilyn Thibeau, & Roger A. Dixon

Index   278

FRONTIERS IN DEVELOPMENTAL SCIENCE
EXECUTIVE FUNCTION VOLUME
(ALL VOLUMES HAVE A LIFE-SPAN FOCUS)

Theme	*Chapters Focused on That Theme*
Atypical development	Chapter 9 Finch & Obradović Chapter 14 Josev & Anderson Chapter 15 Gelardi, Vilgis, & Guyer Chapter 16 Luciana & Ewan Chapter 17 McFall, Sapkota, Thibeau, & Dixon
Cognition	All chapters
Communication/language	Chapter 11 Poarch & van Hell
Computational modeling	Chapter 8 Buss
Continuity/discontinuity	None
Cross species	None
Cultural context	Chapter 11 Poarch & van Hell
Developmental robotics	None
Emotion/affect	Chapter 9 Finch & Obradović Chapter 10 Hughes & Devine Chapter 15 Gelardi, Vilgis, & Guyer
Family/parenting	Chapter 1 Cuevas, Rajan, & Bryant Chapter 2 Chevalier & Clark Chapter 7 Li & Roberts Chapter 9 Finch & Obradović Chapter 10 Hughes & Devine
Gene-environment	Chapter 1 Cuevas, Rajan, & Bryant Chapter 7 Li & Roberts Chapter 16 Luciana & Ewan Chapter 17 McFall, Sapkota, Thibeau, & Dixon
Individual differences	Chapter 1 Cuevas, Rajan, & Bryant Chapter 2 Chevalier & Clark Chapter 3 Crone, Peters, & Steinbeis Chapter 4 Li, Vadaga, Bruce, & Lai Chapter 7 Li & Roberts Chapter 9 Finch & Obradović Chapter 10 Hughes & Devine Chapter 16 Luciana & Ewan Chapter 17 McFall, Sapkota, Thibeau, & Dixon
Intergenerational transmission	Chapter 1 Cuevas, Rajan, & Bryant Chapter 7 Li & Roberts Chapter 9 Finch & Obradović Chapter 10 Hughes & Devine

Theme	Chapters Focused on That Theme
Mechanisms of developmental change	Chapter 1 Cuevas, Rajan, & Bryant Chapter 2 Chevalier & Clark Chapter 3 Crone, Peters, & Steinbeis Chapter 4 Li, Vadaga, Bruce, & Lai Chapter 5 Moriguchi Chapter 6 West Chapter 7 Li & Roberts Chapter 8 Buss Chapter 9 Finch & Obradović Chapter 10 Hughes & Devine Chapter 11 Poarch & van Hell Chapter 15 Gelardi, Vilgis, & Guyer Chapter 16 Luciana & Ewan Chapter 17 McFall, Sapkota, Thibeau, & Dixon
Neuroscience	Chapter 1 Cuevas, Rajan, & Bryant Chapter 2 Chevalier & Clark Chapter 3 Crone, Peters, & Steinbeis Chapter 4 Li & Vadaga, Bruce, & Lai Chapter 5 Moriguchi Chapter 6 West Chapter 7 Li & Roberts Chapter 9 Finch & Obradović Chapter 11 Poarch & van Hell Chapter 12 Berryman, Pothier, & Bherer Chapter 14 Josev & Anderson Chapter 15 Gelardi, Vilgis, & Guyer Chapter 16 Luciana & Ewan Chapter 17 McFall, Sapkota, Thibeau, & Dixon
Ontogeny	Chapter 1 Cuevas, Rajan, & Bryant Chapter 2 Chevalier & Clark Chapter 3 Crone, Peters, & Steinbeis Chapter 5 Moriguchi Chapter 9 Finch & Obradović Chapter 10 Hughes & Devine Chapter 11 Poarch & van Hell Chapter 14 Josev & Anderson Chapter 15 Gelardi, Vilgis, & Guyer Chapter 16 Luciana & Ewan
Plasticity/repair	Chapter 2 Chevalier & Clark Chapter 3 Crone, Peters, & Steinbeis Chapter 4 Li, Vadaga, Bruce, & Lai Chapter 10 Hughes & Devine Chapter 11 Poarch & van Hell Chapter 12 Berryman, Pothier, & Bherer Chapter 13 Kliegel, Hering, Ihle, & Zuber

Theme	Chapters Focused on That Theme
Sensory/Motor	Chapter 1 Cuevas, Rajan, & Bryant
	Chapter 4 Li, Vadaga, Bruce, & Lai
	Chapter 8 Buss
	Chapter 12 Berryman, Pothier, & Bherer
Social	Chapter 9 Finch & Obradović
	Chapter 10 Hughes & Devine

CONTRIBUTORS

Peter J. Anderson, Murdoch Children's Research Institute and the University of Melbourne, Australia.

Nicolas Berryman, Bishop's University and Institut Universitaire de Gériatrie de Montréal, Canada.

Louis Bherer, Institut Universitaire de Gériatrie de Montréal, Concordia University, University de Montréal, and Montreal Heart Institute, Canada.

Halina Bruce, Concordia University, Canada.

Lauren J. Bryant, University of Connecticut, USA.

Aaron T. Buss, University of Tennessee-Knoxville, USA.

Nicolas Chevalier, University of Edinburgh, UK.

Caron A. C. Clark, University of Nebraska-Lincoln, USA.

Eveline A. Crone, Leiden University, the Netherlands.

Kimberly Cuevas, University of Connecticut, USA.

Rory T. Devine, University of Cambridge, UK.

Roger A. Dixon, University of Alberta, Canada.

Emily Ewan, University of Minnesota, USA.

Jenna E. Finch, Stanford University, USA.

Kristina L. Gelardi, University of California, Davis, USA.

Amanda E. Guyer, University of California, Davis, USA.

Alexandra Hering, University of Geneva, Switzerland.

Claire Hughes, University of Cambridge, UK.

Andreas Ihle, University of Geneva and Swiss National Centre of Competence in Research LIVES, Switzerland.

Elisha K. Josev, Murdoch Children's Research Institute and the University of Melbourne, Australia.

Matthias Kliegel, University of Geneva and Swiss National Centre of Competence in Research LIVES, Switzerland.

Laurence Lai, Concordia University, Canada.

James J. Li, University of Wisconsin-Madison, USA.

Karen Z. H. Li, Concordia University, Canada.

Monica Luciana, University of Minnesota, USA.

G. Peggy McFall, University of Alberta, Canada.

Yusuke Moriguchi, Kyoto University, Japan.

Jelena Obradović, Stanford University, USA.

Sabine Peters, Leiden University, the Netherlands.

Gregory J. Poarch, University of Münster, Germany.

Kristell Pothier, Institut Universitaire de Gériatrie de Montréal and Concordia University, Canada.

Vinaya Rajan, University of the Sciences, USA.

Delanie K. Roberts, University of Wisconsin-Madison, USA.

Shraddha Sapkota, University of Alberta, Canada.

Nikolaus Steinbeis, Leiden University, the Netherlands.

Sherilyn Thibeau, University of Alberta, Canada.

Kiran K. Vadaga, Concordia University, Canada.

Janet G. van Hell, The Pennsylvania State University, USA.

Veronika Vilgis, University of California, Davis, USA.

Robert West, DePauw University, USA.

Sascha Zuber, University of Geneva, Switzerland.

INTRODUCTION

Development and Plasticity of Executive Function Across the Life Span

Sandra A. Wiebe & Julia Karbach

Executive function (EF) encompasses a set of higher-order cognitive processes involved in regulating attention, thoughts, and actions. EF is closely tied to the prefrontal cortex, a region important for top-down control that "is critical in situations when the mappings between sensory inputs, thoughts, and actions either are weakly established relative to other existing ones or are rapidly changing" (Miller & Cohen, 2001, p. 168). Many different models of EF have been proposed, encompassing a variety of processes such as response inhibition, working memory updating, attention shifting, goal monitoring, and action planning, to name just a few. Over the last 15 years perhaps the most prominent model of EF is that put forward by Akira Miyake and his colleagues (Miyake et al., 2000; Miyake & Friedman, 2012), which through several iterations has emphasized the unity of executive processes (common EF) alongside its diversity (processes unique to specific EF components such as working memory updating and shifting).

EF has a protracted developmental trajectory over the life span that parallels the maturation and degeneration of the frontal lobes (Casey, Tottenham, Liston, & Durston, 2005; Hedden & Gabrieli, 2004; Moriguchi, Chapter 5, this volume; West, Chapter 6, this volume), in that EF is among the last cognitive functions to reach adult levels and the first to show decline (Li et al., 2004). Similarly, the structure of EF changes over the life span, becoming more complex through childhood and adolescence (Chevalier & Clark, Chapter 2, this volume; Lee, Bull, & Ho, 2013; Wiebe, Espy, & Charak, 2008), with a corresponding decrease in late adulthood (de Frias, Dixon, & Strauss, 2009; Li, Vadaga, Bruce, & Lai, Chapter 4, this volume).

This protracted developmental course appears to result in increased sensitivity to environmental influences and experience, suggesting that EF development is likely characterized by both increased vulnerability to risk and enhanced opportunities for intervention (Finch & Obradović, Chapter 9, this volume; Hackman, Gallop, Evans, & Farah, 2015; Hughes & Devine, Chapter 10, this volume; Karbach & Unger, 2014; Wass, 2015). This is of particular importance because individual differences in EF have been linked to a number of important life outcomes, such as academic and vocational success, socioemotional development, and mental and physical health (Bull, Espy, & Wiebe, 2008; Crone, Peters, & Steinbeis, Chapter 3, this volume; Gelardi, Vilgis, & Guyer, Chapter 15, this volume; Luciana & Ewan, Chapter 16, this volume; Titz & Karbach, 2014). It is important to note that while environmental factors clearly influence EF, genetic factors also play a role, on their own and in interaction with environment (Friedman et al., 2008; Li & Roberts, Chapter 7, this volume).

Our goal in the present volume is to adopt a life-span perspective to integrate findings from different research areas and provide a state-of-the-art review of EF development across the life span, with a particular focus on how experience and environmental variation shape EF outcomes. To achieve this goal, we have organized this volume into four parts.

The first set of chapters provides a survey of EF development across the life span, intended to provide developmental context for the remainder of the book. In this section, contributors characterize EF development in particular periods of the life span: infancy (Cuevas, Rajan, & Bryant, Chapter 1), childhood (Chevalier & Clark, Chapter 2), adolescence (Crone et al., Chapter 3), and adulthood and aging (Li et al., Chapter 4).

First, Cuevas and colleagues provide an account of EF's emergence in the first few years of life. They conceptualize the control of attention as the foundation on which later developments build. Their account characterizes infant EF in terms of two core abilities: holding information in mind (working memory) vs. overriding or delaying dominant action tendencies (inhibitory control) to support goal-directed behavior. Both of these abilities are argued to improve gradually over the infant period, as shown by behavioral evidence from search paradigms. Furthermore, electrophysiological studies link developmental improvements to changes in frontal lobe activity and connectivity.

Chevalier and Clark trace EF development in childhood, describing a transformation from distractible, fragmented toddlers to goal-directed adolescents. Early childhood is perhaps most impressive in the rate of growth, with transition from perseveration to increased flexibility, and increased abilities to inhibit prepotent responses and resist interference. Middle childhood is marked by pronounced increases in ability to learn from feedback, with more continuous improvements in speed of processing, and other aspects of EF. Chevalier and Clark argue that several mechanisms are critical for EF development in childhood: increases in the strength of mental representation, improvements in goal identification, and better coordination of control processes. The latter two points are particularly useful, as goal identification reflects EF's self-directed nature, which is difficult to measure in the lab but central to EF in day-to-day functioning; coordination becomes more critical as EF differentiates and children add more EF abilities to their "toolkit".

Crone and colleagues describe EF development and its neural correlates in adolescence, which encompasses puberty and the transition to adulthood. This is a period when goal-directed behavior has great significance for long-term life outcomes; hence a key focus of this chapter is the relation between EF and success in school. They also discuss training studies examining plasticity of EF in adolescence, including evidence from behavioral and neuroimaging research, and conclude by highlighting areas where more research would be particularly valuable.

In the final chapter within this section, Li and colleagues discuss EF development in adulthood and aging, when the balance shifts from acquiring and refining new skills to declining capabilities. Echoing patterns earlier in the life span, they identify distinct developmental trajectories for different components of EF, and there appears to be a process of dedifferentiation in the factor structure of EF. With the exception of the ability to divide attention or multi-task, however, it is difficult to separate decline in EF components from cognitive decline in other domains such as processing speed. Li and colleagues also present evidence that EF is critical for daily functioning in aging, both in terms of activities of daily living and motor function.

The second section explores mechanisms underlying EF's development and plasticity, that is, its sensitivity to experiential factors. Despite the vast empirical findings documenting life-span changes in EF and their sensitivity to the effects of experience and environment, the mechanisms driving these changes are often poorly understood. Therefore, contributors to this section discuss evidence on EF plasticity from diverse research methodologies including neuroimaging in childhood (Moriguchi, Chapter 5) and across the adult life span (West, Chapter 6), genetics (Li & Roberts, Chapter 7), and computational modeling (Buss, Chapter 8).

In the chapter focusing on childhood development of EF, Moriguchi provides an overview of studies that have examined the relationship between EF development (cognitive shifting, inhibitory control, and working memory) and fronto-parietal network activity. Taken together, this research reveals that infants and young children activate prefrontal regions during tasks that assess EF while school-aged children rely more on the fronto-parietal network. Focusing on adults, West provides an overview of research on the neural correlates of EF in aging. He concludes that aging is associated with alterations in the neural recruitment related to inhibiting, monitoring, and if–then logical processing, while other aspects of executive processing may be preserved in older age. While both authors succeed at illustrating the tremendous advances in the understanding of the neural mechanisms subserving EF that neuroscientific research has provided over the last decade, they also agree that future research needs to establish a stronger link between neural and behavioral markers of EF.

Taking another approach, Li and Roberts shed light on the role of genetic influences on the development and plasticity of EF. They provide a general overview of how genetic factors contribute to variation in EF over the life span by reviewing twin studies, candidate gene studies, and genome-wide association studies. They conclude that while many studies have confirmed the crucial role of genetic influences for individual differences in EF across development, more research is needed to uncover the specific genes that make up this variation and to understand the gene–environment interplay.

In his chapter on computational models, Buss highlights the value of using mathematically formulated descriptions of the neurocognitive system to model developmental changes and learning processes across development. Based on the examples of standard cognitive tasks he shows how computational models can reproduce patterns of data that help understanding the mechanisms driving developmental and experience-related cognitive changes and inform developmental theories.

The remainder of the book is devoted to surveying specific factors believed to impact EF development. Even though life-span trajectories of different EF domains are well documented in the literature, recent research has identified a number of factors that can impact these changes on the intraindividual and the interindividual levels. Therefore, the third section addresses environmental and lifestyle factors, including contributions focusing on the general effects of adverse economic circumstances (Finch & Obradović, Chapter 9) and the specific effect of individual home environments (Hughes & Devine, Chapter 10). Poarch and van Hell (Chapter 11) provide a critical review of research on the effects of bilingualism and the volume also includes chapters on the effects of physical activity (Berryman, Pothier, & Bherer, Chapter 12) and cognitive training (Kliegel, Hering, Ihle, & Zuber, Chapter 13) on EF.

Finch and Obradović illustrate on a general level how adverse experiences, such as poverty, residential instability, and household chaos are linked to deficits in executive functions. They detail how this association is related to neurobiological mechanisms, including genetics, stress physiology, as well as brain structure and functionality, and identify possible protective factors and early interventions for children at risk. In the following chapter, Hughes and Devine take a more specific approach by reviewing the role of the home environment and parenting in children's EF development. They report studies on the home learning environment, parental well-being, and family chaos as well as parent–child interactions and the mediation of the intergenerational association in EF by parenting behaviors. One question that has gained more and more interest over the years is whether bilingualism modulates the development of EF. Poarch and van Hell review research on EF in bilinguals and monolinguals, with a specific focus on children, and explore the underlying variables that may attenuate the differences between bilinguals and monolinguals when performing EF tasks and may explain heterogeneous findings of previous studies.

The remaining two chapters in this section address the influence of lifestyle choices on the development and plasticity of EF, namely the active engagement in physical activity and cognitive training. Berryman et al. provide an overview of studies supporting the view that an active lifestyle supports EF throughout the life span, whether it is long-term physical training or short bouts of physical activity. They review findings from aerobic and strength training as well as from other modalities such as gross motor skills development and tai chi. In a similar vein, Kliegel et al. address the question whether cognitive training can result in gains beyond the trained task (transfer). They review the literature on EF training, concluding that EF can be trained, that training and particularly transfer effects are critically moderated, and that these moderators are largely unknown. Consequently, they call for further systematic research testing possible moderators of training and transfer effects across the life span.

The fourth and final section deals with contributions on atypical trajectories of EF development in relation to risk factors and clinical outcomes, including preterm birth (Josev & Anderson, Chapter 14), mental health (Gelardi, Vilgis, & Guyer, Chapter 15), substance abuse (Luciana & Ewan, Chapter 16), and neurodegenerative disorders (McFall, Sapkota, Thibeau, & Dixon, Chapter 17).

Josev and Anderson illustrate how early risk factors can alter trajectories of EF development, focusing on the example of children born very preterm, before 32 weeks gestation. They describe patterns of brain pathology related to prematurity and how they relate to deficits in attentional control, information processing, cognitive flexibility, and goal setting. They also discuss the impact of executive dysfunction on children's outcomes in academic and social-emotional domains, relative to their peers born at term.

In their chapter on EF and developmental psychopathology, Gelardi and colleagues focus on emotion dysregulation as a risk factor for anxiety and depression. After a brief overview of typical and atypical trajectories of emotion regulation development, they discuss the links between poor emotion regulation, altered function in frontal-limbic brain networks, and internalizing symptoms. They argue for EF (particularly cognitive control) as a target for intervention in at-risk youth, while calling for more research on possible moderators such as the social context.

Luciana and Ewan describe evidence for a robust correlation between substance misuse and EF dysfunction, but also argue compellingly that the causal pathways underlying this association are hard to disentangle. On the one hand, many substances of misuse are neurotoxic and use often starts in adolescence when the brain is still developing, but on the other hand, there is longitudinal evidence that individual differences in EF early on are predictive of later drug use. Luciana and Ewan propose a developmental cascade model where early difficulties in EF lead to experimentation with and dependence on substances, which in turn exacerbate EF difficulties. They conclude with a call for more longitudinal research to provide strong evidence on this question.

In their chapter, McFall, Sapkota, Thibeau, and Dixon describe EF development in neurodegenerative diseases relative to healthy aging, focusing on Alzheimer's disease and vascular dementia. They emphasize the importance of examining within-person trajectories of EF change (e.g., rates of decline) rather than simply current levels of performance. They also discuss a broad set of risk and protective factors including genetics, lifestyle, and physical health, and consider how these factors act and interact to alleviate or exacerbate rate of decline, and potential implications for early identification and intervention in individuals at higher risk for neurodegenerative disease.

Taken together, the chapters in this book show how much we have learned about EF and its development across the life span. However, they also highlight remaining challenges to surmount. The most fundamental challenge is defining what we mean by EF – this is a persistent challenge for researchers studying it in its "mature" state, in adulthood, and even more so for developmental scientists, when the cognitive skills individuals have at their disposal to work toward goals vary considerably across the life span. Further progress is also needed in linking EF to its neural substrates, which despite the common equation of "executive" and "frontal" go beyond the frontal lobes. There is ample evidence that EF is related to particular experiences and environmental factors, but there are more questions about the underlying reasons for these associations: it is important for researchers to consider potential confounding factors or alternate pathways of causation. However, the contributions in this volume make useful suggestions for next steps to overcome these challenges, and advance our understanding, hinting at breakthroughs in the near future.

We hope that this volume serves as a useful resource for a range of readers, in keeping with the broad importance of EF as a construct. For students, educators, clinicians, and researchers, this book provides an overview of state-of-the-art research on EF from infancy through late adulthood. Furthermore, researchers may be inspired to take some of the next steps highlighted within this book, thereby moving the field forward.

References

Bull, R., Espy, K. A., & Wiebe, S. A. (2008). Short-term memory, working memory, and executive functioning in preschoolers: Longitudinal predictors of mathematical achievement at age 7 years. *Developmental Neuropsychology, 33*, 205–228.

Casey, B. J., Tottenham, N., Liston, C., & Durston, S. (2005). Imaging the developing brain: What have we learned about cognitive development? *Trends in Cognitive Sciences, 9*, 104–110.

de Frias, C. M., Dixon, R. A., & Strauss, E. (2009). Characterizing executive functioning in older special populations: From cognitively elite to cognitively impaired. *Neuropsychology, 23*, 778–791.

Friedman, N. P., Miyake, A., Young, S. E., DeFries, J. C., Corley, R. P., & Hewitt, J. K. (2008). Individual differences in executive functions are almost entirely genetic in origin. *Journal of Experimental Psychology: General, 137*, 201–225.

Hackman, D. A., Gallop, R., Evans, G. W., & Farah, M. J. (2015). Socioeconomic status and executive function: Developmental trajectories and mediation. *Developmental Science, 18*, 686–702.

Hedden, T., & Gabrieli, J. D. E. (2004). Insights into the ageing mind: A view from cognitive neuroscience. *Nature Reviews: Neuroscience, 5*, 87–96.

Karbach, J., & Unger, K. (2014). Executive control training from middle childhood to adolescence. *Frontiers in Psychology, 5*, 390.

Lee, K., Bull, R., & Ho, R. M. H. (2013). Developmental changes in executive functioning. *Child Development, 84*, 1933–1953.

Li, S.-C., Lindenberger, U., Hommel, B., Aschersleben, G., Prinz, W., & Baltes, P. B. (2004). Transformations in the couplings among intellectual cognitive processes across the life span. *Psychological Science, 15*, 155–163.

Miller, E. K., & Cohen, J. D. (2001). An integrative theory of prefrontal cortex function. *Annual Review of Neuroscience, 24*, 167–202.

Miyake, A., & Friedman, N. P. (2012). The nature and organization of individual differences in executive functions: Four general conclusions. *Current Directions in Psychological Science, 21*, 8–14.

Miyake, A., Friedman, N. P., Emerson, M. J., Witzki, A. H., Howerter, A., & Wager, T. D. (2000). The unity and diversity of executive functions and their contributions to complex "frontal lobe" tasks: A latent variable analysis. *Cognitive Psychology, 41*, 49–100.

Titz, C., & Karbach, J. (2014). Working memory and executive functions: Effects of training on academic achievement. *Psychological Research, 78*, 852–868.

Wass, S. V. (2015). Applying cognitive training to target executive functions during early development. *Child Neuropsychology, 21*, 150–166.

Wiebe, S. A., Espy, K. A., & Charak, D. (2008). Using confirmatory factor analysis to understand executive control in preschool children: I. Latent structure. *Developmental Psychology, 44*, 575–587.

Part I

Characterizing Executive Function Development Across the Life Span

1

EMERGENCE OF EXECUTIVE FUNCTION IN INFANCY

Kimberly Cuevas, Vinaya Rajan, & Lauren J. Bryant

The early building blocks of executive functions (EFs) are acquired during infancy and undergo remarkable improvement during this period. In the span of a few short years, infants become capable of sustaining, shifting, and inhibiting their attention, and eventually use these gains in attentional control to hold and update information in mind and exert control over their behavior in the presence of interfering thoughts and actions. These rudimentary EF skills in infancy likely represent developmental precursors to more complex EFs. In this chapter, we focus on the emergence and developmental progression of EF during infancy. We begin by describing our conceptualization of EF and its foundational components. Next, we detail the tasks commonly used to assess infant EF, highlighting our research and selectively reviewing other work examining the biopsychosocial mechanisms that impact EF development and individual variation. We conclude by proposing future directions for the field, emphasizing both functional and integrative analyses.

Conceptualization of EF

EFs are a set of higher-order cognitive processes involved in coordinating, planning, and completing goal-directed actions (Miyake et al., 2000). Varying theoretical perspectives of EF have emphasized a unitary construct (Duncan, Johnson, Swales, & Freer, 1997), dissociable components (Stuss & Alexander, 2000), or an integrative framework underscoring both "unity and diversity" (Garon, Bryson, & Smith, 2008; Miyake et al., 2000). The structure and organization of EF may exhibit a developmental shift from a single latent EF factor during early childhood (Wiebe, Espy, & Charak, 2008; Wiebe et al., 2011) to separate component processes when children are older (Huizinga, Dolan, & van der Molen, 2006; see Chevalier & Clark, Chapter 2, this volume). Despite these differing perspectives, there is agreement that working memory (WM), inhibitory control (IC), and cognitive flexibility are core EF skills (Diamond, 2016).

Attention also plays an integral role in the development of EF (Garon et al., 2008), and attentional control is viewed as a core component of WM (Kane & Engle, 2002; Reynolds & Romano, 2016). According to Engle and colleagues' conceptualization of WM (Engle, Kane, & Tuholski, 1999; Kane & Engle, 2002), there is a limited capacity, domain-free executive attention component capable of maintaining short-term memory representations in situations involving interference or response competition. Individual differences in executive attention or "WM capacity" (Kane & Engle, 2002) are predictive of a wide variety of cognitive abilities (Engle et al., 1999). For a more detailed description of the links between executive attention and WM, see Bell and Deater-Deckard (2007).

Our conceptualization of EF is based on Roberts and Pennington's (1996) framework that emphasizes WM and IC as two core EFs that best represent the role of the prefrontal cortex. WM involves holding information in mind for a brief period of time while actively manipulating that information (Baddeley, 1992). When infants are capable of retrieving a toy from a hiding location after a delay or in the presence of distraction, they are relying on this ability to represent objects in memory. IC involves overriding or delaying a dominant response in order to achieve a goal (Roberts & Pennington, 1996). When infants can forgo reaching for a reward immediately and withhold their response, they are relying on IC to suppress a prepotent response. These two core EF processes work in an interactive function to support goal-directed behavior (Roberts & Pennington, 1996).

Behavioral Development of Infant EF

The measurement of EF during infancy offers particular challenges as tasks must be sensitive to limitations in young participants' motor skills as well as their receptive and productive language abilities. To meet this challenge, researchers have developed a variety of EF tasks that focus on infants' behavioral responses (e.g., looking, reaching) during engaging "games". Table 1.1 provides an overview of some of the most common paradigms used to measure infant EF. In what follows, we provide detailed information regarding delayed-memory search tasks – the most extensively used infant EF paradigm and the focus of our behavioral and psychophysiological research.

Infant Delayed-Memory Search Tasks

The delayed response (DR) task is a widely used nonverbal WM task in human and nonhuman animals (see Pelphrey & Reznick, 2004, for review) that is dependent on the dorsolateral prefrontal cortex (Diamond, 1990b). The Piagetian

TABLE 1.1 Selected Overview of Infant Executive Function Tasks

Task (Examples)	EF Component	Description	Age	Selected References
Antisaccade (*Freeze Frame*)	IC	Infants are required to inhibit looking to peripheral distractor cues	4 months +	Holmboe et al., 2008, 2010; Johnson, 1995
Delayed-memory search (*DR; A-not-B*)	WM+IC	Infant must find object in one of multiple locations after a delay, and inhibit searching in previously rewarded locations	4.5 months +	Cuevas & Bell, 2010; Diamond & Doar, 1989; Diamond et al., 1997; Pelphrey et al., 2004; Wiebe et al., 2010
Object retrieval (*Barrier Detour*)	WM+IC	Infants have to inhibit reaching directly toward object (visible through transparent barrier), reaching instead for side opening while holding the object and alternate route in mind	6 months +	Bell & Fox, 1992; Diamond, 1990a; Diamond et al., 1997; Matthews et al., 1996; Noland, 2008
Prohibition (*Forbidden Toy*)	IC	Infants are instructed to withhold a prepotent response (e.g., touching desirable object)	8 months +	Kochanska et al., 1998; Miller & Marcovitch, 2015
N boxes/pots (*Stationary vs. Scrambled*)	WM	N objects are hidden in N containers (stationary or scrambled between trials★). Infant must find each object	12 months +	Bernier, 2010; Diamond et al., 1997; Miller & Marcovitch, 2015; Wiebe et al., 2010

Notes
DR = delayed response; IC = inhibitory control; WM = working memory.
★ Only the scrambled version is related to dorsolateral prefrontal cortex function (for review, see Diamond et al., 1997).

infant A-not-B task is a variant of the DR task with longitudinal work indicating comparable developmental progression in performance across tasks (Diamond, 1985, 1990b; Diamond & Doar, 1989). In these tasks, participants watch as a desirable object is hidden in one of multiple locations, a brief delay is imposed while participants' visual attention to the hiding location is distracted (i.e., reducing likelihood of low-level marking strategies), and then participants search (i.e., look, reach) for the object. In the DR task, the hiding location is predetermined (independent of responding); however, in the A-not-B task, after successful search in one location (A; typically multiple times), the hiding location is switched (B). The A-not-B error refers to continued search at the incorrect location (A) on reversal trials (B). Thus, based on the broader animal and neuroscience literature, we conceptualize successful task performance as requiring WM and IC (Diamond, 1990b): maintaining the object location, updating its location between trials, and inhibiting previously rewarded responses. A variety of computational models (see Buss, Chapter 8, this volume) have been developed to understand the processes underlying A-not-B task performance, including different types of memories/processes (active–latent: Munakata, 1998; fast–slow: Thelen, Schöner, Scheier, & Smith, 2001) and conscious representational systems (Marcovitch & Zelazo, 2009). Although a memory process (akin to WM) is included in most accounts, there is debate regarding the need for separate inhibitory processes (Diamond, 2001; Munakata, 1998; Smith, Thelen, Titzer, & McLin, 1999).

EF demands of infant delayed-memory search tasks can be heightened by increasing the number of hiding objects/locations, the number of A trials prior to reversal, and the delay between hiding and searching (Cuevas & Bell, 2010; Marcovitch & Zelazo, 1999; Miller & Marcovitch, 2015). Likewise, the A-not-B task with invisible displacement has been used to extend this task into the second year (Diamond, Prevor, Callender, & Druin, 1997; Wiebe, Lukowski, & Bauer, 2010). The procedures are similar, but invisible displacement is considered more challenging because infants do not directly see the object hidden at either location. Additional non-EF factors (e.g., distance between hiding locations; simultaneous covering of hiding locations) can also affect task performance; thus, methodological rigor is critical (Diamond, Cruttenden, & Neiderman, 1994; Marcovitch & Zelazo, 1999). Further, from a dynamic systems perspective, perseverative reaching (i.e., reaching to the incorrect, previously rewarded location) involves "complex interactions of visual input, direction of gaze, posture, and memory" (Smith et al., 1999, p. 253), emphasizing the role of multiple factors in reaching behavior, especially motor planning.

Emergence and Development of EF

Early evidence regarding the emergence and developmental progression of EF came from a series of longitudinal delayed-memory search investigations, in which Diamond tested infants bi-weekly, beginning when they could first reach for a hidden object (Diamond, 1985; Diamond & Doar, 1989). Although there were individual differences in the maximum delay for accurate searches in the A-not-B and DR tasks, there was a general linear trend in performance between 7.5 and 12 months with an increase of 2–3 s per month in the delay required to produce errors in either task. Furthermore, a 2–3 s increase/decrease of the delay for individual infants resulted in perseverative or successful responding.

Looking versions of delayed-memory tasks have been used to extend downward the age of initial assessment. The earliest data come from 4.5–5 months of age, at which time the majority of infants perform poorly (Baird et al., 2002; Cuevas & Bell, 2010; Reznick, Morrow, Goldman, & Snyder, 2004); however, infants older than 5.5 months can maintain the hiding location for 1–2 s delays in a social "peek-a-boo" DR task (Reznick et al., 2004). Within-subjects comparisons of looking and reaching versions of delayed-memory search tasks reveal comparable performance (Bell & Adams, 1999; Matthews, Ellis, & Nelson, 1996; Pelphrey et al., 2004). Figure 1.1 displays the month-to-month performance of a group of infants assessed on both versions of the A-not-B task between 5 and 10 months of age (Cuevas & Bell, 2010).

As can be seen, once infants have developed the motoric skills requisite for the reaching response (8–9 months), performance is overlapping across response modality. Taken together, these findings indicate that by around 8 months, there are substantial improvements in A-not-B performance with about half of infants being able to succeed on reversal trials with a minimal delay (Bell & Fox, 1992; Diamond, 1985). Near 10 months, a basic level of WM and IC is demonstrated with most infants succeeding on reversal trials with minimal delay and some succeeding with longer delays (Diamond, 1985; Matthews et al., 1996). In sum, these findings indicate that the same executive demands are likely required for looking and reaching versions of delayed-memory search tasks.

During the second year, infants continue to exhibit improvements in more advanced versions of the A-not-B task as well as other EF tasks, such as 3-boxes (WM) and forbidden toys (IC) tasks (see Table 1.1; Diamond et al., 1997; Miller & Marcovitch, 2015; Wiebe et al., 2010). However, a recurring finding is the lack of longitudinal stability for individual EF tasks. In addition, unlike findings with children and adults, there is often minimal evidence of consistency across EF measures. As a notable exception, however, Diamond et al. (1997) found similar patterns of age-related change in A-not-B and object retrieval (WM + IC) tasks (see Table 1.1), with performance across tasks being modestly correlated from 8 to 11 months. These findings highlight some of the challenges in measuring EF during infancy, and are potentially related to early emerging executive skills often varying between contexts, yet providing the foundations for the "unitary" structure of EF identified by factorial models during the preschool-age period (Miller & Marcovitch, 2015). Orthogonal to theoretical analysis of the underlying structure of emerging EF, there is consensus that

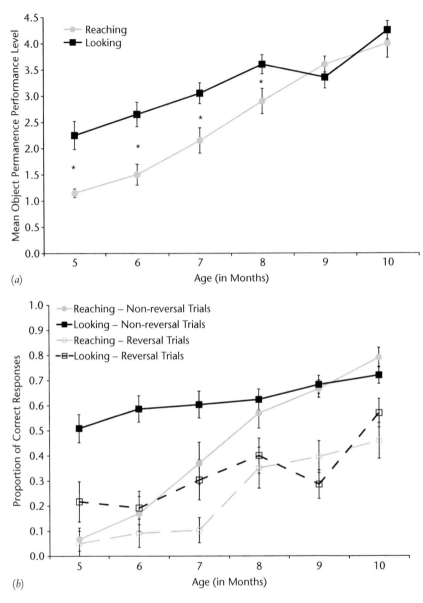

FIGURE 1.1a–b Mean performance (and SE) on the looking and reaching versions of the A-not-B task with incremental delay from 5 to 10 months of age. (*a*) Mean Object Permanence Scale score. 1–3 = Finding an object partially (1) or completely covered with one cloth (2) or two identical cloths (3). 4–5 = Finding an object in A-not-B procedure with a 0–slight delay (4) or 2-s delay (5). (*b*) Proportion of correct responses on non-reversal (A) and reversal (B) trials. Error bars represent one standard error. (From Cuevas & Bell, 2010. Copyright 2010 by American Psychological Association. Adapted with permission.)

further work is needed to analyze the psychometric properties (i.e., validity, reliability) and increase the precision of measurement of EF throughout early development.

Developmental antecedents. The trajectory of latent EF appears to begin as a unitary, domain-general process that becomes more differentiated with development (Best & Miller, 2010; Garon et al., 2008). The exact nature of this unitary component of EF in early development is unclear. Garon et al. (2008) posit that EFs share a common executive attention component that serves as a precursor for WM to develop initially, followed by IC, and then cognitive flexibility. In a longitudinal study, we found that individual differences in infant attention at 5 months (thought to reflect information processing efficiency) were associated with better EF at 24, 36, and 48 months (Cuevas & Bell, 2014), and others have found that measures of processing speed and recognition memory in infancy were related to EF at 11 years (Rose, Feldman, & Jankowski, 2012). If the roots of more complex executive skills are found in infancy, it may be promising to determine whether these rudimentary skills can be improved. Indeed, there is preliminary evidence that attentional control training leads to brief improvement in some aspects of EF (cognitive control but not WM) and sustained attention in 11-month-olds (Wass, Porayska-Pomsta, & Johnson, 2011).

In adults, many studies have failed to show evidence for improved cognitive function or distal transfer following training (Owen et al., 2010; Shipstead, Redick, & Engle, 2012; see Kliegel, Hering, Ihle, & Zuber, Chapter 13, this volume). Training studies with children have been more successful (Rueda, Rothbart, McCandliss, Saccomanno, & Posner, 2005; Thorell, Lindqvist, Nutley, Bohlin, & Klingberg, 2009), which may be indicative of increased behavioral and brain plasticity early in development (Huttenlocher, 2002). Given that EF deficits manifest in the context of early adversity (Hackman, Farah, & Meaney, 2010) but may be amenable to training during this period of increased plasticity, early behavioral interventions may prevent a cascade of EF failure in at-risk children. If early intervention is key, then determining the specific cognitive abilities in infancy that serve as developmental antecedents of childhood EF is of critical importance.

A Biopsychosocial Perspective of Infant EF

Our work has used a developmental psychobiological approach to examine individual differences and the developmental progression of EF. However, to understand the mechanisms underlying age-related changes and variations in EF, it is important to consider the multiple levels at which biology, behavior, and environment interact to contribute to frontal lobe and EF development (Calkins, 2015; Rueda, Posner, & Rothbart, 2004; Sameroff, 2010). Although this section primarily focuses on our brain electrical activity examinations of frontal lobe development, we briefly review contributions of other biological variables (temperament, genetics) as well as co-occurring socialization experiences with the family to provide a more integrative biopsychosocial perspective of infant EF.

Frontal Lobe Development

Current theory proposes that maturation of the frontal cortex and associated neural circuitry during the latter half of the first year supports the emergence of higher-order cognitive processes, including EFs (Colombo & Cheatham, 2006; Diamond, 1990b). The frontal cortex exhibits protracted development, and during the first postnatal year, there are substantial changes in frontal maturation, including increases in glucose metabolism, dendritic and axonal growth, synaptogenesis, and myelination (Chugani, Phelps, & Mazziotta, 1987; Deoni et al., 2011; Huttenlocher & Dabholkar, 1997; Tsekhmistrenko, Vasil'eva, Shumeiko, & Vologirov, 2004). In this section, we review converging biobehavioral evidence of frontal cortex contributions to early WM and IC as assessed via infant delayed-memory search tasks.

As reviewed by Diamond (1990b), findings from a wide array of biobehavioral techniques have established that performance on delayed-memory search tasks is dependent on the involvement of the dorsolateral prefrontal cortex. For instance, studies with adult and infant nonhuman primates reveal that lesions to the dorsolateral prefrontal cortex disrupts A-not-B/DR performance, resulting in perseverative errors after brief delays (Diamond, 1990b; Diamond & Goldman-Rakic, 1989). Most infant EF research examining brain–behavior associations has used electroencephalography (EEG), which measures the electrical activity on the scalp that is considered to be indicative of underlying cortical activity (i.e., the summation of postsynaptic potentials). This work has investigated frequency-dependent changes in neural activity (EEG power: excitability of group of neurons) and neural synchrony (EEG coherence: functional connectivity between two electrode sites) as a function of executive processing during the A-not-B task. As can be seen in Table 1.2, measures of EEG power and coherence in the 6–9 Hz frequency band are functionally related to EF during infancy. Unless otherwise specified, all findings reviewed below are for the 6–9 Hz frequency band.

Initial infant EF investigations used resting baseline measures of EEG activity (often associated with brain maturation) to provide information about the organization and activation of the brain when not involved in active cognitive processing. Cross-sectional and longitudinal data from 7- to 12-month-olds revealed that baseline measures of frontal and occipital EEG power and frontal-posterior EEG coherence were related to performance on the reaching A-not-B task (Bell & Fox, 1992, 1997). Likewise, recent diffusion tensor imaging (DTI) research with 12-month-olds indicates that white matter tract microstructure

TABLE 1.2 Overview of Infant Executive Function and 6–9 Hz EEG Findings

Finding	EEG Power	EEG Coherence
Resting baseline EEG is associated with WM+IC (*A-not-B task performance*)	Bell & Fox, 1992, 1997	Bell & Fox, 1992
Changes in EEG during executive processing [*baseline-to-task related changes during delay (WM+IC)*]	Bell, 2001, 2002; Bell & Wolfe, 2007; Cuevas & Bell, 2011; Cuevas et al., 2012a	Bell & Wolfe, 2007; Cuevas & Bell, 2011; Cuevas et al., 2012a; Cuevas et al., 2012b
Changes in EEG during increased inhibitory demands [*nonreversal (WM) vs. reversal (WM+IC) trials*]		Cuevas et al., 2012c
Changes in EEG during executive processing are associated with WM+IC (*A-not-B task performance*)	Bell, 2001, 2002; Cuevas et al., 2012a	Bell, 2001, 2012; Cuevas et al., 2012b★
Resting baseline EEG is associated with future EF	Kraybill & Bell, 2013	

Note
★ = 10–13 Hz EEG.

connecting frontal, parietal, and temporal regions is associated with DR task performance (Short et al., 2013). Taken together, these findings suggest that measures of brain maturation in regions that support EF in older children and adults (Jonides et al., 1993; Klingberg, 2006) are associated with individual differences in delayed-memory search task performance during infancy.

With the development of the looking A-not-B task, subsequent work has examined task-related changes in EEG measures during the delay portion of the task (requiring WM and IC); thus providing information about changes in brain activation and organization as a function of mental activity (e.g., Sauseng, Klimesch, Schabus, & Doppelmayr, 2005). Longitudinal and cross-sectional work has revealed that 5- to 10-month-olds exhibit task-related changes in EEG power (e.g., Bell, 2001; Cuevas & Bell, 2011) and frontal EEG coherence (e.g., Bell & Wolfe, 2007; Cuevas, Bell, Marcovitch, & Calkins, 2012). Subsequent analyses have examined how these changes are associated with task demands and performance. In terms of task difficulty, 10-month-olds exhibit increases in frontal EEG coherence in response to added inhibitory demands (i.e., non-reversal [A: WM] vs. reversal [B: WM+IC] trials; Cuevas, Swingler, Bell, Marcovitch, & Calkins, 2012). Likewise, 8-month-olds display higher EEG power (6–9 Hz) and frontal functional connectivity (10–13 Hz) during trials with correct as opposed to incorrect responses (Bell, 2002; Cuevas, Raj, & Bell, 2012). These findings are consistent with the results of individual differences analyses of EF.

Eight-month-old "high performers" exhibited a similar pattern as above – task-related increases in EEG coherence (medial frontal-parietal) and power (frontal, parietal, occipital); whereas "low performers" displayed task-related decreases in EEG coherence (medial frontal-parietal, frontal pole-medial frontal) and no task-related changes in EEG power (Bell, 2001, 2012). Furthermore, converging evidence from functional near-infrared spectroscopy research demonstrates that increases in frontal brain activity are associated with 5- to 12-month-olds' object permanence (Baird et al., 2002). Together, these findings suggest that increases in frontal neural activity, EEG power, and functional connectivity (i.e., integration of function between frontal and other cortical areas, especially parietal) underlie successful executive processing and higher levels of performance, with greater frontal synchrony also supporting increased EF demands. Our EEG findings likely reflect the emergence of the frontal-parietal network and its engagement during key aspects of EF processing – supporting the ability to focus attention, maintain information over longer delays, and withhold prepotent responses.

We have also used regression analyses to investigate whether frontal-parietal networks make unique contributions to individual differences in infant EF. Although electrocardiogram and EEG measures fail to account for variability in 5-month-olds' task performance (i.e., when most infants perform poorly; Cuevas, Bell et al., 2012), by 8 months of age (i.e., when there is a high level of variability in performance), medial frontal-parietal coherence and heart rate are unique predictors of task performance (Bell, 2012). Likewise, medial and lateral frontal EEG power and heart rate are unique predictors of 10-month-olds' task performance (Cuevas, Bell et al., 2012). Although rudimentary forms of WM and IC are emerging during infancy, our EEG data reveal a role of frontal-parietal networks in early EF in terms of individual differences, age-related changes, and executive processing demands – consistent with evidence from older children and adults (e.g., Sauseng et al., 2005; see Moriguchi, Chapter 5, this volume).

Genetic Influences

Although behavioral genetic studies of childhood EF are limited, there is moderate (Polderman et al., 2006; Wang, Deater-Deckard, Cutting, Thompson, & Petrill, 2012) to high heritability (Engelhardt, Briley, Mann, Harden, & Tucker-Drob, 2015; see Li & Roberts, Chapter 7, this volume) of childhood EF. Other approaches, examining the parent–child "familial" resemblance, found that mother and child EF was modestly correlated from 24 to 48 months (Cuevas, Deater-Deckard, Kim-Spoon, Wang et al., 2014). One potential mechanism underlying genetic contributions to individual variation in EF involves variations in genes that affect dopamine production and utilization (Casey, Tottenham, & Fossella, 2002). Evidence from electrophysiological studies in nonhuman primates (Sawaguchi & Goldman-Rakic, 1991) and neuropsychological studies on children treated for phenylketonuria (Diamond et al., 1997) reveal that dopamine plays a key role in frontal cortex functioning. Furthermore, Holmboe et al. (2010) found that 9-month-olds with the Met/Met and Val/Met COMT genotype (associated with higher levels of prefrontal dopamine) performed better on a saccadic inhibition task than infants with the Val/Val genotype (associated with the lowest levels of prefrontal dopamine). These findings suggest that polymorphisms in dopamine system genes can contribute to variation in EF ability as early as infancy.

Temperament

Rothbart and Bates (2006) privilege a biological approach to temperament as relating to individual differences in motor, emotional, and attentional reactivity and the subsequent self-regulation processes that modulate reactivity. In infancy, the temperamental aspect of attentional regulation is conceptually similar to the construct of executive attention; both are hypothesized characteristics of the individual rather than based on experience (Kane & Engle, 2002; Rothbart & Bates, 2006). Moreover, the executive attention aspects of temperament and cognitive aspects of EF share common neurobiological substrates, such as the frontal lobe architecture of the anterior cingulate cortex (Bell & Deater-Deckard, 2007; Posner & Rothbart, 2000). Indeed, laboratory-based measures of WM and IC are associated with temperament during infancy (Bell, 2012; Morasch & Bell, 2011). Thus, the conceptualization and neurobiological underpinnings of the executive attention aspects of temperament and cognitive-based aspects of EF share many commonalities. Future research should examine the degree to which these constructs are related and distinct (Liew, 2012; Zhou, Chen, & Main, 2012).

Environmental Factors

The effects of adverse economic circumstances (see Finch & Obradović, Chapter 9, this volume) on brain maturation can be detected as early as infancy (Hanson et al., 2013; Tomalski et al., 2013) and may impact the development of EF through the effects of stress on the brain, altering the structure and activity of the prefrontal cortex (Hackman et al., 2010). Variations in parenting (see Hughes & Devine, Chapter 10, this volume) across different familial contexts may be one potential pathway through which the environmental context affects stress physiology and neurocognitive development. It may be the *absence* of nurturing behavior and reduced positive parenting in low socioeconomic status homes that increases children's cortisol levels (Blair et al., 2011), which can affect EF development. Even in the context of non-adverse home environments, negative maternal caregiving behaviors (assessed at 10, 24, and 36 months) are negatively associated with child EF at 3 and 4 years (Cuevas, Deater-Deckard, Kim-Spoon, Watson et al., 2014). Conversely, positive parenting practices during infancy, such as maternal sensitivity, positive affect, and autonomy support, favor the development of later childhood EF (Bernier, Carlson, & Whipple, 2010; Kraybill & Bell, 2013) and brain maturation by contributing to increases in resting frontal EEG power during infancy (Bernier, Calkins, & Bell, 2016). These findings underscore the critical role that the environmental context and parent–child relationships play in fostering optimal EF and brain development.

Summary and Future Directions

Throughout this chapter we have highlighted some of the important contributions to our understanding of infant EF, with a focus on core processes of WM and IC. Examination of infant EF development through a biopsychosocial lens has greatly informed our knowledge of age-related change and individual variation. With respect to its early behavioral development, evidence from infant delayed-memory search tasks reveals substantial improvement from 8 to 12 months: infants make gains in (1) representing objects in memory as they tolerate increasing delays and (2) inhibiting previously rewarded responses as accuracy of reversal trials increases. Converging neuroscience investigations reveal a role of frontal-parietal networks in infant executive processing, including individual variation, age-related changes, and task demands – consistent with evidence from older children and adults. Since the development of EF is multi-faceted, early biological (e.g., brain development, genetics, temperament) and environmental factors (e.g., caregiving, financial resources) may aid or hinder its development.

Although we have previously identified particular areas in which future research is needed (e.g., measurement, conceptualization, developmental antecedents), we conclude by taking a more global perspective regarding future directions for the field. In Rovee-Collier's (1996) International Society on Infant Studies presidential address, she called for "shifting the focus from what to why" (p. 385); this was a call for examining the *function of infant behavior* – the perspective that infants are perfectly adapted for their ecological niche as opposed to incomplete adults who get better with age – as well as *integration between fractionized, highly specialized subfields* of infancy. Twenty years later, her "call to arms" resonates and is relevant to the study of infant EF. With respect to her latter point, recent paradigm shifts toward a biopsychosocial perspective are critical to integrative analysis of the complex mechanisms that underlie the emergence and variation in early EF development. For instance, examinations of epigenetic changes, such as DNA methylation (for review, see Szyf & Bick, 2013), provide one potential mechanism for the influence of early life experiences on EF development (for review of rodent work, see Jensen Peña &

Champagne, 2012). Likewise, computational modeling (see Buss, Chapter 8, this volume) and structural equation modeling offer promising methods for testing theoretical predictions derived from biopsychosocial models of EF. To our knowledge, the broader, evolutionary questions regarding function *"why they do it when they do"* (Rovee-Collier, 1996, p. 385) remain largely unanswered in the field of EF. *Are there adaptive advantages for young infants to not withhold prepotent responses or to actively maintain information regarding nonvisible stimuli for very brief periods of time?* Perseveration, for instance, has been conceptualized as "a sign of developmental achievement on the path to stable and flexible behavior" by dynamic systems accounts (Clearfield, Diedrich, Smith, & Thelen, 2006, p. 435). Further, in terms of the infant's ecological niche, perhaps aspects of executive processing emerge earlier and are facilitated by social stimuli/contexts (e.g., Reznick et al., 2004). The novel ideas and hypotheses driven by *functional* and *integrative* biopsychosocial lines of research can lead to entirely new areas of inquiry in analysis of the *"why"* and *"how"* of infant EF.

Author Note

Much of our research highlighted in this chapter was supported by grants HD049878 and HD043057 (PI: M. A. Bell) from the Eunice Kennedy Shriver National Institute of Child Health and Human Development (NICHD). The content of this chapter is solely the responsibility of the authors and does not necessarily represent the official views of the NICHD or the National Institutes of Health.

References

Baddeley, A. (1992). Working memory. *Science, 255*, 556–559.
Baird, A. A., Kagan, J., Gaudette, T., Walz, K. A., Hershlag, N., & Boas, D. A. (2002). Frontal lobe activation during object permanence: Data from near-infrared spectroscopy. *NeuroImage, 16*, 1120–1126.
Bell, M. A. (2001). Brain electrical activity associated with cognitive processing during a looking version of the A-not-B task. *Infancy, 2*, 311–330.
Bell, M. A. (2002). Power changes in infant EEG frequency bands during a spatial working memory task. *Psychophysiology, 39*, 450–458.
Bell, M. A. (2012). A psychobiological perspective on working memory performance at 8 months of age. *Child Development, 83*, 251–265.
Bell, M. A., & Adams, S. E. (1999). Comparable performance on looking and reaching versions of the A-not-B task at 8 months of age. *Infant Behavior and Development, 22*, 221–235.
Bell, M. A., & Deater-Deckard, K. (2007). Biological systems and the development of self-regulation: Integrating behavior, genetics, and psychophysiology. *Journal of Developmental & Behavioral Pediatrics, 28*, 409–420.
Bell, M. A., & Fox, N. A. (1992). The relations between frontal brain electrical activity and cognitive development during infancy. *Child Development, 63*, 1142–1163.

Bell, M. A., & Fox, N. A. (1997). Individual differences in object permanence performance at 8 months: Locomotor experience and brain electrical activity. *Developmental Psychobiology, 31*, 287–297.

Bell, M. A., & Wolfe, C. D. (2007). Changes in brain functioning from infancy to early childhood: Evidence from EEG power and coherence during working memory tasks. *Developmental Neuropsychology, 31*, 21–38.

Bernier, A., Calkins, S. D., & Bell, M. A. (2016). Longitudinal associations between the quality of mother–infant interactions and brain development across infancy. *Child Development, 87*, 1159–1174.

Bernier, A., Carlson, S. M., & Whipple, N. (2010). From external regulation to self-regulation: Early parenting precursors of young children's executive functioning. *Child Development, 81*, 326–339.

Best, J. R., & Miller, P. H. (2010). A developmental perspective on executive function. *Child Development, 81*, 1641–1660.

Blair, C., Raver, C. C., Granger, D., Mills-Koonce, R., Hibel, L., & the Family Life Project Key Investigators. (2011). Allostasis and allostatic load in the context of poverty in early childhood. *Development and Psychopathology, 23*, 845–857.

Calkins, S. D. (2015). Introduction to the volume: Seeing infant development through a biopsychosocial lens. In S. D. Calkins (Ed.), *Handbook of infant biopsychosocial development* (pp. 3–10). New York: The Guilford Press.

Casey, B. J., Tottenham, N., & Fossella, J. (2002). Clinical, imaging, lesion, and genetic approaches toward a model of cognitive control. *Developmental Psychobiology, 40*, 237–254.

Chugani, H. T., Phelps, M. E., & Mazziotta, J. C. (1987). Positron emission tomography study of human brain functional development. *Annals of Neurology, 22*, 487–497.

Clearfield, M. W., Diedrich, F. J., Smith, L. B., & Thelen, E. (2006). Young infants reach correctly in A-not-B tasks: On the development of stability and perseveration. *Infant Behavior and Development, 29*, 435–444.

Colombo, J., & Cheatham, C. L. (2006). The emergence and basis of endogenous attention in infancy and early childhood. In R. V. Kail (Ed.), *Advances in child development and behavior* (Vol. 34, pp. 283–322). Amsterdam: Elsevier.

Cuevas, K., & Bell, M. A. (2010). Developmental progression of looking and reaching performance on the A-not-B task. *Developmental Psychology, 46*, 1363–1371.

Cuevas, K., & Bell, M. A. (2011). EEG and ECG from 5 to 10 months of age: Developmental changes in baseline activation and cognitive processing during a working memory task. *International Journal of Psychophysiology, 80*, 119–128.

Cuevas, K., & Bell, M. A. (2014). Infant attention and early childhood executive function. *Child Development, 85*, 397–404.

Cuevas, K., Bell, M. A., Marcovitch, S., & Calkins, S. D. (2012). Electroencephalogram and heart rate measures of working memory at 5 and 10 months of age. *Developmental Psychology, 48*, 907–917.

Cuevas, K., Deater-Deckard, K., Kim-Spoon, J., Wang, Z., Morasch, K. C., & Bell, M. A. (2014). A longitudinal intergenerational analysis of executive functions during early childhood. *British Journal of Developmental Psychology, 32*, 50–64.

Cuevas, K., Deater-Deckard, K., Kim-Spoon, J., Watson, A. J., Morasch, K. C., & Bell, M. A. (2014). What's mom got to do with it? Contributions of maternal executive function and caregiving to the development of executive function across early childhood. *Developmental Science, 17*, 224–238.

Cuevas, K., Raj, V., & Bell, M. A. (2012). Functional connectivity and infant spatial working memory: A frequency band analysis. *Psychophysiology, 49*, 271–280.

Cuevas, K., Swingler, M. M., Bell, M. A., Marcovitch, S., & Calkins, S. D. (2012). Measures of frontal functioning and the emergence of inhibitory control processes at 10 months of age. *Developmental Cognitive Neuroscience, 2*, 235–243.

Deoni, S. C. L., Mercure, E., Blasi, A., Gasston, D., Thomson, A., Johnson, M., et al. (2011). Mapping infant brain myelination with magnetic resonance imaging. *The Journal of Neuroscience, 31*, 784–791.

Diamond, A. (1985). Development of the ability to use recall to guide action, as indicated by infants' performance on AB. *Child Development, 56*, 868–883.

Diamond, A. (1990a). Developmental time course in human infants and infant monkeys and the neural basis of inhibitory control in reaching. In A. Diamond (Ed.), *The development and neural bases of higher cognitive functions* (pp. 637–676). New York: New York Academy of Sciences Press.

Diamond, A. (1990b). The development and neural bases of memory function as indexed by the AB and delayed response tasks in human infants and infant monkeys. In A. Diamond (Ed.), *The development and neural bases of higher cognitive functions* (pp. 267–317). New York: New York Academy of Sciences Press.

Diamond, A. (2001). Looking closely at infants' performance and experimental procedures in the A-not-B task. *Behavioral and Brain Sciences, 24*, 38–41.

Diamond, A. (2016). Why improving and assessing executive functions early in life is critical. In J. A. Griffin, P. McCardle, & L. S. Freund (Eds.), *Executive function in preschool-age children: Integrating measurement, neurodevelopment, and translational research* (pp. 11–43). Washington, DC: American Psychological Association.

Diamond, A., Cruttenden, L., & Neiderman, D. (1994). AB with multiple wells: 1. Why are multiple wells sometimes easier than two wells? 2. Memory or memory + inhibition? *Developmental Psychology, 30*, 192–205.

Diamond, A., & Doar, B. (1989). The performance of human infants on a measure of frontal cortex function, the delayed response task. *Developmental Psychobiology, 22*, 271–294.

Diamond, A., & Goldman-Rakic, P. S. (1989). Comparison of human infants and rhesus monkeys on Piaget's AB task: Evidence for dependence on dorsolateral prefrontal cortex. *Experimental Brain Research, 74*, 24–40.

Diamond, A., Prevor, M. B., Callender, G., & Druin, D. P. (1997). Prefrontal cortex cognitive deficits in children treated early and continuously for PKU. *Monographs of the Society for Research in Child Development, 62*(4, Serial No. 252), 1–208.

Duncan, J., Johnson, R., Swales, M., & Freer, C. (1997). Frontal lobe deficits after head injury: Unity and diversity of function. *Cognitive Neuropsychology, 14*, 713–741.

Engelhardt, L. E., Briley, D. A., Mann, F. D., Harden, K. P., & Tucker-Drob, E. M. (2015). Genes unite executive functions in childhood. *Psychological Science, 26*, 1151–1163.

Engle, R. W., Kane, M. J., & Tuholski, S. W. (1999). Individual differences in working memory capacity and what they tell us about controlled attention, general fluid intelligence, and functions of the prefrontal cortex. In A. Miyake & P. Shah (Eds.), *Models of working memory: Mechanisms of active maintenance and executive control* (pp. 102–134). New York: Cambridge University Press.

Garon, N., Bryson, S. E., & Smith, I. M. (2008). Executive function in preschoolers: A review using an integrative framework. *Psychological Bulletin, 134*, 31–60.

Hackman, D. A., Farah, M. J., & Meaney, M. J. (2010). Socioeconomic status and the brain: Mechanistic insights from human and animal research. *Nature Reviews Neuroscience, 11*, 651–659.

Hanson, J. L., Hair, N., Shen, D. G., Shi, F., Gilmore, J. H., Wolfe, B. L., & Pollak, S. D. (2013). Family poverty affects the rate of human infant brain growth. *PLoS ONE, 8*, e80954.

Holmboe, K., Nemoda, Z., Pasco Fearon, R. M., Csibra, G., Sasvari-Szekely, M., & Johnson, M. H. (2010). Polymorphisms in dopamine system genes are associated with individual differences in attention in infancy. *Developmental Psychology, 46*, 404–416.

Holmboe, K., Pasco Fearon, R. M., Csibra, G., Tucker, L. A., & Johnson, M. H. (2008). Freeze-frame: A new infant inhibition task and its relation to frontal cortex tasks during infancy and early childhood. *Journal of Experimental Child Psychology, 100*, 89–114.

Huizinga, M., Dolan, C. V., & van der Molen, M. W. (2006). Age-related change in executive function: Developmental trends and a latent variable analysis. *Neuropsychologia, 44*, 2017–2036.

Huttenlocher, P. R. (2002). *Neural plasticity: The effects of environment on the development of the cerebral cortex.* Cambridge, MA: Harvard University Press.

Huttenlocher, P. R., & Dabholkar, A. S. (1997). Regional differences in synaptogenesis in human cerebral cortex. *Journal of Comparative Neurology, 387*, 167–178.

Jensen Peña, C. L., & Champagne, F. A. (2012). Epigenetic and neurodevelopmental perspectives on variation in parenting behavior. *Parenting: Science and Practice, 12*, 202–211.

Johnson, M. H. (1995). The inhibition of automatic saccades in early infancy. *Developmental Psychobiology, 28*, 281–291.

Jonides, J., Smith, E. E., Koeppe, R. A., Awh, E., Minoshima, S., & Mintun, M. A. (1993). Spatial working memory in humans as revealed by PET. *Nature, 363*, 623–625.

Kane, M. J., & Engle, R. W. (2002). The role of prefrontal cortex in working-memory capacity, executive attention, and general fluid intelligence: An individual-differences perspective. *Psychonomic Bulletin & Review, 9*, 637–671.

Klingberg, T. (2006). Development of a superior frontal-intraparietal network for visuospatial working memory. *Neuropsychologia, 44*, 2171–2177.

Kochanska, G., Tjebkes, T. L., & Forman, D. R. (1998). Children's emerging regulation of conduct: Restraint, compliance, and internalization from infancy to the second year. *Child Development, 69*, 1378–1389.

Kraybill, J. H., & Bell, M. A. (2013). Infancy predictors of preschool and postkindergarten executive function. *Developmental Psychobiology, 55*, 530–538.

Liew, J. (2012). Effortful control, executive functions, and education: Bringing self-regulatory and social-emotional competencies to the table. *Child Development Perspectives, 6*, 105–111.

Marcovitch, S., & Zelazo, P. D. (1999). The A-not-B error: Results from a logistic meta-analysis. *Child Development, 70*, 1297–1313.

Marcovitch, S., & Zelazo, P. D. (2009). A hierarchical competing systems model of the emergence and early development of executive function. *Developmental Science, 12*, 1–18.

Matthews, A., Ellis, A. E., & Nelson, C. A. (1996). Development of preterm and full-term infant ability on AB, recall memory, transparent barrier detour, and means-end tasks. *Child Development, 67*, 2658–2676.

Miller, S. E., & Marcovitch, S. (2015). Examining executive function in the second year of life: Coherence, stability, and relations to joint attention and language. *Developmental Psychology, 51*, 101–114.

Miyake, A., Friedman, N. P., Emerson, M. J., Witzki, A. H., Howerter, A., & Wager, T. D. (2000). The unity and diversity of executive functions and their contributions to complex "frontal lobe" tasks: A latent variable analysis. *Cognitive Psychology, 41,* 49–100.

Morasch, K. C., & Bell, M. A. (2011). The role of inhibitory control in behavioral and physiological expressions of toddler executive function. *Journal of Experimental Child Psychology, 108,* 593–606.

Munakata, Y. (1998). Infant perseveration and implications for object permanence theories: A PDP model of the AB task. *Developmental Science, 1,* 161–184.

Noland, J. S. (2008). Executive functioning demands of the object retrieval task for 8-month-old infants. *Child Neuropsychology, 14,* 504–509.

Owen, A. M., Hampshire, A., Grahn, J. A., Stenton, R., Dajani, S., Burns, A. S., et al. (2010). Putting brain training to the test. *Nature, 465,* 775–778.

Pelphrey, K. A., & Reznick, J. S. (2004). Working memory in infancy. In R. V. Kail (Ed.), *Advances in child development and behavior* (Vol. 31, pp. 173–227). San Diego, CA: Elsevier.

Pelphrey, K. A., Reznick, J. S., Goldman, B. D., Sasson, N., Morrow, J., Donahoe, A., & Hodgson, K. (2004). Development of visuospatial short-term memory in the second half of the 1st year. *Developmental Psychology, 40,* 836–851.

Polderman, T. J. C, Gosso, M. F., Posthuma, D., Van Beusterveldt, T. C. E. M., Heutink, P., Verhulsi, F. C., & Boomsma, D. I. (2006). A longitudinal twin study on IQ, executive functioning, and attention problems during childhood and early adolescence. *Acta Neurologica Belgica, 106,* 191–207.

Posner, M. I., & Rothbart, M. K. (2000). Developing mechanisms of self-regulation. *Development and Psychopathology, 12,* 427–441.

Reynolds, G. D., & Romano, A. C. (2016). The development of attention systems and working memory in infancy. *Frontiers in Systems Neuroscience, 10,* 15.

Reznick, J. S., Morrow, J. D., Goldman, B. D., & Snyder, J. (2004). The onset of working memory in infants. *Infancy, 6,* 145–154.

Roberts, R. J., Jr., & Pennington, B. F. (1996). An interactive framework for examining prefrontal cognitive processes. *Developmental Neuropsychology, 12,* 105–126.

Rose, S. A., Feldman, J. F., & Jankowski, J. J. (2012). Implications of infant cognition for executive functions at age 11. *Psychological Science, 23,* 1345–1355.

Rothbart, M. K., & Bates, J. E. (2006). Temperament. In N. Eisenberg (Vol. Ed.), & W. Damon & R. M. Lerner (Eds.), *Handbook of child psychology: Vol. 3. Social, emotional, and personality development* (6th ed., pp. 99–166). Hoboken, NJ: Wiley.

Rovee-Collier, C. (1996). Shifting the focus from what to why. *Infant Behavior and Development, 19,* 385–400.

Rueda, M. R., Posner, M. I., & Rothbart, M. K. (2004). Attentional control and self-regulation. In R. F. Baumeister & K. D. Vohs (Eds.), *Handbook of self-regulation: Research, theory, and applications* (pp. 283–300). New York: Guilford Press.

Rueda, M. R., Rothbart, M. K., McCandliss, B. D., Saccomanno, L., & Posner, M. I. (2005). Training, maturation, and genetic influences on the development of executive attention. *Proceedings of the National Academy of Sciences, 102,* 14931–14936.

Sameroff, A. (2010). A unified theory of development: A dialectic integration of nature and nurture. *Child Development, 81,* 6–22.

Sauseng, P., Klimesch, W., Schabus, M., & Doppelmayr, M. (2005). Fronto-parietal EEG coherence in theta and upper alpha reflect central executive functions of working memory. *International Journal of Psychophysiology, 57,* 97–103.

Sawaguchi, T., & Goldman-Rakic, P. S. (1991). D1 dopamine receptors in prefrontal cortex: Involvement in working memory. *Science, 251*, 947–950.

Shipstead, Z., Redick, T. S., & Engle, R. W. (2012). Is working memory training effective? *Psychological Bulletin, 138*, 628–654.

Short, S. J., Elison, J. T., Goldman, B. D., Styner, M., Gu, H., Connelly, M., et al. (2013). Associations between white matter microstructure and infants' working memory. *NeuroImage, 64*, 156–166.

Smith, L. B., Thelen, E., Titzer, R., & McLin, D. (1999). Knowing the context of acting: The task dynamics of the A-not-B error. *Psychological Review, 106*, 235–260.

Stuss, D. T., & Alexander, M. P. (2000). Executive functions and the frontal lobes: A conceptual view. *Psychological Research, 63*, 289–298.

Szyf, M., & Bick, J. (2013). DNA methylation: A mechanism for embedding early life experiences in the genome. *Child Development, 84*, 49–57.

Thelen, E., Schöner, G., Scheier, C., & Smith, L. B. (2001). The dynamics of embodiment: A field theory of infant perseverative reaching. *Behavioral and Brain Sciences, 24*, 1–86.

Thorell, L. B., Lindqvist, S., Nutley, S. B., Bohlin, G., & Klingberg, T. (2009). Training and transfer effects of executive functions in preschool children. *Developmental Science, 12*, 106–113.

Tomalski, P., Moore, D. G., Ribiero, H., Axelsson, E. L., Murphy, E., Karmiloff-Smith, A., et al. (2013). Socioeconomic status and functional brain development: Associations in early infancy. *Developmental Science, 16*, 676–687.

Tsekhmistrenko, T. A., Vasil'eva, V. A., Shumeiko, N. S., & Vologirov, A. S. (2004). Quantitative changes in the fibroarchitectonics of the human cortex from birth to the age of 12 years. *Neuroscience and Behavioral Physiology, 34*, 983–988.

Wang, Z., Deater-Deckard, K., Cutting, L., Thompson, L. A., & Petrill, S. A. (2012). Working memory and parent-rated components of attention in middle childhood: A behavioral genetic study. *Behavior Genetics, 42*, 199–208.

Wass, S., Porayska-Pomsta, K., & Johnson, M. H. (2011). Training attentional control in infancy. *Current Biology, 21*, 1543–1547.

Wiebe, S. A., Espy, K. A., & Charak, D. (2008). Using confirmatory factor analysis to understand executive control in preschool children: I. Latent structure. *Developmental Psychology, 44*, 575–587.

Wiebe, S. A., Lukowski, A. F., & Bauer, P. J. (2010). Sequence imitation and reaching measures of executive control: A longitudinal examination in the second year of life. *Developmental Neuropsychology, 35*, 522–538.

Wiebe, S. A., Sheffield, T., Nelson, J. M., Clark, C. A. C., Chevalier, N., & Espy, K. A. (2011). The structure of executive function in 3-year-olds. *Journal of Experimental Child Psychology, 108*, 436–452.

Zhou, Q., Chen, S. H., & Main, A. (2012). Commonalities and differences in the research on children's effortful control and executive function: A call for an integrated model of self-regulation. *Child Development Perspectives, 6*, 112–121.

2
EXECUTIVE FUNCTION IN EARLY AND MIDDLE CHILDHOOD

Nicolas Chevalier & Caron A. C. Clark

If you have watched a 1-year-old play with a shape sorter, you know infant attention is quite fragile and stimulus-driven, a characteristic that parents often take advantage of when they distract their newly crawling 9-month-old with a toy to avoid broken ornaments. Yet it will also be clear to you that infants are capable of engaging in simple forms of goal-directed behavior (as illustrated in the preceding chapter), providing a critical platform for the protracted and gradual process of executive control development that continues through childhood and into adolescence (e.g., Cuevas & Bell, 2014; Cuevas, Rajan, & Bryant, Chapter 1, this volume). Resting on this platform of rudimentary attentional control, executive function (EF) undergoes some of its most profound changes in early and middle childhood. During this period, children grow from young preschoolers who are "all over the place", struggle to resist impulses, and follow social rules, to young adolescents who already demonstrate most foundations of adult-like thinking and behavior. Emerging EF supports increasingly flexible and adaptive behavior, and promotes greater autonomy. It plays a central role in cognitive and social development, attention in the classroom and academic achievement, and problem behavior (e.g., Clark, Sheffield, Wiebe, & Espy, 2013; Clark & Woodward, 2014; Hughes, 1998).

As EF is supported by higher-order association cortices, including prefrontal and parietal cortices (Moriguchi, Chapter 5, this volume; West, Chapter 6, this volume), which are among the slowest to develop (Gogtay et al., 2004), it shows a protracted developmental trajectory from birth to early childhood. Yet, individual differences in EF are amazingly stable through the preschool years and middle childhood, with childhood EF scores predicting EF performance in adolescence (Friedman, Miyake, Robinson, & Hewitt, 2011) and up to 40 years later (Casey et al., 2011). The origin of these individual differences has been argued to be almost entirely genetic in nature (Friedman et al., 2008; see also Li & Roberts, Chapter 7, this volume). Yet, EF is largely influenced by various environmental factors (Poarch & van Hell, Chapter 11, this volume; Berryman, Pothier, & Bherrer, Chapter 12, this volume), such as poverty (Blair et al., 2011; Finch & Obradović, Chapter 9, this volume) and culture (Sabbagh, Xu, Carlson, Moses, & Lee, 2006). In the rest of this chapter, we chart its major milestones and progressive differentiation with age, then review some of the mechanisms driving this development.

EF Transitions Occurring in Early and Middle Childhood

One marked characteristic of young children's behavior is their inability to overcome well-rehearsed or prepotent response tendencies, as illustrated by their seeming inability to stop themselves from blurting out information they have been told to keep secret. The A-not-B task represents one such experimental example of a perseverative tendency to reach toward the previously reinforced A location during the B trial (Diamond, 1990). During the toddler and preschool years, such perseverative behaviors are perhaps most clearly evident on sorting tasks such as the Dimensional Change Card Sort (DCCS) (Doebel & Zelazo, 2015). By 3 years of age, children are able to sort bivalent DCCS stimulus cards (e.g., a blue dog, a red dog, a blue boat, a red boat) according to a single dimension (e.g. by color). However, when they are asked to sort by a new dimension (e.g., by shape), the majority of 3-year-olds continue to sort by the pre-switch dimension (Zelazo, Müller, Frye, & Marcovitch, 2003). Perseveration on previously relevant information, which has traditionally been thought of as poor control, may indeed reflect a stepping stone toward flexible switching, showing that children have transitioned away from an earlier inability to consistently apply a single rule across trials and phases (Blakey, Visser, & Carroll, 2016). By 4 years of age, the majority of children are able to pass the standard version of the DCCS (Doebel & Zelazo, 2015).

Concomitant with this ability to switch between two different rule sets, children show an increasing ability to inhibit prepotent behavioral response tendencies through the preschool period. Between 1 and 2 years of age, children are better able to resist distraction in the face of short-term memory task delays and they can wait for short periods of time before touching a forbidden toy (Wiebe, Lukowski, & Bauer, 2010). Between 3 and 5 years of age, children show dramatic increases in the ability to make a correct verbal or motor response that is counter-intuitive to a well-learned or well-rehearsed response (Carlson, 2005). Likewise, they are better able to suppress motor responses to no-go trials of button-press tasks (Wiebe, Sheffield, & Espy, 2012); inhibit previously relevant verbal responses (Clark, Sheffield, Chevalier, et al., 2013); stop themselves from pointing to a larger reward in order to obtain that reward (Müller, Zelazo, Hood, Leone, & Rohrer, 2004); and delay gratification (Carlson, Moses, &

Claxton, 2004). Notably, 3-year-old children who show the quickest response times on these types of tasks also are the least likely to be able to inhibit motor responses and they have lower levels of working memory, suggesting that a more cautious, reflective approach may signify more efficient executive control at this young age (Clark, Sheffield, Chevalier, et al., 2013; Wiebe et al., 2012).

Growth in more complex forms of inhibitory control continues into adolescence (Romine & Reynolds, 2005). In particular, there are continued changes in middle childhood in the ability to resist interference from irrelevant attentional stimuli (Cragg, 2015). In the early preschool period, children's errors on inhibitory control tasks often are associated with distracting or task-irrelevant information, whereas errors tend to become focused on task-relevant information as children mature and selective attention becomes more refined (Clark, Sheffield, Chevalier, et al., 2013; Fisher, Godwin, & Seltman, 2014). This is in keeping with the idea that children may have limited understanding or appreciation of the goals of the task, instead becoming distracted by irrelevant cues (Chevalier, 2015a).

As illustrated by the DCCS studies, children also show marked advances in the ability to switch flexibly between different response options by age 4, provided that tasks involve exogenous cues or instructions (Clark, Sheffield, Chevalier, et al., 2013). Often in life, however, we are not told of the need to switch and instead must infer this need through cues or feedback in the environment (Snyder & Munakata, 2010). On sorting tasks that involve abstraction of the switching rule based on feedback, children's growth is more gradual, with gains in growth in accuracy occurring between 6 and 10 years of age (e.g., Chelune & Baer, 1986) and speed increasing through middle childhood and adolescence (Anderson, Anderson, Northam, Jacobs, & Catroppa, 2001). Children aged 7–12 years show a slower decline in switch costs over the course of repeated non-switch trials in a switching task relative to older children and adults, indicating that part of the difficulty of switching between strategies in this age group may relate to interference from the previous task set (Cepeda, Kramer, & Gonzalez de Sather, 2001). More complex measures of verbal fluency, which likely involve an array of complex cognitive skills, continue to improve through late adolescence into early adulthood (e.g., Romine & Reynolds, 2005).

A third aspect of executive control that shows steady change over the course of childhood is working memory, defined by the ability to hold a representation online for periods of time in service of goal-directed tasks (Cowan, 2015). Working memory tasks, however, vary in the taxation they place on executive management processes. In early childhood, working memory tasks often involve simple maintenance of information. Examples of these types of tasks include forward digit span and spatial delay tasks. In adulthood, working memory tasks often involve simultaneous manipulation of the information to be maintained through updating, inhibition, or transformation processes (Linares, Bajo, & Pelegrina, 2016). Examples of these types of tasks include counting span, self-ordered pointing tasks, or listening span tasks. Performance on simple measures of spatial memory maintenance increases linearly across early and middle childhood and appears to level off by early adolescence (Gathercole, Pickering, Ambridge, & Wearing, 2004; Luciana, Conklin, Hooper, & Yarger, 2005). Complex working memory updating tasks that require, for example, retrieval of information that was relevant two trials earlier, show particularly pronounced growth in middle childhood and adolescence and continue to develop through late adolescence (e.g., Carriedo, Corral, Montoro, Herrero, & Rucián, 2016).

Together, these findings show major changes in EF performance across early and middle childhood. However, substantial unevenness in performance can be observed across tasks at any given age (e.g., perseveration is no longer typical on the A-not-B task after 12 months, but is observed at 3 years in the DCCS) because performance ultimately depends on the level of interference, cue saliency, and difficulty of the non-executive processes specific to each task. Therefore, no single task is likely to reflect the complexity of EF development.

Differentiation of Executive Control with Age

The above review of developmental changes on different types of EF tasks suggests that EF may be underscored by a variety of component processes, with some EF capacities maturing later than others. Nonetheless, there is continued debate regarding the component processes that fall under the general umbrella of EF and the organization of the EF system over the course of development (Friedman & Miyake, 2016). Several studies using factor analysis to examine the underlying architecture of EF in preschool-aged children have found that a single, unitary factor best models the shared variance of EF measures, as opposed to distinct working memory, cognitive flexibility, or inhibitory control components (Fuhs, Nesbitt, Farran, & Dong, 2014; Wiebe, Espy, & Charak, 2008; Wiebe et al., 2011; Willoughby, Blair, Wirth, & Greenberg, 2012; although see Miller, Giesbrecht, Müller, McInerney, & Kerns, 2012 for a counter-argument). Studies in late childhood, in contrast, have tended to replicate adult findings of correlated and yet distinct latent EF components (Miyake et al., 2000), with some studies showing separate working memory updating and cognitive flexibility factors (St. Clair-Thompson & Gathercole, 2006; van der Sluis, de Jong, & van der Leij, 2007) and a few reporting evidence for a further, distinct inhibition component (Huizinga, Dolan, & van der Molen, 2006; Lehto, Juujärvi, Kooistra, & Pulkkinen, 2003). Further studies that have used the same executive tasks in different age groups have found support for the gradual emergence of distinct EF factors over the course of childhood (Gandolfi, Viterbori, Traverso, & Usai, 2014; Lee, Bull, & Ho, 2013; Shing, Lindenberger, Diamond, Li, & Davidson, 2010; Xu et al., 2013). For instance, in a longitudinal study tracking participants between 6 and 15 years of age, Lee et al. (2013) found that a two-factor structure comprising a working memory updating and a joint cognitive flexibility/inhibitory control factor provided the best model of EF in 6–12-year-olds, but that cognitive flexibility and inhibitory control were best modeled as two distinct factors by age 15 years.

Interestingly, this gradual differentiation of EF components maps on to progressive specialization observed at the neural level. Specifically, the neural networks that support EF show progressive segregation (i.e., reduced local functional connectivity within brain regions such as the prefrontal or anterior cingulate cortex) and integration (i.e., greater distal functional connectivity between brain regions, e.g., prefrontal and parietal cortices) (e.g., Fair et al., 2007), and EF performance is supported by increasingly focal prefrontal activity with age (Durston et al., 2006). These findings suggest that EF development cannot simply be viewed as the growth of performance on EF tasks. Instead, there may be fundamental changes in the organization of cognition and the interrelations among cognitive systems that facilitate the increasing ability to engage in independent, goal-directed, executive behavior.

Mechanisms Supporting Developmental Change

Increases in the Strength of Mental Representations

If EF relies on the maintenance of a goal or representation "online", then the ability to form, process, and reflect on information related to that goal is fundamental. Morton and Munakata's (2002) active vs. latent representation model, for instance, suggests that the ability to build a strong representation of a goal is critical for overcoming interference from well-learned representations that may conflict with that goal. In this and other neural network models (see Buss, Chapter 8, this volume), strong inputs for shape, color, and linguistic information boost excitation of populations of neurons in the PFC, building up more stable representational traces and inspiring lateral inhibition of irrelevant representations (Buss & Spencer, 2014).

The dramatic gains in executive function through the preschool period occur alongside gains in several other cognitive domains, including the ability to reflect on the intentions of the self and others, episodic memory, language, and conceptual development. Because they afford more fluent access to concepts and an increase in the speed at which information can be processed and accessed from long-term memory, these capacities all are likely to provide a scaffold for the formation of strong, active representations. For instance, Miller and Marcovitch (2015) found that, whereas children's performance on EF tasks at 14 months did not predict their performance on EF measures at 18 months, language skills and joint attention, factors that support representational salience, did predict 18-month performance. Others have reported similarly strong links between language ability and EF (Fuhs & Day, 2010; Gooch, Thompson, Nash, Snowling, & Hulme, 2016). Clark et al. (2014) found that children's language skills and the speed with which children could name basic shapes and colors shared variance with children's EF performance at age 3 to such a degree that a separate, unique executive control factor could not be extracted from this variance. By age 5, however, a unique EF factor did emerge, suggesting that early EF tasks are strongly conflated with children's basic fluency in recognizing and processing information (see Cepeda, Blackwell, & Munakata, 2013, for converging evidence).

Apart from the speed and efficiency with which children can build mental representations, however, there are also changes in the resolution of short-term memory, such that the precision of visuospatial representations increases over the course of childhood (Burnett Heyes, Zokaei, van der Staaij, Bays, & Husain, 2012). Moreover, providing children with the experience of sorting colors and shapes aids their ability to switch in the DCCS task (Perone, Molitor, Buss, Spencer, & Samuelson, 2015). These examples lend support to the idea that, although EF is often conceptualized as the organizer and conductor of other cognitive systems, its development may be critically intertwined with the developing efficiency of these systems. Experience-dependent conceptual development may aid children's ability to build up efficient representations that contribute to EF performance.

Cue Processing and Goal Identification

The very function of EF is to ensure goal attainment. Goals maintained in the PFC guide actions by biasing activity in posterior brain regions (Miller & Cohen, 2001; Munakata, Snyder, & Chatham, 2012). Thus, EF can only be implemented if children know what goals (i.e., task, action, or state) they need to reach (e.g., win a game, get the homework done, clean up the room). Goal identification has traditionally been considered trivial, even for young children who were thought to know what they needed to do but failed to do it. Yet, recent evidence suggests that children often struggle to identify goals and goal identification progress is a major force driving EF development (Chevalier, 2015a).

To identify which goal is most relevant, one often needs to process information in the environment (e.g., traffic lights signaling whether or not to cross the street, a glare from a teacher signifying that one should stop talking and focus on one's work). Environmental cues differ in nature and in how strongly or explicitly they signal which task to reach, when and how to do so, making them especially challenging to process. Preschoolers struggle to identify relevant task goals when these are signaled by arbitrary cues bearing no semantic relation to the tasks or actions they signal and show much better performance when additional or more salient cues are provided to facilitate task goal identification (e.g., Chevalier & Blaye, 2009; Chevalier, Huber, Wiebe, & Espy, 2013). Importantly, children's performance progressively becomes less dependent on cue saliency with age, pointing out substantial progress in cue processing and goal identification during middle childhood (Chevalier & Blaye, 2009).

In other words, the ability to engage EF depends on the cues available in the environment and ultimately how easily children can identify what they need or are expected to do. The difficulty of goal identification (also referred to as task selection or inference) is far from anecdotal. First, it varies with age to a much greater extent than the difficulty of switching tasks per se (Chevalier, Dauvier, & Blaye, in press). Second, goal identification seems to drive the increasing relations among the main EF components over the preschool period (Chevalier et al., 2012), hence suggesting it may underpin "common EF" abilities (Friedman & Miyake, 2016). Third, practicing rapid detection of relevant cues subsequently yields better response inhibition performance than the practice of motoric stopping per se in middle childhood (Chevalier, Chatham, & Munakata, 2014). Finally, although cue processing becomes easier with age, it is never trivial even in adulthood (e.g., Chatham et al., 2012).

Control Coordination

EF development does not exclusively reflect more executive resources with age, but also better and more flexible engagement of extant executive resources. Because of ever-changing task demands, the most adaptive way to engage EF varies from moment to moment, hence calling for dynamic adjustment of control. For example, a bike ride to school may need high control engagement to monitor for traffic lights and switch attention between cars and pedestrians, whereas EF can be disengaged on more quiet bike paths while remaining mobilizable in case a dog unexpectedly runs across the path. Increased flexibility in EF engagement relies on acquisition of new control strategies, such as verbal strategies (Fatzer & Roebers, 2012) or proactive control strategies (Munakata et al., 2012), as well as more optimal coordination of this growing repertoire of strategies (Chevalier, 2015b).

Control coordination is essential because different ways of engaging EF have their own benefits and costs. For instance, proactive control, which refers to anticipation and preparation for upcoming task demands (e.g., actively monitoring for traffic light changes), requires substantial cognitive effort to maintain goal-relevant information in working memory through sustained PFC activity, but ensures high performance in situations where task demands can be reliably predicted, by preventing interference before it even occurs. In contrast, reactive control, which corresponds to in-the-moment EF engagement to solve interference that has already arisen (e.g., avoid the dog unexpectedly running across the bike path), is not usually as effective as proactive control but it requires less effort as it relies on goal-information retrieval and transient PFC activity and is especially useful when task demands cannot be predicted (Braver, 2012; Braver, Paxton, Locke, & Barch, 2009). Indeed, children who engage proactive control show greater performance when task preparation is possible, but lower performance than those who engage reactive control in situations where it is especially difficult to prepare (Blackwell & Munakata, 2014).

EF engagement is relatively rigid during the preschool years (despite some evidence of strategy change after initial failure; Dauvier, Chevalier, & Blaye, 2012) but subsequently gains in flexibility. Young children are largely biased to reactive control, whereas older children and adolescents more flexibly engage either type of control as a function of task demands (e.g., Andrews-Hanna et al., 2011; Chatham, Frank, & Munakata, 2009; Chevalier, James, Wiebe, Nelson, & Espy, 2014; Lucenet & Blaye, 2014). Young children's bias to reactive control does not merely reflect inability to engage EF proactively, as shown by the fact that children do engage it when reactive control is more difficult (Chevalier, Martis, Curran, & Munakata, 2015).

Improvement in control coordination may be related to metacognition development, which shows close links with EF progress more broadly (Roebers & Feurer, 2015). Control coordination may be achieved by weighing the respective cost (i.e., mental effort) and benefit (i.e., reward) of each task and strategy, and selecting those with the most advantageous ratios, a process that may be supported by the anterior cingulate cortex (Shenhav, Botvinick, & Cohen, 2013). Greater metacognitive abilities may support better appreciation of task demands, representation of strategy costs and benefits, and performance monitoring. As they grow older, children get better at monitoring online how well they are engaging EF (Chevalier & Blaye, 2016) as well as processing feedback on their behaviors (e.g., DuPuis et al., 2015). More optimal control coordination may not always lead to more EF engagement, but also in less EF engagement in situations where EF may be detrimental, potentially resulting in more economic cognitive functioning with age. Consistently, performance on EF tasks is increasingly supported by posterior regions specialized in task-specific processes across adolescence (Luna, Padmanabhan, & O'Hearn, 2010).

Concluding Remarks

Together, the growing literature of EF paints a picture of increasingly efficient, differentiated, and flexible engagement of cognitive control toward more easily identified goals. This development is supported by increasingly specialized neural networks and intermingled with other cognitive acquisitions during childhood. Because of its central role in child development, "too little" EF is clearly a major risk factor for academic failure and cascading negative outcomes (Daly, Delaney, Egan, & Baumeister, 2015; Moffitt et al., 2011), and is also associated with developmental disorders such as autism and ADHD (Barkley, 1997; Rosenthal et al., 2013). Poor environmental circumstances, such as low care quality, environmental stress or poverty, can have detrimental influences on children's EF, which can in turn exert a negative influence on the environment a child grows up in (e.g., children with lower EF elicit more negative, punitive behavior from their parents; Blair, Raver, & Berry, 2014; Clark & Woodward, 2014; Hughes & Devine, Chapter 10, this volume). This vicious circle may be broken through early interventions to support EF during childhood (see Kliegel, Hering, Ihle, & Zuber, Chapter 13, this volume). For instance, working memory and attention can be enhanced through computer-based training in typically developing children as well as those with ADHD or intellectual disabilities, although effects have been inconsistent with respect to far-transfer domains (Kirk, Gray, Riby, & Cornish, 2015; Rueda et al., 2005; Titz & Karbach, 2014). In younger children, training with games like "Simon Says" and preschool programs that integrate training in self-control also show promise (Bierman, Nix, Greenberg, Blair, & Domitrovich, 2008; Diamond, 2012; Halperin et al., 2013), with perhaps greater far-transfer effects on other cognitive domains in this early childhood period, when cognitive abilities rely on less differentiated and less specialized neural networks (Wass, Scerif, & Johnson, 2012).

Although such interventions are needed for children with lower EF for their age group, they probably should not be over-relied on for other children, as "too much" EF may not be any more optimal than "too little" EF. For instance, excessively high EF could be associated with poor social interactions, with over-controlled behaviors contributing to social reticence and anxiety in middle childhood (Lamm et al., 2014). Healthy development likely relies on an optimal balance between controlled and non-controlled behaviors. This optimal balance may vary across childhood, progressively favoring more control with age (Thompson-Schill, Ramscar, & Chrysikou, 2009). Immature EF may be adaptive in early childhood, because without the constraints of a specific goal, young children can explore their environments more freely, promoting learning of statistical regularities (see Gopnik, Griffiths, & Lucas, 2015) and, in turn, other important cognitive acquisitions, such as language and social conventions, which, as we have seen, may ultimately aid effective executive control. With advancing age and growing expectation for autonomy, the optimal balance shifts toward greater EF as it sharpens behavioral efficiency.

References

Anderson, V. A., Anderson, P., Northam, E., Jacobs, R., & Catroppa, C. (2001). Development of executive functions through late childhood and adolescence in an Australian sample. *Developmental Neuropsychology, 20*, 385–406.

Andrews-Hanna, J. R., Mackiewicz Seghete, K. L., Claus, E. D., Burgess, G. C., Ruzic, L., & Banich, M. T. (2011). Cognitive control in adolescence: Neural underpinnings and relation to self-report behaviors. *PloS One, 6*, e21598.

Barkley, R. A. (1997). Behavioral inhibition, sustained attention, and executive functions: Constructing a unifying theory of ADHD. *Psychological Bulletin, 121*, 65–94.

Bierman, K. L., Nix, R. L., Greenberg, M. T., Blair, C., & Domitrovich, C. E. (2008). Executive functions and school readiness intervention: Impact, moderation, and mediation in the Head Start REDI program. *Development and Psychopathology, 20*, 821–843.

Blackwell, K. A., & Munakata, Y. (2014). Costs and benefits linked to developments in cognitive control. *Developmental Science, 17*, 203–211.

Blair, C., Granger, D. A., Willoughby, M., Mills-Koonce, R., Cox, M., Greenberg, M. T., et al. (2011). Salivary cortisol mediates effects of poverty and parenting on executive functions in early childhood. *Child Development, 82*, 1970–1984.

Blair, C., Raver, C. C., & Berry, D. J. (2014). Two approaches to estimating the effect of parenting on the development of executive function in early childhood. *Developmental Psychology, 50*, 554–565.

Blakey, E., Visser, I., & Carroll, D. J. (2016). Different executive functions support different kinds of cognitive flexibility: Evidence from 2-, 3-, and 4-year-olds. *Child Development, 87*, 513–526.

Braver, T. S. (2012). The variable nature of cognitive control: A dual mechanisms framework. *Trends in Cognitive Sciences, 16*, 106–113.

Braver, T. S., Paxton, J. L., Locke, H. S., & Barch, D. M. (2009). Flexible neural mechanisms of cognitive control within human prefrontal cortex. *Proceedings of the National Academy of Sciences of the United States of America, 106*, 7351–7356.

Burnett Heyes, S., Zokaei, N., van der Staaij, I., Bays, P. M., & Husain, M. (2012). Development of visual working memory precision in childhood. *Developmental Science, 15*, 528–539.

Buss, A. T., & Spencer, J. P. (2014). The emergent executive: A dynamic field theory of the development of executive function. *Monographs of the Society for Research in Child Development, 79*(2), 1–132.

Carlson, S. M. (2005). Developmentally sensitive measures of executive function in preschool children. *Developmental Neuropsychology, 28*, 595–616.

Carlson, S. M., Moses, L. J., & Claxton, L. J. (2004). Individual differences in executive functioning and theory of mind: An investigation of inhibitory control and planning ability. *Journal of Experimental Child Psychology, 87*, 299–319.

Carriedo, N., Corral, A., Montoro, P. R., Herrero, L., & Rucián, M. (2016). Development of the updating executive function: From 7-year-olds to young adults. *Developmental Psychology, 52*, 666–678.

Casey, B. J., Somerville, L. H., Gotlib, I. H., Ayduk, O., Franklin, N. T., Askren, M. K., et al. (2011). Behavioral and neural correlates of delay of gratification 40 years later. *Proceedings of the National Academy of Sciences, 108*, 14998–15003.

Cepeda, N. J., Blackwell, K. A., & Munakata, Y. (2013). Speed isn't everything: Complex processing speed measures mask individual differences and developmental changes in executive control. *Developmental Science, 16*, 269–286.

Cepeda, N. J., Kramer, A. F., & Gonzalez de Sather, J. C. M. (2001). Changes in executive control across the life span: Examination of task-switching performance. *Developmental Psychology, 37*, 715–729.

Chatham, C. H., Claus, E. D., Kim, A., Curran, T., Banich, M. T., & Munakata, Y. (2012). Cognitive control reflects context monitoring, not motoric stopping, in response inhibition. *PloS One, 7*, e31546.

Chatham, C. H., Frank, M. J., & Munakata, Y. (2009). Pupillometric and behavioral markers of a developmental shift in the temporal dynamics of cognitive control. *Proceedings of the National Academy of Sciences of the United States of America, 106*, 5529–5533.

Chelune, G. J., & Baer, R. A. (1986). Developmental norms for the Wisconsin Card Sorting test. *Journal of Clinical and Experimental Neuropsychology, 8*, 219–228.

Chevalier, N. (2015a). Executive function development: Making sense of the environment to behave adaptively. *Current Directions in Psychological Science, 24*, 363–368.

Chevalier, N. (2015b). The development of executive function: Toward more optimal coordination of control with age. *Child Development Perspectives, 9*, 239–244.

Chevalier, N., & Blaye, A. (2009). Setting goals to switch between tasks: Effect of cue transparency on children's cognitive flexibility. *Developmental Psychology, 45*, 782–797.

Chevalier, N., & Blaye, A. (2016). Metacognitive monitoring of executive control engagement during childhood. *Child Development, 87*, 1264–1276.

Chevalier, N., Chatham, C. H., & Munakata, Y. (2014). The practice of going helps children to stop: The importance of context monitoring in inhibitory control. *Journal of Experimental Psychology: General, 143*, 959–965.

Chevalier, N., Dauvier, B., & Blaye, A. (in press). From prioritizing objects to prioritizing cues: A developmental shift for cognitive control. *Developmental Science*.

Chevalier, N., Huber, K. L., Wiebe, S. A., & Espy, K. A. (2013). Qualitative change in executive control during childhood and adulthood. *Cognition, 128*, 1–12.

Chevalier, N., James, T. D., Wiebe, S. A., Nelson, J. M., & Espy, K. A. (2014). Contribution of reactive and proactive control to children's working memory performance: Insight from item recall durations in response sequence planning. *Developmental Psychology, 50*, 1999–2008.

Chevalier, N., Martis, S. B., Curran, T., & Munakata, Y. (2015). Metacognitive processes in executive control development: The case of reactive and proactive control. *Journal of Cognitive Neuroscience, 27*, 1125–1136.

Chevalier, N., Sheffield, T. D., Nelson, J. M., Clark, C. A. C., Wiebe, S. A., & Espy, K. A. (2012). Underpinnings of the costs of flexibility in preschool children: The roles of inhibition and working memory. *Developmental Neuropsychology, 37*, 99–118.

Clark, C. A. C., Nelson, J. M., Garza, J., Sheffield, T. D., Wiebe, S. A., & Espy, K. A. (2014). Gaining control: Changing relations between executive control and processing speed and their relevance for mathematics achievement over course of the preschool period. *Frontiers in Psychology, 5*, 107.

Clark, C. A. C., Sheffield, T. D., Chevalier, N., Nelson, J. M., Wiebe, S. A., & Espy, K. A. (2013). Charting early trajectories of executive control with the Shape School. *Developmental Psychology, 49*, 1481–1493.

Clark, C. A. C., Sheffield, T. D., Wiebe, S. A., & Espy, K. A. (2013). Longitudinal associations between executive control and developing mathematical competence in preschool boys and girls. *Child Development, 84*, 662–677.

Clark, C. A. C., & Woodward, L. J. (2014). Relation of perinatal risk and early parenting to executive control at the transition to school. *Developmental Science, 18*, 525–542.

Cowan, N. (2015). Working memory maturation: Can we get at the essence of cognitive growth? *Perspectives on Psychological Science, 11*, 239–264.

Cragg, L. (2015). The development of stimulus and response interference control in mid-childhood. *Developmental Psychology, 52*, 242–252.

Cuevas, K., & Bell, M. A. (2014). Infant attention and early childhood executive function. *Child Development, 85*, 397–404.

Daly, M., Delaney, L., Egan, M., & Baumeister, R. F. (2015). Childhood self-control and unemployment throughout the life span: Evidence from two British cohort studies. *Psychological Science, 26*, 709–723.

Dauvier, B., Chevalier, N., & Blaye, A. (2012). Using finite mixture of GLMs to explore variability in children's flexibility in a task-switching paradigm. *Cognitive Development, 27*, 440–454.

Diamond, A. (1990). The development and neural bases of memory functions as indexed by the AB and delayed response tasks in human infants and infant monkeys. *Annals of the New York Academy of Sciences, 608*, 267–309.

Diamond, A. (2012). Activities and programs that improve children's executive functions. *Current Directions in Psychological Science, 21*, 335–341.

Doebel, S., & Zelazo, P. D. (2015). A meta-analysis of the Dimensional Change Card Sort: Implications for developmental theories and the measurement of executive function in children. *Developmental Review, 38*, 241–268.

DuPuis, D., Ram, N., Willner, C. J., Karalunas, S., Segalowitz, S. J., & Gatzke-Kopp, L. M. (2015). Implications of ongoing neural development for the measurement of the error-related negativity in childhood. *Developmental Science, 18*, 452–468.

Durston, S., Davidson, M. C., Tottenham, N., Galvan, A., Spicer, J., Fossella, J. A., & Casey, B. J. (2006). A shift from diffuse to focal cortical activity with development. *Developmental Science, 9*, 1–8.

Fair, D. A., Dosenbach, N. U. F., Church, J. A., Cohen, A. L., Brahmbhatt, S., Miezin, F. M., et al. (2007). Development of distinct control networks through segregation and integration. *Proceedings of the National Academy of Sciences of the United States of America, 104*, 13507–13512.

Fatzer, S. T., & Roebers, C. M. (2012). Language and executive functions: The effect of articulatory suppression on executive functioning in children. *Journal of Cognition and Development, 13*, 454–472.

Fisher, A. V., Godwin, K. E., & Seltman, H. (2014). Visual environment, attention allocation, and learning in young children: When too much of a good thing may be bad. *Psychological Science, 25*, 1362–1370.

Friedman, N. P., & Miyake, A. (2016). Unity and diversity of executive functions: Individual differences as a window on cognitive structure. *Cortex, 86*, 186–204.

Friedman, N. P., Miyake, A., Robinson, J. L., & Hewitt, J. K. (2011). Developmental trajectories in toddlers' self-restraint predict individual differences in executive functions 14 years later: A behavioral genetic analysis. *Developmental Psychology, 47*, 1410–1430.

Friedman, N. P., Miyake, A., Young, S. E., Defries, J. C., Corley, R. P., & Hewitt, J. K. (2008). Individual differences in executive functions are almost entirely genetic in origin. *Journal of Experimental Psychology. General, 137*, 201–225.

Fuhs, M. W., & Day, J. D. (2010). Verbal ability and executive functioning development in preschoolers at head start. *Developmental Psychology, 47*, 404–416.

Fuhs, M. W., Nesbitt, K. T., Farran, D. C., & Dong, N. (2014). Longitudinal associations between executive functioning and academic skills across content areas. *Developmental Psychology, 50*, 1698–1709.

Gandolfi, E., Viterbori, P., Traverso, L., & Usai, M. C. (2014). Inhibitory processes in toddlers: A latent-variable approach. *Frontiers in Psychology, 5*, 381.

Gathercole, S. E., Pickering, S. J., Ambridge, B., & Wearing, H. (2004). The structure of working memory from 4 to 15 years of age. *Developmental Psychology, 40*, 177–190.

Gogtay, N., Giedd, J. N., Lusk, L., Hayashi, K. M., Greenstein, D., Vaituzis, A. C., et al. (2004). Dynamic mapping of human cortical development during childhood through early adulthood. *Proceedings of the National Academy of Sciences of the United States of America, 101*, 8174–8179.

Gooch, D., Thompson, P., Nash, H. M., Snowling, M. J., & Hulme, C. (2016). The development of executive function and language skills in the early school years. *Journal of Child Psychology and Psychiatry, 57*, 180–187.

Gopnik, A., Griffiths, T. L., & Lucas, C. G. (2015). When younger learners can be better (or at least more open-minded) than older ones. *Current Directions in Psychological Science, 24*, 87–92.

Halperin, J. M., Marks, D. J., Bedard, A. C., Chacko, A., Curchack, J. T., Yoon, C. A., & Healey, D. M. (2013). Training executive, attention, and motor skills: A proof-of-concept study in preschool children with ADHD. *Journal of Attention Disorders, 17*, 711–721.

Hughes, C. (1998). Finding your marbles: Does preschoolers' strategic behavior predict later understanding of mind? *Developmental Psychology, 34*, 1326–1339.

Huizinga, M., Dolan, C. V., & van der Molen, M. W. (2006). Age-related change in executive function: Developmental trends and a latent variable analysis. *Neuropsychologia, 44*, 2017–2036.

Kirk, H. E., Gray, K., Riby, D. M., & Cornish, K. M. (2015). Cognitive training as a resolution for early executive function difficulties in children with intellectual disabilities. *Research in Developmental Disabilities, 38*, 145–160.

Lamm, C., Walker, O. L., Degnan, K. A., Henderson, H. A., Pine, D. S., McDermott, J. M., & Fox, N. A. (2014). Cognitive control moderates early childhood temperament in predicting social behavior in 7-year-old children: An ERP study. *Developmental Science, 17*, 667–681.

Lee, K., Bull, R., & Ho, R. M. H. (2013). Developmental changes in executive functioning. *Child Development, 84*, 1933–1953.

Lehto, J. E., Juujärvi, P., Kooistra, L., & Pulkkinen, L. (2003). Dimensions of executive functioning: Evidence from children. *British Journal of Developmental Psychology, 21*, 59–80.

Linares, R., Bajo, M. T., & Pelegrina, S. (2016). Age-related differences in working memory updating components. *Journal of Experimental Child Psychology, 147*, 39–52.

Lucenet, J., & Blaye, A. (2014). Age-related changes in the temporal dynamics of executive control: A study in 5- and 6-year-old children. *Frontiers in Psychology, 5*, 831.

Luciana, M., Conklin, H. M., Hooper, C. J., & Yarger, R. S. (2005). The development of nonverbal working memory and executive control processes in adolescents. *Child Development, 76*, 697–712.

Luna, B., Padmanabhan, A., & O'Hearn, K. (2010). What has fMRI told us about the development of cognitive control through adolescence? *Brain and Cognition, 72*, 101–113.

Miller, E. K., & Cohen, J. D. (2001). An integrative theory of prefrontal cortex function. *Annual Review of Neuroscience, 24*, 167–202.

Miller, M. R., Giesbrecht, G. F., Müller, U., McInerney, R. J., & Kerns, K. A. (2012). A latent variable approach to determining the structure of executive function in preschool children. *Journal of Cognition and Development, 13*, 395–423.

Miller, S. E., & Marcovitch, S. (2015). Examining executive function in the second year of life: Coherence, stability, and relations to joint attention and language. *Developmental Psychology, 51*, 101–114.

Miyake, A., Friedman, N. P., Emerson, M. J., Witzki, A. H., Howerter, A., & Wager, T. D. (2000). The unity and diversity of executive functions and their contributions to complex "frontal lobe" tasks: A latent variable analysis. *Cognitive Psychology, 41*, 49–100.

Moffitt, T. E., Arseneault, L., Belsky, D., Dickson, N., Hancox, R. J., Harrington, H., et al. (2011). A gradient of childhood self-control predicts health, wealth, and public safety. *Proceedings of the National Academy of Sciences of the United States of America, 108*, 2693–2698.

Morton, J. B., & Munakata, Y. (2002). Active versus latent representations: A neural network model of perseveration, dissociation, and decalage. *Developmental Psychobiology, 40*, 255–265.

Müller, U., Zelazo, P. D., Hood, S., Leone, T., & Rohrer, L. (2004). Interference control in a new rule use task: Age-related changes, labeling, and attention. *Child Development, 75*, 1594–1609.

Munakata, Y., Snyder, H. R., & Chatham, C. H. (2012). Developing cognitive control: Three key transitions. *Current Directions in Psychological Science, 21*, 71–77.

Perone, S., Molitor, S. J., Buss, A. T., Spencer, J. P., & Samuelson, L. K. (2015). Enhancing the executive functions of 3-year-olds in the Dimensional Change Card Sort task. *Child Development, 86*, 812–827.

Roebers, C. M., & Feurer, E. (2015). Linking executive functions and procedural metacognition. *Child Development Perspectives, 10*, 39–44.

Romine, C. B., & Reynolds, C. R. (2005). A model of the development of frontal lobe functioning: Findings from a meta-analysis. *Applied Neuropsychology, 12*, 190–201.

Rosenthal, M., Wallace, G. L., Lawson, R., Wills, M. C., Dixon, E., Yerys, B. E., & Kenworthy, L. (2013). Impairments in real-world executive function increase from childhood to adolescence in autism spectrum disorders. *Neuropsychology, 27*, 13–18.

Rueda, M. R., Rothbart, M. K., McCandliss, B. D., Saccomanno, L., & Posner, M. I. (2005). Training, maturation, and genetic influences on the development of executive attention. *Proceedings of the National Academy of Sciences of the United States of America, 102*, 14931–14936.

Sabbagh, M. A., Xu, F., Carlson, S. M., Moses, L. J., & Lee, K. (2006). The development of executive functioning and theory of mind: A comparison of Chinese and U.S. preschoolers. *Psychological Science, 17*, 74–81.

Shenhav, A., Botvinick, M. M., & Cohen, J. D. (2013). The expected value of control: An integrative theory of anterior cingulate cortex function. *Neuron, 79*, 217–240.

Shing, Y. L., Lindenberger, U., Diamond, A., Li, S.-C., & Davidson, M. C. (2010). Memory maintenance and inhibitory control differentiate from early childhood to adolescence. *Developmental Neuropsychology, 35*, 679–697.

Snyder, H. R., & Munakata, Y. (2010). Becoming self-directed: Abstract representations support endogenous flexibility in children. *Cognition, 116*, 155–167.

St. Clair-Thompson, H. L., & Gathercole, S. E. (2006). Executive functions and achievements in school: Shifting, updating, inhibition, and working memory. *Quarterly Journal of Experimental Psychology, 59*, 745–759.

Thompson-Schill, S. L., Ramscar, M., & Chrysikou, E. G. (2009). Cognition without control: When a little frontal lobe goes a long way. *Current Directions in Psychological Science, 18*, 259–263.

Titz, C., & Karbach, J. (2014). Working memory and executive functions: Effects of training on academic achievement. *Psychological Research, 78*, 852–868.

van der Sluis, S., de Jong, P. F., & van der Leij, A. (2007). Executive functioning in children, and its relations with reasoning, reading, and arithmetic. *Intelligence, 35*, 427–449.

Wass, S. V., Scerif, G., & Johnson, M. H. (2012). Training attentional control and working memory: Is younger, better? *Developmental Review, 32*, 360–387.

Wiebe, S. A., Espy, K. A., & Charak, D. (2008). Using confirmatory factor analysis to understand executive control in preschool children: I. Latent structure. *Developmental Psychology, 44*, 575–587.

Wiebe, S. A., Lukowski, A. F., & Bauer, P. J. (2010). Sequence imitation and reaching measures of executive control: A longitudinal examination in the second year of life. *Developmental Neuropsychology, 35*, 522–538.

Wiebe, S. A., Sheffield, T. D., & Espy, K. A. (2012). Separating the fish from the sharks: A longitudinal study of preschool response inhibition. *Child Development, 83*, 1245–1261.

Wiebe, S. A., Sheffield, T., Nelson, J. M., Clark, C. A. C., Chevalier, N., & Espy, K. A. (2011). The structure of executive function in 3-year-olds. *Journal of Experimental Child Psychology, 108*, 436–452.

Willoughby, M. T., Blair, C. B., Wirth, R. J., & Greenberg, M. (2012). The measurement of executive function at age 5: Psychometric properties and relationship to academic achievement. *Psychological Assessment, 24*, 226–239.

Xu, F., Han, Y., Sabbagh, M. A., Wang, T., Ren, X., & Li, C. (2013). Developmental differences in the structure of executive function in middle childhood and adolescence. *PloS One, 8*, e77770.

Zelazo, P. D., Müller, U., Frye, D., & Marcovitch, S. (2003). The development of executive function in early childhood. *Monographs of the Society for Research in Child Development, 68*(3), 1–136.

3

EXECUTIVE FUNCTION DEVELOPMENT IN ADOLESCENCE

Eveline A. Crone, Sabine Peters, & Nikolaus Steinbeis

Introduction

Executive function is defined as the ability to control thoughts and actions for the purpose of achieving future goals (Diamond, 2013). It relies on our ability to keep information in mind (working memory), to respond to a changing environment (cognitive flexibility), and to stop inappropriate actions or impulses (inhibition). When these functions work well in concert, this allows individuals to plan ahead and respond flexibly to a changing environment (Miyake et al., 2000). In this chapter we will focus on three questions: (1) how do executive functions develop during adolescence, (2) can executive functions be trained in adolescence, and (3) what are the implications of executive functions for school settings such as reading and mathematics performance?

The focus is on adolescence, which is a key transition period in development. Adolescence is defined broadly as the age range of 10–22 years, during which children show changes in physical, cognitive, affective, and social domains (Steinberg, 2008). The first phase of adolescence is referred to as puberty, which spans approximately ages 10–14 years. Puberty starts roughly 1.5 years earlier for girls than for boys, and varies in onset between individuals (Shirtcliff, Dahl, & Pollak, 2009). Puberty is associated with extensive hormonal changes, which affects the physical appearance of adolescents, but also has effects on social-affective processes. Studies have shown that hormones have an impact on brain development, such that adolescents who are more advanced in puberty have more mature brain connectivity patterns (Ladouceur, Peper, Crone, & Dahl, 2012). The second phase of adolescence is defined as the age period of approximately 16–22 years, during which individuals develop mature cognitive and social goals and become independent members of society (Crone & Dahl, 2012).

The Development of Executive Functions in Adolescence

Whereas much is known about the development of executive functions in early and mid-childhood (Chevalier & Clark, Chapter 2, this volume; Cuevas, Rajan, & Bryant, Chapter 1, this volume; Davidson, Amso, Anderson, & Diamond, 2006), relatively less attention has been devoted to the development of executive functions in adolescence. One of the seminal articles on executive function development across childhood and adolescence distinguished between three basic executive functions: working memory, shifting, and inhibition (Huizinga, Dolan, & van der Molen, 2006), following Miyake et al.'s (2000) model on unity and diversity of executive functions in adults. The researchers estimated latent factors based on the shared variance between tasks that tapped into each of the three basic functions. A battery of tasks was administered to participants of the following ages: 7 years (n=71), 11 years (n=101), 15 years (n=111), and 21 years (n=94). Latent factors were obtained for working memory and shifting, and the results showed that working memory developed until the age of 21, whereas shifting was mature at age 15 (Huizinga et al., 2006). These findings are consistent with other studies that have demonstrated that working memory performance increases over the whole period of adolescent development (Asato, Sweeney, & Luna, 2006; Luna, Garver, Urban, Lazar, & Sweeney, 2004). In this same study, no common latent factor could be derived for the inhibition tasks (Huizinga et al., 2006). Inhibition was assessed using a stop-signal task, which requires the inhibition of a prepotent response; a flanker task, which requires inhibition of irrelevant information; and a Stroop task, which requires inhibition of a conflicting stimulus property. Apparently, inhibition is a multi-faceted construct and different types of inhibition do not necessarily correlate with each other. The developmental patterns were also distinct. Stop-signal inhibition and flanker inhibition developed until age 15, consistent with several other studies that reported that inhibition improves until early adolescence (Prencipe et al., 2011).

Recent research showed that from late adolescence to early adulthood the three-factor structure of working memory, cognitive flexibility, and inhibition is quite stable. This was evident in a 6-year longitudinal study that followed 840 individuals (Friedman et al., 2016). These findings suggest that when entering adulthood, executive functions are no longer changing.

An important question concerns whether these relatively basic executive functions have predictive value for situations in which multiple executive functions need to be combined. For this purpose, basic executive functions tasks, or the latent factors of these functions, can be used to test how well these basic executive functions predicted performance on more complex executive function tasks, such as the Wisconsin Card Sorting Task (WCST) and the Tower of London (ToL) task. The WCST requires participants to sort cards according to a certain response rule (e.g., match on color), and participants need to flexibly use feedback to switch between sorting rules (Milner, 1963). The ToL task requires response planning by reordering visuospatial stimuli (Asato et al., 2006). Prior studies reported that working memory is a significant predictor of WCST performance (Huizinga et al., 2006), and that flanker inhibition is a predictor for performance on the ToL, showing that this basic executive function has explanatory power for complex executive function tasks (see also Asato et al., 2006).

Taken together, working memory shows a protracted development, which occurs over the whole period of adolescence, whereas inhibition and shifting show most pronounced improvement in early adolescence after which it stabilizes (Huizinga et al., 2006; Luna et al., 2004). These functions in turn predict performance on more complex flexibility and planning tasks.

Four recent studies tested how executive functions develop from early to late adolescence and seem to point toward optimal performance and more flexibility in adolescence relative to other ages (see Figure 3.1). One study examined cognitive learning based on performance feedback and working memory development in 208 participants between ages 8 and 27 who were tested longitudinally across two measurements that were separated by 2 years. The researchers found that both cognitive learning and working memory peaked around 17 years of age, after which there was stable performance for some individuals and declines for other individuals (Peters, Van Duijvenvoorde, Koolschijn, & Crone, 2016). A second study examined divergent thinking and creative insight in children, adolescents, and adults using a battery of verbal and spatial tasks. Whereas creative insight increased during early adolescence and stabilized in mid-adolescence, spatial divergent thinking was best in 15–16-year-old adolescents, after which it decreased in young adults (Kleibeuker, De Dreu, & Crone, 2013). Comparable findings were obtained for another spatial divergent task where 16–17-year-olds slightly outperformed 25–30-year-old participants (Kleibeuker, Koolschijn, et al., 2013). A fourth study examined age differences in a range of cognitive functions, including executive functions, across the whole age range of adolescence (ages 8–21, n = 9,138). This study had a slightly different focus by testing age differences in within-subject variability in performance. They found that within-subject variability decreased between ages 8 and 13, after which it gradually increased again from age 17 to 21. Possibly this suggests increased flexibility in whether or not executive functions are recruited for specific tasks. In other words, increased instability of cognitive functioning may indicate heightened amenability to environmental influences, because behavior is less stable. Interestingly, after age 13 gender differences emerged, with larger variability in males than in females (Roalf et al., 2014).

The findings indicative of optimal performance and increased variability during late adolescence suggest that this developmental stage may be an optimal period for the development of executive functions, marking this time as particularly receptive for input. As such it may be a tipping point for positive vs.

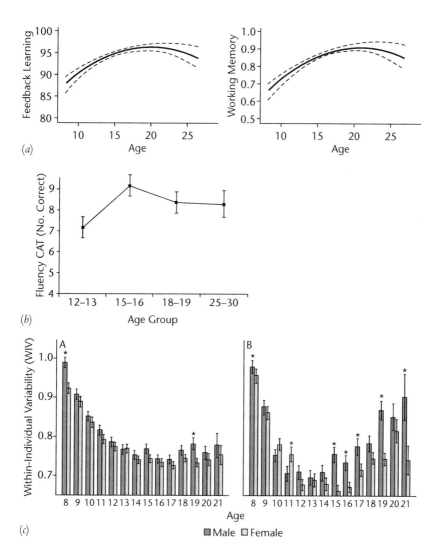

FIGURE 3.1 (*a*) Predicted patterns (solid lines, and 95% confidence intervals in dashed lines) of cognitive learning and working memory based on 2-year longitudinal data from 208 participants between ages 8 and 27 (reprinted with permission from Peters et al., 2016); (*b*) 4 age groups compared cross-sectionally on visual-spatial fluency showing best performance at age 15–16 (reprinted with permission from Kleibeuker et al., 2013); (*c*) within-individual variability in accuracy (left) and speed (right) decreases and increases between ages 8 and 21 (reprinted with permission from Roalf et al., 2014).

negative developmental trajectories (Crone & Dahl, 2012). This leads to the question whether adolescence is also a sensitive period in which executive functions can be enhanced.

Can Executive Functions Be Trained in Adolescence?

Whether, and the extent to which, executive functions can be trained has been the subject of intense empirical research (Kliegel, Hering, Ihle, & Zuber, Chapter 13, this volume). The key question in these studies is not only whether cognitive skills can be trained, but also whether such training improves not only the trained task (near transfer), but also cognitive functioning in daily life (e.g., in school; far transfer). Initially, researchers presumed that transfer effects were limited. These effects were based, for example, on a large study including 11,430 healthy adult participants that examined the effects of 6-week cognitive training in reasoning, memory, planning, visuospatial skills, and attention. Whereas participants improved considerably on the trained task, the evidence for transfer to other domains (such as domains important for school performance) was limited (Owen et al., 2010). Furthermore, a developmental study that used a working memory practice task with varying difficulty levels also found significant practice benefits in the trained domain in 13-year-old adolescents, but no transfer effects to other domains (Jolles, van Buchem, Rombouts, & Crone, 2012). Similar effects were obtained for practice tasks in 7–9-year-old children (Dunning, Holmes, & Gathercole, 2013).

However, recent insights show that whether far-transfer effects can be observed after cognitive training depends on what is actually being trained. Improved performance during training can be attributed to improvements in an underlying cognitive skill or merely reflect the use of an effective strategy for the specific task being trained (Diamond & Ling, 2016). Seeing that strategies are developed in response to meet the specific demands of a certain task, the likelihood of a far-transfer effect is much smaller than in the case of training an underlying cognitive skill (see also Figure 3.2). It has recently been argued that training needs to be continually novel, complex, and diverse and be motivating to maximize transfer to other domains (Moreau & Conway, 2014). Indeed, several studies found support for transfer effects of training in adults as long as the training was adaptive to the performance level of the participants (i.e., the task becomes more difficult when performance is improving) (Constantinidis & Klingberg, 2016; Corbett et al., 2015).

Given that executive functions are still improving considerably and reaching stability across adolescence, this may be a period where adaptive training has most benefits. In terms of studies on adolescence, studies have mostly focused on the domain of working memory, and observed that adaptive training of working memory was associated with performance improvements on the trained task and showed transfer to reasoning skills in participants aged 7–15 with attention deficit

FIGURE 3.2 Schematic overview of outcome, transfer type, and neural mechanisms of training properties. Whereas repetitive training schedules will likely only train task-related strategies, this can lead to near transfer effects as long as other tasks require similar strategies. Such training effects will likely result in the recruitment of additional strategy specific brain regions. When training schedules are adaptive, novel, complex, and diverse there is a greater likelihood of an underlying skill being trained, with more extensive effects for far transfer. Such training effects will likely result in strengthening of functional networks required for training.

hyperactivity disorder (ADHD) (Klingberg, Forssberg, & Westerberg, 2002), and in participants aged 7–12 with ADHD (Klingberg et al., 2005). Training effects were also found for adaptive training programs in 7–9-year-old children with poor working memory with transfer to other working memory tasks (Dunning et al., 2013; Holmes, Gathercole, & Dunning, 2009).

It has recently been suggested that training and transfer effects might be enhanced during specific developmental periods (Kliegel et al., Chapter 13, this volume). A recent meta-analysis on working memory training indeed showed that age was a significant moderator variable of the extent of training benefits, with larger effects observed at younger compared to older ages (Melby-Lervag & Hulme, 2013). Another meta-analysis of studies into training executive functions and their transfer effects also looked at far-transfer effects with age in an age range from 1 to 90 years (Wass, Scerig, & Johnson, 2012). The authors showed that transfer effects decline with age and are largest during childhood and early adolescence. None of these studies, however, included mid- to late adolescents, so it remains to be determined how training is beneficial in this specific age range.

What Are the Implications of Executive Function Training for Scholarly Tasks?

Probably one of the most important forms of transfer for children and adolescents is transfer to school performance. It may be beneficial to train executive functions to improve general abilities such as reading and mathematics. Several studies have examined the relation between executive functions and scholastic performance measures such as reading and mathematics as outcomes measures (Alloway & Alloway, 2010; Blair & Razza, 2007). One of the executive functions that plays arguably the most vital role in academic learning is working memory. Children and adolescents who have poor working memory are at risk for poor school performance, as these children often also perform poorly on reading and mathematics (Blair & Razza, 2007; Gathercole, Pickering, Knight, & Stegmann, 2004; Geary, Hoard, Byrd-Craven, & DeSoto, 2004; Raghubar, Barnes, & Heckt, 2010; Swanson & Jerman, 2006).

In recent years the question has been addressed whether adaptive training of executive functions also has benefits for academic performance. One study showed that 9–10-year-old children who participated in a working memory training study showed benefits in the trained domain, and additionally showed better mathematics performance than the control group 6 months later (Holmes et al., 2009). It should be noted that the control group was not included in the 6-month follow-up, so it is possible that some of the effects could be related to test–retest effects. Another recent study examined if these training effects were also observed in more ecologically valid settings. In this study, 9–11-year-old children followed teacher-administered working memory training. Improvements were observed in both working memory performance and academic performance including mathematics (Holmes & Gathercole, 2014). A study that included children aged 8–11 observed that working memory training was associated with improvement in reading (Loosli, Buschkuehl, Perrig, & Jaeggi, 2012). Similar effects were found for younger children aged 7–9 following 14 sessions of adaptive working memory training (Karbach, Strobach, & Schubert, 2015). A training study with children with special educational needs and attention problems aged 9–12 also showed benefits of reading comprehension that persisted until at least 6 months after the training had ended (Dahlin, 2011). The same author found improved performance in mathematics following 5-week working memory training in boys with attentional problems aged 9–12 (Dahlin, 2013).

These findings suggest that working memory training can be beneficial in a school context. It is yet to be determined if the same training benefits are also observed when working memory is trained in adolescents rather than children, and there is still debate about effects sizes of training outcomes (Gathercole, Dunning, & Holmes, 2012; Hulme & Melby-Lervag, 2012; Roberts et al., 2016), but this research provides promising starting points.

The Neural Correlates of Executive Functions

One way to better understand which mechanisms develop in which order and how they contribute to real-life cognitive functioning is to study neural development (Moriguchi, Chapter 5, this volume; West, Chapter 6, this volume). The brain matures considerably during child and adolescent development, in terms of its structure, connections, and functions (Mills & Tamnes, 2014). When it is shown that training enhances cognitive functioning, neuroimaging may help to determine exactly which process is being improved by the training, such as through increasing working memory capacity, switching to a better strategy, or becoming better at suppressing interfering stimuli. This might also help in determining which types of training will result in far-transfer effects. Brain imaging and patient studies have consistently observed that executive functions put demands on the frontal and parietal cortices (Miller & Cohen, 2001).

Executive functions such as working memory, inhibition, and switching put demands on overlapping regions within the frontal cortex (specifically dorsolateral and ventrolateral prefrontal cortex), suggesting that despite the differences between these processes, they also rely on overlapping neural areas (Kim, Cilles, Johnson, & Gold, 2012; Niendam et al., 2012).

Developmental comparison studies have provided consistent evidence for protracted development of brain activity across adolescence in the frontal-parietal network for several executive functions including working memory (Satterthwaite et al., 2013) and inhibition (Rubia et al., 2013; Vink et al., 2014). This pattern was also observed for higher-order cognitive learning tasks that placed demands on several executive functions (Peters et al., 2016). Neural responses in the prefrontal and parietal cortex during executive functioning tasks were found to be predictive for academic tasks such as mathematics in participants aged 6–16. Notably, neural activity predicted future mathematical performance even better than behavioral testing alone (Dumontheil & Klingberg, 2012). Similar results were previously found in a study showing that structural and functional brain imaging measures were additional predictors of children's gains in reading decoding ability over the school year beyond behavioral measures alone (Hoeft et al., 2007). Together, these studies suggest that the core network in the brain that is important for executive functions in adulthood is developing until late adolescence and seems relevant for predicting school performance.

The theory of interactive specialization describes the way in which these neurodevelopmental changes may take place (Johnson, 2011). According to this theory, at a young age children use cortical regions for a variety of cognitive functions. Over the course of child and adolescent development, brain regions become more specialized to perform a specific cognitive function. This specialization occurs by a process of activity-dependent interaction and competition between regions. This theory makes important predictions for developmental changes in the influence of training on neural functions and resulting transfer effects. Thus, the wider the set of cognitive functions associated with a brain region (as is the case in younger children and adolescents), the more extensive the transfer should be following training.

Brain Plasticity and Training Executive Functions

Insight into the mechanisms of training executive functions has been obtained by examining changes in brain responses before and after training. It is thought that in adults training of working memory increases connections between these brain regions, which results in more efficient processing (Constantinidis & Klingberg, 2016).

Very few studies have examined neural responses before and after training in children and adolescents. One prior study used a working memory practice paradigm (non-adaptive) with several hours of training during 6 weeks, which was administered in 12–13-year-old adolescents (Jolles, Grol, Van Buchem, Rombouts, & Crone, 2010). Before training, participants showed less activity in dorsolateral prefrontal cortex than adults. After 6 weeks of training, activity in dorsolateral prefrontal cortex was enhanced in adolescents and they no longer differed in activity patterns and behavioral performance from adults. It should be noted that there was no evidence for transfer to other domains, so possibly this study was more successful in training a strategy suitable to the specific trained task, rather than a broader underlying skill.

A recent study examined effects of working memory training on brain connectivity in children aged 8–11 by studying resting state magnetoencephalography (MEG) changes, a direct measure of neural activity. In this study, 13 children received 20–25 sessions of adaptive working memory training over a period of 4–5 weeks, and 14 children received placebo (non-adaptive) training. The adaptive training group increased more in performance compared to the placebo group on the trained task, and moreover showed transfer to a non-trained working memory task. The trained group also showed increases in connectivity patterns between right lateral prefrontal cortex and left occipital cortex. The extent of connectivity changes was predictive of training outcomes (Astle, Barnes, Baker, Colclough, & Woolrich, 2015). This study provides the first direct measure of training effects on the neural level in children and early adolescents.

Longitudinal studies can also provide important insight into mechanisms that predict change over time. A longitudinal imaging study examined how neural activity during a working memory task was related to changes in working memory capacity 2 years later. This study found that activity in and white matter connectivity with the striatum was predictive of future working memory capacity (Darki & Klingberg, 2015). Interestingly, a study in adults also found that training was associated with increases in striatum activity (Jolles et al., 2010). An important neurotransmitter that is often associated with working memory training, dopamine, has its main receptors in the striatum (Darki &

Klingberg, 2015; Ullman, Almeida, & Klingberg, 2014), so possibly this implies that the striatum is an important region in the development and plasticity of working memory development.

One possible explanation for these developmental shifts in brain activity patterns is that in the developing brain, children rely on a wider network of brain regions to improve in certain cognitive functions. Mere practice may result in more activity in regions such as dorsolateral prefrontal cortex (Jolles et al., 2012). Larger executive control improvements over time may be associated with reliance on compensatory brain regions such as the striatum (Backman et al., 2011; Darki & Klingberg, 2015) and connections with regions outside prefrontal cortex (Astle et al., 2015). Following predictions from the interactive specialization theory, adaptive training in the developing brain may result in activity changes in a wider network than in adults, or, alternatively, in more specialized activity (reflecting more adult-like activity) in a localized network of brain regions (Johnson, 2011). These are all questions that should be addressed in future studies.

Future Directions

Several compelling questions arise from these studies that should be addressed in future research. First, even though many behavioral studies have focused on development and plasticity in children, early adolescents, and adults, there are few studies that examined training effects in mid- to late adolescence. Yet, developmental studies suggest that this may be an optimal window for working memory training, given possible enhanced cognitive performance (Peters et al., 2016), variability (Roalf et al., 2014), and flexibility (Crone & Dahl, 2012). Future studies are needed to examine training effects in this important transition period in life (Hughes & Devine, Chapter 10, this volume; Berryman, Pothier, & Bherer, Chapter 12, this volume).

Second, little is yet understood about the commonalities between different executive functions and whether there are benefits to training them in concert rather than separately (Ang & Lee, 2010), even though neuroscience studies have shown that many of these functions put demands on the same neural regions (Niendam et al., 2012). To date, most studies have focused on working memory training, but possibly far-transfer effects to, for example, school performance will only be found using more complex executive functioning training. Previous work suggests that complexity is an integral part of the success of cognitive training (Diamond & Ling, 2016). While researchers tend to isolate specific components of executive functions, future studies ought to assess the interactive and super-additive components of combining training across executive function domains. These questions can be addressed by examining training effects of single domain and multiple domain programs and their effects on behavior and brain activity.

Conclusions

In the past decades the development of executive functions has been extensively studied, especially in young children and early adolescence, and to a lesser extent in mid- to late adolescence. These studies were consistent in showing that working memory improves over the teenage years, whereas inhibition and shifting develop before mid-adolescence (Huizinga et al., 2006; Luna et al., 2004; Roalf et al., 2014). Neuroscience studies have elaborated on the neural mechanisms underlying these developmental improvements. These studies found that the prefrontal and parietal cortices, which are important for executive functions in adults, show a protracted development during adolescence (Luna, Padmanabhan, & O'Hearn, 2010), which are correlated with performance changes (Peters et al., 2016; Satterthwaite et al., 2013). However, activities in other regions of the brain, such as the striatum, are important for performance improvements as well (Darki & Klingberg, 2015). Possibly, these changes can be understood in terms of the interactive specialization theory (Johnson, 2011). That is to say, children and adolescents may rely on a wider network of brain regions than adults during cognitive processes.

Important starting points have been made for understanding executive function development in terms of increased potential for flexibility and plasticity in adolescence. Using training paradigms, it was found that extensive and adaptive training enhances executive functions. The training is considered successful in case executive functions are trained as a skill, such that training transfers to other domains that make use of the same resources (Klingberg, 2010, 2014; Moreau & Conway, 2014).

Many of the studies to date have focused on children and adolescents who have poor working memory, such as children with learning disabilities or children with ADHD, and these groups show pronounced benefits of training (Spencer-Smith & Klingberg, 2015). An important question concerns whether training only leads to enhancements in children who are behind in development, or if executive functions can be trained in everybody. Also known as compensation vs. magnification accounts of the effects of training (Titz & Karbach, 2014), this calls for addressing questions of who benefits most from cognitive training and why. Unraveling these individual sensitivities will most likely be crucial to understand why there is variability in training outcomes and which type of training works for whom.

References

Alloway, T. P., & Alloway, R. G. (2010). Investigating the predictive roles of working memory and IQ in academic attainment. *Journal of Experimental Child Psychology, 106*(1), 20–29.

Ang, S. Y., & Lee, K. (2010). Exploring developmental differences in visual short-term memory and working memory. *Developmental Psychology, 46*(1), 279–285.

Asato, M. R., Sweeney, J. A., & Luna, B. (2006). Cognitive processes in the development of ToL performance. *Neuropsychologia, 44*(12), 2259–2269.

Astle, D. E., Barnes, J. J., Baker, K., Colclough, G. L., & Woolrich, M. W. (2015). Cognitive training enhances intrinsic brain connectivity in childhood. *Journal of Neuroscience, 35*(16), 6277–6283.

Backman, L., Nyberg, L., Soveri, A., Johansson, J., Andersson, M., Dahlin, E., et al. (2011). Effects of working-memory training on striatal dopamine release. *Science, 333*(6043), 718.

Blair, C., & Razza, R. P. (2007). Relating effortful control, executive function, and false belief understanding to emerging math and literacy ability in kindergarten. *Child Development, 78*(2), 647–663.

Constantinidis, C., & Klingberg, T. (2016). The neuroscience of working memory capacity and training. *Nature Review Neuroscience, 17*, 438–449.

Corbett, A., Owen, A., Hampshire, A., Grahn, J., Stenton, R., Dajani, S., et al. (2015). The effect of an online cognitive training package in healthy older adults: An online randomized controlled trial. *Journal of the American Medical Directors Association, 16*(11), 990–997.

Crone, E. A., & Dahl, R. E. (2012). Understanding adolescence as a period of social-affective engagement and goal flexibility. *Nature Review Neuroscience, 13*(9), 636–650.

Dahlin, K. I. (2011). Effects of working memory training on reading in children with special needs. *Reading and Writing, 24*, 479–491.

Dahlin, K. I. (2013). Working memory training and the effect of mathematical achievement in children with Attention Deficit and Special Needs. *Journal of Education and Learning, 2*, 118–133.

Darki, F., & Klingberg, T. (2015). The role of fronto-parietal and fronto-striatal networks in the development of working memory: A longitudinal study. *Cerebral Cortex, 25*(6), 1587–1595.

Davidson, M. C., Amso, D., Anderson, L. C., & Diamond, A. (2006). Development of cognitive control and executive functions from 4 to 13 years: Evidence from manipulations of memory, inhibition, and task switching. *Neuropsychologia, 44*(11), 2037–2078.

Diamond, A. (2013). Executive functions. *Annual Review of Psychology, 64*, 135–168.

Diamond, A., & Ling, D. S. (2016). Conclusions about interventions, programs, and approaches for improving executive functions that appear justified and those that, despite much hype, do not. *Developmental Cognitive Neuroscience, 18*, 34–48.

Dumontheil, I., & Klingberg, T. (2012). Brain activity during a visuospatial working memory task predicts arithmetical performance 2 years later. *Cerebral Cortex, 22*(5), 1078–1085.

Dunning, D. L., Holmes, J., & Gathercole, S. E. (2013). Does working memory training lead to generalized improvements in children with low working memory? A randomized controlled trial. *Developmental Science, 16*(6), 915–925.

Friedman, N. P., Miyake, A., Altamirano, L. J., Corley, R. P., Young, S. E., Rhea, S. A., & Hewitt, J. K. (2016). Stability and change in executive function abilities from late adolescence to early adulthood: A longitudinal twin study. *Developmental Psychology, 52*(2), 326–340.

Gathercole, S. E., Dunning, D. L., & Holmes, J. (2012). Cogmed training: Let's be realistic about intervention research. *Journal of Applied Research in Memory and Cognition, 1*(3), 201–203.

Gathercole, S. E., Pickering, S. J., Knight, C., & Stegmann, Z. (2004). Working memory skills and educational attainment: Evidence from national curriculum assessments at 7 and 14 years of age. *Applied Cognitive Psychology, 18*, 1–16.

Geary, D. C., Hoard, M. K., Byrd-Craven, J., & DeSoto, M. C. (2004). Strategy choices in simple and complex addition: Contributions of working memory and counting knowledge for children with mathematical disability. *Journal of Experimental Child Psychology, 88*(2), 121–151.

Hoeft, F., Ueno, T., Reiss, A. L., Meyler, A., Whitfield-Gabrieli, S., Glover, G. H., et al. (2007). Prediction of children's reading skills using behavioral, functional, and structural neuroimaging measures. *Behavioral Neuroscience, 121*(3), 602–613.

Holmes, J., & Gathercole, S. E. (2014). Taking working memory training from the laboratory into schools. *Educational Psychology, 34*(4), 440–450.

Holmes, J., Gathercole, S. E., & Dunning, D. L. (2009). Adaptive training leads to sustained enhancement of poor working memory in children. *Developmental Science, 12*(4), F9–15.

Huizinga, M., Dolan, C. V., & van der Molen, M. W. (2006). Age-related change in executive function: Developmental trends and a latent variable analysis. *Neuropsychologia, 44*(11), 2017–2036.

Hulme, C., & Melby-Lervag, M. (2012). Current evidence does not support the claims made for CogMed working memory training. *Journal of Applied Research in Memory and Cognition, 1*, 197–200.

Johnson, M. H. (2011). Interactive specialization: A domain-general framework for human functional brain development? *Developmental Cognitive Neuroscience, 1*(1), 7–21.

Jolles, D. D., Grol, M. J., Van Buchem, M. A., Rombouts, S. A., & Crone, E. A. (2010). Practice effects in the brain: Changes in cerebral activation after working memory practice depend on task demands. *NeuroImage, 52*(2), 658–668.

Jolles, D. D., van Buchem, M. A., Rombouts, S. A., & Crone, E. A. (2012). Practice effects in the developing brain: A pilot study. *Developmental Cognitive Neuroscience, 2 Suppl 1*, S180–191.

Karbach, J., Strobach, T., & Schubert, T. (2015). Adaptive working-memory training benefits reading, but not mathematics in middle childhood. *Child Neuropsychology, 21*, 285–301.

Kim, C., Cilles, S. E., Johnson, N. F., & Gold, B. T. (2012). Domain general and domain preferential brain regions associated with different types of task switching: A meta-analysis. *Human Brain Mapping, 33*(1), 130–142.

Kleibeuker, S. W., De Dreu, C. K., & Crone, E. A. (2013). The development of creative cognition across adolescence: Distinct trajectories for insight and divergent thinking. *Developmental Science, 16*(1), 2–12.

Kleibeuker, S. W., Koolschijn, P. C., Jolles, D. D., Schel, M. A., De Dreu, C. K., & Crone, E. A. (2013). Prefrontal cortex involvement in creative problem solving in middle adolescence and adulthood. *Developmental Cognitive Neuroscience, 5*, 197–206.

Klingberg, T. (2010). Training and plasticity of working memory. *Trends in Cognitive Sciences, 14*(7), 317–324.

Klingberg, T. (2014). Childhood cognitive development as a skill. *Trends in Cognitive Sciences, 18*(11), 573–579.

Klingberg, T., Fernell, E., Olesen, P. J., Johnson, M., Gustafsson, P., Dahlstrom, K., et al. (2005). Computerized training of working memory in children with ADHD: A randomized, controlled trial. *Journal of the American Academy of Child and Adolescent Psychiatry, 44*(2), 177–186.

Klingberg, T., Forssberg, H., & Westerberg, H. (2002). Training of working memory in children with ADHD. *Journal of Clinical Experimental Neuropsychology, 24*(6), 781–791.

Ladouceur, C. D., Peper, J. S., Crone, E. A., & Dahl, R. E. (2012). White matter development in adolescence: The influence of puberty and implications for affective disorders. *Developmental Cognitive Neuroscience, 2*(1), 36–54.

Loosli, S. V., Buschkuehl, M., Perrig, W. J., & Jaeggi, S. M. (2012). Working memory training improves reading processes in typically developing children. *Child Neuropsychology, 18*(1), 62–78.

Luna, B., Garver, K. E., Urban, T. A., Lazar, N. A., & Sweeney, J. A. (2004). Maturation of cognitive processes from late childhood to adulthood. *Child Development, 75*(5), 1357–1372.

Luna, B., Padmanabhan, A., & O'Hearn, K. (2010). What has fMRI told us about the development of cognitive control through adolescence? *Brain and Cognition, 72*(1), 101–113.

Melby-Lervag, M., & Hulme, C. (2013). Is working memory training effective? A meta-analytic review. *Developmental Psychology, 49*(2), 270–291.

Miller, E. K., & Cohen, J. D. (2001). An integrative theory of prefrontal cortex function. *Annual Review of Neuroscience, 24*, 167–202.

Mills, K. L., & Tamnes, C. K. (2014). Methods and considerations for longitudinal structural brain imaging analysis across development. *Developmental Cognitive Neuroscience, 9*, 172–190.

Milner, B. (1963). Effects of different brain lesions on card sorting. *Archives of Neurology, 9*, 100–110.

Miyake, A., Friedman, N. P., Emerson, M. J., Witzki, A. H., Howerter, A., & Wager, T. D. (2000). The unity and diversity of executive functions and their contributions to complex "Frontal Lobe" tasks: A latent variable analysis. *Cognitive Psychology, 41*(1), 49–100.

Moreau, D., & Conway, A. R. (2014). The case for an ecological approach to cognitive training. *Trends in Cognitive Sciences, 18*(7), 334–336.

Niendam, T. A., Laird, A. R., Ray, K. L., Dean, Y. M., Glahn, D. C., & Carter, C. S. (2012). Meta-analytic evidence for a superordinate cognitive control network subserving diverse executive functions. *Cognitive, Affective & Behavioral Neuroscience, 12*(2), 241–268.

Owen, A. M., Hampshire, A., Grahn, J. A., Stenton, R., Dajani, S., Burns, A. S., et al. (2010). Putting brain training to the test. *Nature, 465*(7299), 775–778.

Peters, S., Van Duijvenvoorde, A. C., Koolschijn, P. C., & Crone, E. A. (2016). Longitudinal development of frontoparietal activity during feedback learning: Contributions of age, performance, working memory and cortical thickness. *Developmental Cognitive Neuroscience, 19*, 211–222.

Prencipe, A., Kesek, A., Cohen, J., Lamm, C., Lewis, M. D., & Zelazo, P. D. (2011). Development of hot and cool executive function during the transition to adolescence. *Journal of Experimental Child Psychology, 108*(3), 621–637.

Raghubar, K. P., Barnes, M. A., & Heckt, S. A. (2010). Working memory and mathematics: A review of developmental, individual difference, and cognitive approaches. *Learning and Individual Differences, 20*(2), 110–122.

Roalf, D. R., Gur, R. E., Ruparel, K., Calkins, M. E., Satterthwaite, T. D., Bilker, W. B., et al. (2014). Within-individual variability in neurocognitive performance: Age- and sex-related differences in children and youths from ages 8 to 21. *Neuropsychology, 28*(4), 506–518.

Roberts, G., Quach, J., Spencer-Smith, M., Anderson, P. J., Gathercole, S., Gold, L., et al. (2016). Academic outcomes 2 years after working memory training for children with low working memory: A randomized clinical trial. *JAMA Pediatrics, 170*(5), e154568.

Rubia, K., Lim, L., Ecker, C., Halari, R., Giampietro, V., Simmons, A., et al. (2013). Effects of age and gender on neural networks of motor response inhibition: From adolescence to mid-adulthood. *NeuroImage, 83*, 690–703.

Satterthwaite, T. D., Wolf, D. H., Erus, G., Ruparel, K., Elliott, M. A., Gennatas, E. D., et al. (2013). Functional maturation of the executive system during adolescence. *Journal of Neuroscience, 33*(41), 16249–16261.

Shirtcliff, E. A., Dahl, R. E., & Pollak, S. D. (2009). Pubertal development: Correspondence between hormonal and physical development. *Child Development, 80*(2), 327–337.

Spencer-Smith, M., & Klingberg, T. (2015). Benefits of a working memory training program for inattention in daily life: A systematic review and meta-analysis. *PLoS One, 10*(3), e0119522.

Steinberg, L. (2008). *Adolescence*. New York: McGraw-Hill.

Swanson, H. L., & Jerman, O. (2006). Math disabilities: A selective meta-analysis of the literature. *Review of Educational Research, 76*(2), 249–274.

Titz, C., & Karbach, J. (2014). Working memory and executive functions: Effects of training on academic achievement. *Psychological Research, 78*(6), 852–868.

Ullman, H., Almeida, R., & Klingberg, T. (2014). Structural maturation and brain activity predict future working memory capacity during childhood development. *Journal of Neuroscience, 34*(5), 1592–1598.

Vink, M., Zandbelt, B. B., Gladwin, T., Hillegers, M., Hoogendam, J. M., van den Wildenberg, W. P., et al. (2014). Frontostriatal activity and connectivity increase during proactive inhibition across adolescence and early adulthood. *Human Brain Mapping, 35*(9), 4415–4427.

Wass, S., Scerig, G., & Johnson, M. H. (2012). Training attentional control and working memory: Is younger better? *Developmental Review, 32*, 360–387.

4

EXECUTIVE FUNCTION DEVELOPMENT IN AGING

Karen Z. H. Li, Kiran K. Vadaga, Halina Bruce, & Laurence Lai

Overview

In old age, a major hallmark of independent functioning is the ability to efficiently carry out specific tasks, be they physical or cognitive. This is achieved by planning and maintaining goals, flexibly coordinating or regulating component actions, and, in some cases, avoiding competing actions or thoughts (Baddeley, 1986; Norman & Shallice, 1986). These cognitive processes comprise what has been termed executive functions (EFs). The topic of age-related changes in EF has been an area of vigorous investigation for the past several decades (Luszcz, 2011; Rabbitt, 1997; Reuter-Lorenz, Festini, & Jantz, 2015).

Prior to explicit models of EF and aging, Hasher and Zacks (1979) proposed that controlled, effortful, or resource-limited cognitive processes are subject to age-related declines, whereas automatic processes are relatively preserved in later life. Similar ideas of age-related decline in effortful processing have been echoed in broader models of cognitive aging such as the Processing Resource Deficit Model (Craik & Byrd, 1982), Prefrontal Cortex Function Theory of Cognitive Aging (West, 1996), and Context Processing Theory of Cognitive Control (Braver et al., 2001).

Contemporary models of EF posit a multi-factorial construct (e.g., Engle, Tuholski, Laughlin, & Conway, 1999; Miyake et al., 2000) that includes several distinct but moderately correlated factors: for example, memory updating, task shifting/switching, response inhibition, and dual-task coordination. This diversity perspective is well-supported by lesion studies and brain imaging work showing distinct brain regions and neural circuits associated with each executive control process (e.g., Stuss, 2011). With the development of newer neuroscientific techniques, convergent evidence links volumetric loss and cortical thinning in prefrontal regions with chronological age, and with behavioral tests of EFs (e.g., Yuan & Raz, 2014). Indices of white matter degradation such as white matter hyperintensities (e.g., Gunning-Dixon & Raz, 2000) and fractional anisotropy (e.g., Grieve, Williams, Paul, Clark, & Gordon, 2007) are also associated with poor EFs, and underscore the known links between prefrontal regions and other cortical and subcortical regions (e.g., striatum, thalamus, cerebellum) in EF (Duncan & Owen, 2000).

On a behavioral level, the concept of EFs as unitary or multi-faceted has been informed by cross-sectional and longitudinal age studies (de Frias, Dixon, & Strauss, 2006, 2009; McFall, Sapkota, Thibeau, & Dixon, Chapter 17, this volume). In line with the principle of dedifferentiation (Garrett, 1938), or the increased covariation among previously independent factors in young adulthood, de Frias and colleagues considered multiple measures of inhibition and shifting in a sample of middle-aged to old adults, and arrived at a single-factor EF model (de Frias et al., 2006). Follow-up work revealed that cognitively elite older adults displayed a multi-factor structure resembling that of young adults (Miyake et al., 2000), whereas cognitively normal and impaired older adults displayed a more dedifferentiated, single-factor pattern of EFs (de Frias et al., 2009). Using a collection of experimental EF measures, Verhaeghen (2011) employed Brinley plot analyses to assess age-related deficits in selection, shifting, and divided attention with multiple experimental indices of each. In support of a diversity viewpoint, he found that only dual-tasking and global task shifting exhibited unique age-related variance above and beyond age effects on the baseline conditions (i.e., single-task and non-switch conditions), suggesting that age-related declines in EFs may be more limited than originally thought.

A methodological point that has been explored in the context of aging and EFs is the question of whether EF should be considered as a construct that uniquely explains age-related variance in other cognitive abilities such as reasoning, memory, or general intelligence. In a series of studies (Salthouse, 2005; Salthouse, Atkinson, & Berish, 2003), experimental tests of EFs (inhibition, updating, dual-tasking) and measures of other cognitive abilities were given to adults spanning young to old-old adulthood. Across data sets, the latent EF factors did not fully account for age-related variance in the cognitive abilities and showed poor discriminant validity. Verhaeghen (2011) used a similar logic to examine the EF constructs shifting and resistance to interference in relation to complex cognitive abilities (e.g., episodic memory), and demonstrated that the two EF factors explained significant proportions of the age-related variance, in addition to other cognitive processes such as processing speed. We note that the results will be influenced by the level of measurement resolution, given that clinical tests of EFs are typically less process-pure and more likely to correlate with other cognitive measures or with chronological age than experimental measures that control for other age-sensitive task components (Luszcz, 2011).

A final trend in current EF and aging research is to examine prefrontal activity as a compensatory response to brain aging and cognitive decline. For example, the Scaffolding Theory of Aging Cognition (STAC; Park & Reuter-Lorenz, 2009) links the over-activation of prefrontal brain regions associated with EFs to more successful cognitive performance in older adults, although over-activation has limits in its efficacy depending on task demands. Similar patterns of EF involvement and compensatory prefrontal brain activity have been found in sensory and sensorimotor aging research (Li & Bruce, 2016). We next turn to more detailed reviews of four EFs and age findings, followed by a discussion of the broader relevance of EFs in everyday functioning, and mobility in aging.

Working Memory, Updating, and Aging

Working memory (WM) is a "mental workspace" comprising representations and processes that, at any given moment, are the focus of attention. This mental workspace serves to store and actively manipulate information. The earliest notion that WM is a proxy to EF comes from Baddeley and Hitch's (1974) seminal WM model that included domain-specific passive storage buffers (i.e., verbal and visual) and the domain-general central executive. A crucial feature of this multi-component WM model is that there are capacity limitations on both the storage and the executive processing components.

The capacity limitations of WM are often measured by "complex span" tasks such as reading span (Daneman & Carpenter, 1980), operation span (Turner & Engle, 1989), counting span (Case, Kurland, & Goldberg, 1982), and spatial rotation span (e.g., Shah & Miyake, 1996). These span measures share features with dual tasks, requiring information storage in the context of simultaneous processing of other information. For example, in the reading span task, participants read a series of short sentences for comprehension and then recall the sentence-ending words from the series. The sets vary in size and the largest set-size that a participant can correctly recall (all the items in that set) is commonly used as an index of WM capacity.

The cumulative evidence from cross-sectional data suggests that older adults show capacity declines in both verbal and spatial WM tasks (e.g., Hale et al., 2011). In a meta-analysis, Bopp and Verhaeghen (2005) found that older adults' WM capacity on complex span tasks was 74% that of younger adults. In another meta-analysis, focusing on studies using a full adult age range, Verhaeghen (2014) found an age–WM correlation of −0.42.

It is widely accepted that complex span measures are not a process-pure measure of EF and reflect the contributions of rehearsal, coding, and strategies that are domain-specific (Kane, Conway, Hambrick, & Engle, 2007). Therefore, global causes for age-related declines in WM capacity are commonly explained by task-general influences such as general processing speed

(Salthouse, 1996; Verhaeghen, 2014), inhibitory processes (e.g., Hasher, Lustig, & Zacks, 2007), and dual-task coordination (Braver & West, 2008). With respect to inhibition and WM, May, Hasher, and Kane (1999) manipulated the testing procedures of complex span measures and found that resistance to proactive interference or "Deletion-type inhibition" was critical in explaining age-related differences in WM capacity. These behavioral findings converge with neuroscientific findings in which WM-related brain regions such as dorsolateral prefrontal cortex show both structural and functional changes in older adults. For instance, in addition to white matter and gray matter shrinkage, older adults tend to show bilateral activation of WM-related prefrontal areas, presumably to compensate for the capacity limitations (e.g., Reuter-Lorenz & Sylvester, 2005).

Task Switching and Aging

Task switching (TS) is considered a prototypical form of EF as it involves both cognitive stability and flexibility in a dynamic task environment. In experimental contexts, TS is assessed using two principle methods. In the alternating run paradigms (AABB) the task sequences are predictable (e.g., Rogers & Monsell, 1995), whereas in the task-cueing paradigm, the task sequences are random, and the upcoming task is explicitly signaled by a task cue (e.g., Meiran, 1996). In both these methods, a "task set" refers to a given stimulus–response contingency. For example, in a typical two-task design, the bivalent stimulus could be a colored numeral (e.g., 6), and the task sets could be in the form of odd/even number judgments (Task A) and magnitude judgments; greater or less than 5 (Task B). Switching efficiency is reflected by two types of switching costs. First, "local" switch costs, which is the difference in response time (RT) between task switches and task repetitions in a mixed block (i.e., BBAB). This "local" switch cost is an indication of the executive process associated with switching from one task set to the next. Second, the "global" switch cost is the difference in RT between task switches in a mixed block trial (i.e., BBAB) and task repetitions in a single block trial (i.e., BBB). The global switch cost is considered to indicate the set-up cost associated with maintaining and scheduling two task sets. The evidence to date suggests local switch costs are immune to aging, whereas older adults show larger global switch costs compared to younger adults (e.g., Kray & Lindenberger, 2000). Using three tasks, Mayr (2001) replicated Kray and Lindenberger's (2000) findings, but also explained the source of age effects in global switch costs by manipulating the degree of overlap between stimuli and between responses of the competing task sets. Specifically, in task conditions in which there was a high degree of stimulus ambiguity and response overlap, older adults showed larger global switch costs, presumably because of updating and WM demands imposed in this condition (Mayr, 2001). Consistent with these findings, using meta-analysis of 26 published studies Wasylyshyn,

Verhaeghen, and Sliwinski (2011) found that local switch costs are age-invariant, whereas global switch costs are age-sensitive. These behavioral dissociations are echoed in functional neuroimaging studies showing, for example, that global switch costs are associated with age-sensitive right anterior prefrontal cortex, whereas local switch costs are associated with non-frontal regions (e.g., right superior parietal cortex) (Braver, Reynolds, & Donaldson, 2003).

Inhibition and Aging

Inhibition within the context of EF models is defined as the deliberate suppression of dominant or prepotent responses (Miyake et al., 2000), although, more broadly, other inhibitory processes such as selecting out distraction, or down-regulating representations that are held in WM, have also received extensive examination in cognitive aging research (Gorfein & MacLeod, 2007). A multi-component view is supported by finding differential aging effects across inhibitory processes and across tasks (Hasher, Zacks, & May, 1999; Kramer, Humphrey, Larish, Logan, & Strayer, 1994).

Response inhibition has commonly been assessed using the Stroop interference paradigm (Stroop, 1935), which requires participants to indicate the ink color of words printed in incongruent ink colors (e.g., blue in green ink). Relative to the baseline condition (stating the color of neutral stimuli such as asterisks), the interference effect is reliably larger in older adults than younger adults, although Brinley plot analyses have revealed no additional age-related variance beyond that captured by the baseline condition (Verhaeghen, 2011).

Another neuropsychological test of response inhibition showing age-sensitivity is the Hayling Sentence Completion task (Burgess & Shallice, 1997), in which participants are required to complete high-context sentences with unrelated words and avoid the logical, high-frequency word. Across a wide age range of healthy older adults, response times for unrelated word production increase substantially and disproportionately compared to the high-frequency baseline condition (Bielak, Mansueti, Strauss, & Dixon, 2006).

Experimental tests of response inhibition include the stop-signal paradigm (Logan, 1994), in which participants perform a simple RT task but are instructed to withhold responding when a stop signal (e.g., tone) is occasionally presented shortly after the stimulus onset. Variable delays between the stimulus onset and stop-signal onset are used to determine the participant's stop-signal RT (i.e., time required to effectively inhibit the response). Age-related increases in stop-signal RT above and beyond general slowing were reported by Kramer et al. (1994). However, a comparable study using a simpler primary task showed negligible slowing of the stop-signal RT and significant slowing of the go-signal RTs as a function of aging (Williams, Ponesse, Schachar, Logan, & Tannock, 1999), suggesting that overall task demands may impinge on the resources available for response inhibition. Similarly, the Go-No-Go task, involving occasional lures to which participants should withhold responding in an otherwise rapid RT task, has shown marked age-related increases in commission errors. Functional neuroimaging of older adults performing the Go-No-Go task has revealed compensatory brain activity in contralateral prefrontal regions as compared to younger adults (Nielson, Langenecker, & Garavan, 2002).

Dividing Attention and Aging

Dividing attention, or dual-tasking, has been conceptualized as a complex EF task due to its association with the central executive component of working memory (e.g., Baddeley, 1986). Miyake's latent variable analysis revealed that dual-task performance was not well predicted by any of the identified EF factors (shifting, inhibition, updating), suggesting that it is relatively independent of other EFs (Miyake et al., 2000).

Divided attention tests require the concurrent performance of two cognitive tasks, which may or may not share common characteristics at the stimulus or response levels. Divided attention proficiency is commonly measured in experimental designs by comparing component task performance (i.e., single-task/full attention conditions) to dual-task performance (divided attention conditions). Task costs are typically expressed either as absolute costs (e.g., dual-task minus single-task) or proportional costs (e.g., [dual minus single-task]/single-task). Less commonly reported is a global time sharing measure, derived by dividing the dual–single absolute cost by the single-task standard deviation for both tasks, and averaging the two standardized scores (Salthouse & Miles, 2002).

With respect to behavioral evidence of age effects, dividing attention appears to be reliably age-sensitive, commonly showing interactions of age group and attentional load (e.g., McDowd & Craik, 1988; Salthouse, Fristoe, Lineweaver, & Coon, 1995). Using a typical experimental dual-task paradigm (visual RT task paired with an auditory RT task), Verhaeghen's (2011) Brinley plot analyses of divided attention performance revealed a reliable but modest age effect independent of the age-related variance observed under single-task conditions. This finding suggested that dividing attention and task coordination may be more age-sensitive than other EFs beyond the effects of general slowing. Age differences in underlying neural correlates of dividing attention reveal patterns of neural recruitment in older adults under divided attention conditions compared to younger adults (e.g., Fernandes, Pacurar, Moscovitch, & Grady, 2006), in line with models of increased prefrontal brain activation as compensatory scaffolding (Grady, McIntosh, & Craik, 2005).

A related experimental paradigm used to study dual-task costs is the psychological refractory period (PRP) paradigm (Welford, 1952), devised to further identify the stage of processing at which dual-task interference occurs. This is achieved by presenting Task 1 and Task 2 with a variable stimulus onset asynchrony (SOA) and observing a slowed Task 2 response when a processing bottleneck is encountered. In two experiments Allen, Smith, Vires-Collins, and Sperry (1998) showed that older adults had a larger PRP effect than young adults. However, older adults also exhibited a strategy difference by withholding their Task 1 response until the Task 2 response was prepared. Experimental manipulations of stimulus and response characteristics have led to differing views on the nature of the age-related differences, such as generalized and task-specific slowing, cautiousness (Glass et al., 2000), input interference (Hein & Schubert, 2004), or response interference (Hartley, 2001). On balance, the available data show that older adults exhibit a greater time sharing decrement compared to younger adults, even after extensive training or removal of structural interference (Hartley & Maquestiaux, 2007; Maquestiaux, Laguë-Beauvais, Ruthruff, Hartley, & Bherer, 2010). The persistence of age effects is similar in dual-task training studies without such strict timing constraints and equal emphasis instructions (Bherer et al., 2005).

Complex EF and Activities of Daily Living

Departing from our discussion of age-comparative studies of EFs, another topic of growing interest is the examination of how EFs play a role in everyday functioning in older adults (Grigsby, Kaye, Baxter, Shetterly, & Hamman, 1998). Among the EFs, standardized clinical measures of cognitive flexibility/switching have been significantly associated with instrumental activities of daily living (iADL) (Katz, 1983). For example, Trail Making Test-B (TMT-B) and Verbal Fluency significantly predict caregiver-rated iADL in an older community sample (Cahn-Weiner, Boyle, & Malloy, 2002). Similarly, TMT-B significantly predicted subjective iADL and mortality rate in older adults (Johnson, Lui, & Yaffe, 2007). Objective, performance-based measures of iADL were significantly predicted by the combination of TMT-B and Wisconsin Card Sort after controlling for age, sex, and education (Bell-McGinty, Podell, Franzen, Baird, & Williams, 2002). We note that the above studies did not compute a subtraction score (TMT-B minus TMT-A), thus a caveat is that the correlations may reflect switching proficiency as well as sensory-motor ability.

EFs and Mobility in Aging

A newer avenue of research on EFs and everyday functioning links EFs such as inhibition, planning, and divided attention with postural control and gait (Yogev-Seligmann, Hausdorff, & Giladi, 2008). This is in line with theories of cognitive compensation and scaffolding (Park & Reuter-Lorenz, 2009) and ability dedifferentiation (Lindenberger & Baltes, 1994). Good performance on standardized clinical measures such as the Trail Making Test (e.g., Ble et al., 2005), Stroop Color Word Test (Hausdorff, Yogev, Springer, Simon, & Giladi, 2004), Verbal Fluency (e.g., van Iersel, Kessels, Bloem, Verbeek, & Rikkert,

2008), and Digit Span task (e.g., Buchman, Boyle, Leurgans, Barnes, & Bennett, 2011) correlate with gait speed and stride regularity. Moreover, lower levels of inhibition and WM ability correlated with the risk of developing a mobility impairment (Buchman et al., 2011).

Experimental manipulations of cognitive load, using the dual-task paradigm, have also been used to demonstrate the increased reliance on high-level cognitive processes during motor performance with aging. Researchers have shown that older adults demonstrate a drop in performance on measures of memory (Lindenberger, Marsiske, & Baltes, 2000) and verbal abstraction (van Iersel et al., 2008) under dual-task conditions relative to younger adults, and that cognitive status is correlated with changes in gait speed under dual-task conditions (Al-Yahya et al., 2011). Similarly, in response to an unpredictable platform perturbation performed concurrently with WM, attention, or inhibition tasks, older adults exhibit greater dual-task costs in terms of cognitive performance and/or postural sway and recovery, particularly in the later, less automatic stages of the postural response (Woollacott & Shumway-Cook, 2002). The link between EFs and motor functioning in old age has been corroborated using mobile neuroimaging techniques such as functional near infrared spectroscopy, showing that older adults exhibited increased levels of oxygenation during walking in the prefrontal cortex, premotor cortex, and supplementary motor area (Holtzer, Epstein, Mahoney, Izzetoglu, & Blumen, 2014).

Training EFs

A rapidly growing topic concerns the improvement of EFs (see Kliegel, Hering, Ihle, & Zuber, Chapter 13, this volume; Berryman, Pothier, & Bherer, Chapter 12, this volume). Relative to training of other cognitive skills such as memory encoding or strategy use, process-based EF training (e.g., working memory) has shown relatively greater transfer to untrained cognitive tasks (Lustig, Shah, Seidler, & Reuter-Lorenz, 2009). EF training has also shown some promise in improving everyday functioning outside the laboratory. The Advanced Cognitive Training for Independent and Vital Elderly (ACTIVE) trial involved the assignment of older adults to EF (inductive reasoning), speed, or memory training. The EF training group not only improved on EF measures over time, but reported significantly higher scores on self-reported iADL than the other training groups 5 and 10 years after training (Rebok et al., 2014; Willis et al., 2006). Similar to the intervention work on iADL, an exciting new domain of transfer effects involves dual-task training and gains in balance and global mobility (e.g., Li et al., 2010), which underscores the potential for far-transfer effects in older adults who undergo EF training (Hertzog, Kramer, Wilson, & Lindenberger, 2008).

Conclusions

To summarize the current status of EF development in aging research, several notable trends emerge. First, the question of whether EFs are age-sensitive has been replaced by theoretical questions such as whether age-related declines in EF are independent of the aging of basic processes such as cognitive speed. There is relatively more consensus in demonstrating age-related dedifferentiation of the factor structure of EFs. As Luszcz (2011) has discussed, the choice of measurement precision can influence the predictive power of EFs vis-à-vis other constructs such as complex cognition and behavior. Finally, the potential for broad transfer from training EFs, owing to their modality non-specific nature, is an exciting new direction of work in aging and EFs.

References

Allen, P. A., Smith, A. F., Vires-Collins, H., & Sperry, S. (1998). The psychological refractory period: Evidence for age differences in attentional time-sharing. *Psychology and Aging, 13,* 218–229.

Al-Yahya, E., Dawes, H., Smith, L., Dennis, A., Howells, K., & Cockburn, J. (2011). Cognitive motor interference while walking: A systematic review and meta-analysis. *Neuroscience & Biobehavioral Reviews, 35,* 715–728.

Baddeley, A. D. (1986). *Working memory.* New York, NY: Oxford University Press.

Baddeley, A. D., & Hitch, G. J. (1974). Working memory. In G. H. Bower (Ed.), *The psychology of learning and motivation* (Vol. 8, pp. 47–89). New York, NY: Academic Press.

Bell-McGinty, S., Podell, K., Franzen, M., Baird, A. D., & Williams, M. J. (2002). Standard measures of executive function in predicting instrumental activities of daily living in older adults. *International Journal of Geriatric Psychiatry, 17,* 828–834.

Bherer, L., Kramer, A. F., Peterson, M. S., Colcombe, S., Erickson, K., & Becic, E. (2005). Training effects on dual-task performance: Are there age-related differences in plasticity of attentional control? *Psychology and Aging, 20,* 695–709.

Bielak, A. A. M., Mansueti, L., Strauss, E., & Dixon, R. A. (2006). Performance on the Hayling and Brixton tests in older adults: Norms and correlates. *Archives of Clinical Neuropsychology, 21,* 141–149.

Ble, A., Volpato, S., Zuliani, G., Guralnik, J. M., Bandinelli, S., Lauretani, F., et al. (2005). Executive function correlates with walking speed in older persons: The InCHIANTI study. *Journal of the American Geriatrics Society, 53,* 410–415.

Bopp, K. L., & Verhaeghen, P. (2005). Aging and verbal memory span: A meta-analysis. *The Journals of Gerontology: Series B: Psychological Sciences and Social Sciences, 60B,* P223–P233.

Braver, T. S., Barch, D. M., Keys, B. A., Carter, C. S., Cohen, J. D., Kaye, J. A., et al. (2001). Context processing in older adults: Evidence for a theory relating cognitive control to neurobiology in healthy aging. *Journal of Experimental Psychology: General, 130,* 746–763.

Braver, T. S., Reynolds, J. R., & Donaldson, D. I. (2003). Neural mechanisms of transient and sustained cognitive control during task switching. *Neuron, 39,* 713–726.

Braver, T. S., & West, R. (2008). Working memory, executive control, and aging. In F. M. Craik & T. A. Salthouse (Eds.), *The handbook of aging and cognition* (3rd ed., pp. 311–372). New York, NY: Psychology Press.

Buchman, A. S., Boyle, P. A., Leurgans, S. E., Barnes, L. L., & Bennett, D. A. (2011). Cognitive function is associated with the development of mobility impairments in community-dwelling elders. *The American Journal of Geriatric Psychiatry, 19*, 571–580.

Burgess, P. W., & Shallice, T. (1997). *The Hayling and Brixton tests.* Thurston, Suffolk: Thames Valley Test Company.

Cahn-Weiner, D. A., Boyle, P. A., & Malloy, P. F. (2002). Tests of executive function predict instrumental activities of daily living in community-dwelling older individuals. *Applied Neuropsychology, 9*, 187–191.

Case, R., Kurland, D. M., & Goldberg, J. (1982). Operational efficiency and the growth of short-term memory span. *Journal of Experimental Child Psychology, 33*, 386–404.

Craik, F. I. M., & Byrd, M. (1982). Aging and cognitive deficits. In F. I. M. Craik & S. Trehub (Eds.), *Aging and cognitive processes* (pp. 191–211). New York, NY: Plenum Press.

Daneman, M., & Carpenter, P. A. (1980). Individual differences in working memory and reading. *Journal of Verbal Learning & Verbal Behavior, 19*, 450–466.

de Frias, C. M., Dixon, R. A., & Strauss, E. (2006). Structure of four executive functioning tests in healthy older adults. *Neuropsychology, 20*, 206–214.

de Frias, C. M., Dixon, R. A., & Strauss, E. (2009). Characterizing executive functioning in older special populations: From cognitively elite to cognitively impaired. *Neuropsychology, 23*, 778–791.

Duncan, J., & Owen, A. M. (2000). Common regions of the human frontal lobe recruited by diverse cognitive demands. *Trends in Neurosciences, 23*, 475–483.

Engle, R. W., Tuholski, S. W., Laughlin, J. E., & Conway, A. R. A. (1999). Working memory, short-term memory, and general fluid intelligence: A latent-variable approach. *Journal of Experimental Psychology: General, 128*, 309–331.

Fernandes, M. A., Pacurar, A., Moscovitch, M. M., & Grady, C. (2006). Neural correlates of auditory recognition under full and divided attention in younger and older adults. *Neuropsychologia, 44*, 2452–2464.

Garrett, H. E. (1938). Differentiable mental traits. *Psychological Record, 2*, 259–298.

Glass, J. M., Schumacher, E. H., Lauber, E. J., Zurbriggen, E. L., Gmeindl, L., Kieras, D. E., & Meyer, D. E. (2000). Aging and the psychological refractory period: Task-coordination strategies in young and old adults. *Psychology and Aging, 15*, 571–595.

Gorfein, D. S., & MacLeod, C. M. (Eds.). (2007). *Inhibition in cognition.* Washington, DC: American Psychological Association.

Grady, C. L., McIntosh, A. R., & Craik, F. I. M. (2005). Task-related activity in prefrontal cortex and its relation to recognition memory performance in young and old adults. *Neuropsychologia, 43*, 1466–1481.

Grieve, S. M., Williams, L. M., Paul, R. H., Clark, C. R., & Gordon, E. (2007). Cognitive aging, executive function, and fractional anisotropy: A diffusion tensor MR imaging study. *American Journal of Neuroradiology, 28*, 226–235.

Grigsby, J., Kaye, K., Baxter, J., Shetterly, S. M., & Hamman, R. F. (1998). Executive cognitive abilities and functional status among community-dwelling older persons in the San Luis Valley Health and Aging Study. *Journal of the American Geriatrics Society, 46*, 590–596.

Gunning-Dixon, F. M., & Raz, N. (2000). The cognitive correlates of white matter abnormalities in normal aging: A quantitative review. *Neuropsychology, 14*, 224–232.

Hale, S., Rose, N. S., Myerson, J., Strube, M. J., Sommers, M., Tye-Murray, N., & Spehar, B. (2011). The structure of working memory abilities across the adult life span. *Psychology and Aging, 26*, 92–110.

Hartley, A. A. (2001). Age differences in dual-task interference are localized to response generation processes. *Psychology and Aging, 16*, 47–54.

Hartley, A. A., & Maquestiaux, F. (2007). Success and failure at dual-task coordination by younger and older adults. *Psychology and Aging, 22*, 215–222.

Hasher, L., Lustig, C., & Zacks, R. T. (2007). Inhibitory mechanisms and the control of attention. In A. R. A. Conway, C. Jarrold, M. J. Kane, A. Miyake, & J. N. Towse (Eds.), *Variation in working memory* (pp. 227–249). New York, NY: Oxford University Press.

Hasher, L., & Zacks, R. T. (1979). Automatic and effortful processes in memory. *Journal of Experimental Psychology: General, 108*, 356–388.

Hasher, L., Zacks, R. T., & May, C. P. (1999). Inhibitory control, circadian arousal, and age. In D. Gopher & A. Koriat (Eds.), *Attention and performance XVII: Cognitive regulation of performance – Interaction of theory and application* (pp. 653–675). Cambridge, MA: MIT Press.

Hausdorff, J. M., Yogev, G., Springer, S., Simon, E. S., & Giladi, N. (2004). Walking is more like catching than tapping: Gait in the elderly as a complex cognitive task. *Experimental Brain Research, 164*, 541–548.

Hein, G., & Schubert, T. (2004). Aging and input processing in dual-task situations. *Psychology and Aging, 19*, 416–432.

Hertzog, C., Kramer, A. F., Wilson, R. S., & Lindenberger, U. (2008). Enrichment effects on adult cognitive development: Can the functional capacity of older adults be preserved and enhanced? *Psychological Science in the Public Interest, 9*, 1–65.

Holtzer, R., Epstein, N., Mahoney, J. R., Izzetoglu, M., & Blumen, H. M. (2014). Neuroimaging of mobility in aging: A targeted review. *The Journals of Gerontology Series A: Biological Sciences and Medical Sciences, 69*, 1375–1388.

Johnson, J. K., Lui, L. Y., & Yaffe, K. (2007). Executive function, more than global cognition, predicts functional decline and mortality in elderly women. *The Journals of Gerontology Series A: Biological Sciences and Medical Sciences, 62*, 1134–1141.

Kane, M. J., Conway, A. A., Hambrick, D. Z., & Engle, R. W. (2007). Variation in working memory capacity as variation in executive attention and control. In A. A. Conway, C. Jarrold, M. J. Kane, J. N. Towse, A. A. Conway, C. Jarrold, et al. (Eds.), *Variation in working memory* (pp. 21–46). New York, NY: Oxford University Press.

Katz, S. (1983). Assessing self-maintenance: Activities of daily living, mobility, and instrumental activities of daily living. *Journal of the American Geriatrics Society, 31*, 721–727.

Kramer, A. F., Humphrey, D. G., Larish, J. F., Logan, G. D., & Strayer, D. L. (1994). Aging and inhibition: Beyond a unitary view of inhibitory processing in attention. *Psychology and Aging, 9*, 491–512.

Kray, J., & Lindenberger, U. (2000). Adult age differences in task switching. *Psychology and Aging, 15*, 126–147.

Li, K. Z. H., & Bruce, H. (2016). Sensorimotor and cognitive interactions in healthy aging. In S. K. Whitbourne (Ed.), *The encyclopedia of adulthood and aging*. New Malden, MA: Wiley-Blackwell Publishers.

Li, K. Z., Roudaia, E., Lussier, M., Bherer, L., Leroux, A., & McKinley, P. A. (2010). Benefits of cognitive dual-task training on balance performance in healthy older adults. *The Journals of Gerontology Series A: Biological Sciences and Medical Sciences, 65*, 1344–1352.

Lindenberger, U., & Baltes, P. B. (1994). Sensory functioning and intelligence in old age: A strong connection. *Psychology and Aging, 9*, 339–355.

Lindenberger, U., Marsiske, M., & Baltes, P. B. (2000). Memorizing while walking: Increase in dual-task costs from young adulthood to old age. *Psychology and Aging, 15*, 417–436.

Logan, G. D. (1994). On the ability to inhibit thought and action: A user's guide to the stop signal paradigm. In D. Dagenbach & T. H. Carr (Eds.), *Inhibitory processes in attention, memory, and language* (pp. 189–239). San Diego, CA: Academic Press.

Lustig, C., Shah, P., Seidler, R., & Reuter-Lorenz, P. A. (2009). Aging, training, and the brain: A review and future directions. *Neuropsychological Review, 19*, 504–522.

Luszcz, M. (2011). Executive function and cognitive aging. In K. W. Schaie & S. L. Willis (Eds.), *Handbook of the psychology of aging* (7th ed., pp. 59–72). London: Academic Press.

Maquestiaux, F., Laguë-Beauvais, M., Ruthruff, E., Hartley, A., & Bherer, L. (2010). Learning to bypass the central bottleneck: Declining automaticity with advancing age. *Psychology and Aging, 25*, 177.

May, C. P., Hasher, L., & Kane, M. J. (1999). The role of interference in memory span. *Memory & Cognition, 27*, 759–767.

Mayr, U. (2001). Age differences in the selection of mental sets: The role of inhibition, stimulus ambiguity, and response-set overlap. *Psychology and Aging, 16*, 96–109.

McDowd, J. M., & Craik, F. I. M. (1988). Effects of aging and task difficulty on divided attention performance. *Journal of Experimental Psychology: Human Perception and Performance, 14*, 267–280.

Meiran, N. (1996). Reconfiguration of processing mode prior to task performance. *Journal of Experimental Psychology: Learning, Memory, and Cognition, 22*, 1423–1442.

Miyake, A., Friedman, N. P., Emerson, M. J., Witzki, A. H., Howerter, A., & Wager, T. (2000). The unity and diversity of executive functions and their contributions to complex "frontal lobe" tasks: A latent variable analysis. *Cognitive Psychology, 41*, 49–100.

Nielson, K. A., Langenecker, S. A., & Garavan, H. (2002). Differences in functional neuroanatomy of inhibitory control across the adult lifespan. *Psychology and Aging, 17*, 56–71.

Norman, D. A., & Shallice, T. (1986). Attention to action: Willed and automatic control of behavior. In R. J. Davidson (Ed.), *Consciousness and self-regulation* (pp. 1–18). New York, NY: Plenum Press.

Park, D. C., & Reuter-Lorenz, P. (2009). The adaptive brain: Aging and neurocognitive scaffolding. *Annual Review of Psychology, 60*, 173–196.

Rabbitt, P. (1997). Introduction: Methodologies and models in the study of executive function. In P. Rabbitt (Ed.), *Methodology of frontal and executive function* (pp. 1–38). London: Psychology Press.

Rebok, G. W., Ball, K., Guey, L. T., Jones, R. N., Kim, H. Y., King, J. W., & Willis, S. L. (2014). Ten-year effects of the advanced cognitive training for independent and vital elderly cognitive training trial on cognition and everyday functioning in older adults. *Journal of the American Geriatrics Society, 62*, 16–24.

Reuter-Lorenz, P. A., Festini, S. B., & Jantz, T. K. (2015). Executive functions and neurocognitive aging. In K. W. Schaie & S. L. Willis (Eds.), *Handbook of the psychology of aging* (8th ed., pp. 245–262). New York: Elsevier.

Reuter-Lorenz, P. A., & Sylvester, C. C. (2005). The cognitive neuroscience of working memory and aging. In R. Cabeza, L. Nyberg, & D. Park (Eds.), *Cognitive neuroscience of aging: Linking cognitive and cerebral aging* (pp. 186–217). New York: Oxford University Press.

Rogers, R. D., & Monsell, S. (1995). Costs of a predictable switch between simple cognitive tasks. *Journal of Experimental Psychology: General, 124*, 207–231.

Salthouse, T. A. (1996). The processing-speed theory of adult age differences in cognition. *Psychological Review, 103*, 403–428.

Salthouse, T. A. (2005). Relations between cognitive abilities and measures of executive functioning. *Neuropsychology, 19*, 532–545.

Salthouse, T. A., Atkinson, T. M., & Berish, D. E. (2003). Cognitive functioning as a potential mediator of age-related cognitive decline in normal adults. *Journal of Experimental Psychology: General, 132*, 566–594.

Salthouse, T. A., Fristoe, N. M., Lineweaver, T. T., & Coon, V. E. (1995). Aging of attention: Does the ability to divide decline? *Memory & Cognition, 23*, 59–71.

Salthouse, T. A., & Miles, J. D. (2002). Aging and time-sharing aspects of executive control. *Memory & Cognition, 30*, 572–582.

Shah, P., & Miyake, A. (1996). The separability of working memory resources for spatial thinking and language processing: An individual differences approach. *Journal of Experimental Psychology: General, 125*, 4–27.

Smith-Ray, R. L., Hughes, S. L., Prohaska, T. R., Little, D. M., Jurivich, D. A., & Hedeker, D. (2013). Impact of cognitive training on balance and gait in older adults. *The Journals of Gerontology Series B: Psychological Sciences and Social Sciences, 70*, 357–366.

Stroop, J. R. (1935). Studies of interference in serial verbal reactions. *Journal of Experimental Psychology, 18*, 643–662.

Stuss, D. T. (2011). Functions of the frontal lobes: Relation to executive functions. *Journal of the International Neuropsychological Society, 17*, 759–765.

Turner, M. L., & Engle, R. W. (1989). Is working memory capacity task dependent? *Journal of Memory and Language, 28*, 127–154.

van Iersel, M. B., Kessels, R. P., Bloem, B. R., Verbeek, A. L., & Rikkert, M. G. O. (2008). Executive functions are associated with gait and balance in community-living elderly people. *The Journals of Gerontology Series A: Biological Sciences and Medical Sciences, 63*, 1344–1349.

Verghese, J., Mahoney, J., Ambrose, A. F., Wang, C., & Holtzer, R. (2010). Effect of cognitive remediation on gait in sedentary seniors. *The Journals of Gerontology Series A: Biological Sciences and Medical Sciences, 65A*, 1338–1343.

Verhaeghen, P. (2011). Aging and executive control: Reports of a demise greatly exaggerated. *Current Directions in Psychological Science, 20*, 174–180.

Verhaeghen, P. (2014). *The elements of cognitive aging: Meta-analyses of age-related differences in processing speed and their consequences.* New York, NY: Oxford University Press.

Wasylyshyn, C., Verhaeghen, P., & Sliwinski, M. J. (2011). Aging and task switching: A meta-analysis. *Psychology and Aging, 26*, 15–20.

Welford, A. T. (1952). The psychological refractory period and the timing of high-speed performance: A view and a theory. *British Journal of Psychology, 43*, 2–19.

West, R. L. (1996). An application of prefrontal cortex function theory to cognitive aging. *Psychological Bulletin, 120*, 272–292.

Williams, B. R., Ponesse, J. S., Schachar, R. J., Logan, G. D., & Tannock, R. (1999). Development of inhibitory control across the life span. *Developmental Psychology, 35*, 205–213.

Willis, S. L., Tennstedt, S. L., Marsiske, M., Ball, K., Elias, J., Koepke, K. M., & Wright, E. (2006). Long-term effects of cognitive training on everyday functional outcomes in older adults. *Journal of the American Medical Association, 296*, 2805–2814.

Woollacott, M., & Shumway-Cook, A. (2002). Attention and the control of posture and gait: A review of an emerging area of research. *Gait & Posture, 16*, 1–14.

Yogev-Seligmann, G., Hausdorff, J. M., & Giladi, N. (2008). The role of executive function and attention in gait. *Movement Disorders, 23*, 329–342.

Yuan, P., & Raz, N. (2014). Prefrontal cortex and executive functions in healthy adults: A meta-analysis of structural neuroimaging studies. *Neuroscience & Biobehavioral Reviews, 42*, 180–192.

Part II
Understanding Mechanisms of Executive Function Development and Plasticity

Neural Basis of Inhibitory Control during Childhood

Inhibitory control refers to the ability to suppress a dominant response. Research using several specifically designed tasks consistently shows that inhibitory control develops during early childhood. In this section, I introduce studies that have investigated the development of inhibitory control using two tasks: a Go/NoGo task and a Stroop task.

Go/NoGo Task

The Go/NoGo task is one of the major tasks used to investigate inhibitory control. In the Go/NoGo task, participants are asked to respond to targets by pressing a button (Go trials) and to not respond to non-targets. Adult studies using fMRI have shown that the right inferior frontal cortex (BA 44/45/47) is the locus of inhibitory control during Go/NoGo tasks (Aron et al., 2014), although some researchers have proposed that activation in the right inferior frontal regions is not specific to inhibitory control (Hampshire & Sharp, 2015). A recent meta-analysis revealed that the right inferior cortex contributes to inhibitory control whereas the right anterior insula may be related to detecting behaviorally salient events (Cai, Ryali, Chen, Li, & Menon, 2014).

Mehnert et al. (2013) examined brain activation in 4- to 6-year-old children and adults using the Go/NoGo task. Specifically, activity in the PFC, as well as the parietal and temporal regions, was examined using the NIRS system. Results showed that adult participants commit few errors and exhibit activation in the right frontal and parietal regions during NoGo trials when compared to Go trials. On the other hand, children commit more errors than adults, with activation occurring in the right frontal and parietal regions in both Go and NoGo trials. The authors also conducted functional connectivity analyses, and found different patterns between children and adults. Children show stronger partial coherence in short-range connectivity in the right frontal and right parietal cortices, whereas adults exhibit long-range functional connectivity between bilateral frontal and parietal areas. These results suggest that activation in the right frontal and parietal areas plays an important role in the performance of children on the Go/NoGo task, but that the fronto-parietal network may be less efficient and functional in children compared to adults.

Other fNIRS studies have also shown that school-aged children exhibit significant activation in the LPFC during Go/NoGo tasks (Inoue et al., 2012; Xiao et al., 2012). However, activation patterns in children are different from those of adults. For example, research using fMRI has shown that frontal activation is less localized and more bilaterally extended in children compared to adults (Bunge, Dudukovic, Thomason, Vaidya, & Gabrieli, 2002; Durston et al., 2006; Rubia et al., 2006). Indeed, in a longitudinal study, Durston et al. (2006) showed a developmental shift in patterns of cortical activation from diffuse to focal activity during the Go/NoGo task, demonstrating increased VLPFC activation from ages 9 to 11, but decreased activation in other prefrontal regions including the DLPFC.

Stroop Task

Stroop tasks are also used to index the development of inhibitory control (Stroop, 1935). In the classic color-word Stroop task, a color word (e.g., BLUE) appears in an ink color such as yellow. In this task, participants show no difficulty in reading the word and ignoring the color (answering "blue"), but take more time to name the ink color and ignore the word (i.e., answering "yellow"). Adult neuroimaging studies using fMRI have revealed that the anterior cingulate cortex (BA 32), bilateral inferior frontal gyrus (BA 44/45), and left inferior parietal lobule (BA 40) may constitute the core network for the Stroop effect (Derrfuss, et al., 2005; Laird et al., 2005; Roberts & Hall, 2008; West, Chapter 6, this volume).

Few neuroimaging studies have examined the neural basis of the Stroop effect in young children. However, the PFC would be involved in determining the performances of children on Stroop-like tasks during preschool years. Wolfe and Bell (2004) conducted one of the few studies in young children, and examined whether increases in EEG power could be observed in young children during two Stroop-like tasks. In the Day-Night task, children were instructed to say "day" in response to a picture of a moon and "night" in response to a picture of a sun. In order to perform the task correctly, children had to inhibit their dominant response (e.g., children had to inhibit "day" responses when presented with a sun card). In the Yes-No task, children were asked to say "yes" when an experimenter shook their head ("no"), and say "no" when the experimenter nodded their head ("yes"). An EEG assessed activity during the Day-Night and Yes-No tasks, as well as during baseline conditions where children watched a music video. Using this approach, Wolfe and Bell (2004) found significant differences between task and control conditions in the EEG power observed in medial frontal sites. Moreover, they reported that children who perform well on the tasks show stronger 6–9 Hz EEG power than those who perform more poorly on the tasks. However, again, we need to be careful about the interpretation of the spatial information of EEG.

Schroeter, Zysset, Wahl, and von Cramon (2004) gave 7- to 13-year-old children a color-word Stroop task, and examined activation in the LPFC covering F7/8, F3/4, and FC3/4 (BA 9, 10, 45, 47) of the International 10/20 system. This study included a neutral condition and an incongruent condition. In the neutral condition, participants were presented with a colored stimulus "XXXX" (e.g., in red), and asked to judge whether the color of the stimulus matched the words below the stimulus (e.g., "RED"). The incongruent condition was the same as the neutral condition except that stimuli included color words in differently colored ink (e.g., "RED" in blue). Results showed that school-aged children exhibit significant activation in the left LPFC in the incongruent condition compared to those in the neutral condition. Moreover, when comparing data between children and adults, activation in the DLPFC increases as a function of age.

fMRI studies have also examined brain regions underlying the development of the Stroop task. Adleman et al. (2002) compared brain activation in 7- to 11-year-old children to activation in adolescents (aged 12–16) and young adults (aged 18–22) during a color-word Stroop task, and found positive correlations between participant age and activation in the core network (including the left LPFC, left ACC, and left parietal lobe and cortex) during the Stroop task. Such activations pattern in children have been replicated by other researchers (Marsh, Zhu, Wang, Skudlarski, & Peterson, 2007; Peterson et al., 2009). On the other hand, Marsh et al. (2006) reported that the right inferior frontal gyrus (BA 44/45) exhibits increased activation, whereas the ACC shows decreased activation with age. Differences in the activation patterns of the ACC across studies may be due to variability in baseline comparison tasks and/or participant ages; therefore, care should be taken in the interpretation of such findings. Nevertheless, the results suggest that the LPFC plays an important role in the development of Stroop task performance.

In summary, research using both Go/NoGo and Stroop tasks has shown that the LPFC is activated in early childhood during inhibitory control tasks, and that the activations in the regions play an important role in the development of inhibitory control. Other regions such as the ACC and parietal cortex are also important for the development of inhibitory control; however, there is limited evidence to suggest that activation of these regions is essential in preschool-aged children and mixed evidence to suggest a role for these regions in school-aged children.

The Neural Basis of Working Memory during Childhood

Working memory enables the online maintenance and manipulation of information. Baddeley and colleagues proposed an influential model of working memory consisting of distinct storage buffers, with a visuospatial sketchpad, a phonological loop, and a central executive system that acts upon information in the storage buffers (Baddeley, 2000; Baddeley & Hitch, 1974). Previous brain imaging studies have shown that the fronto-parietal network (e.g., BA 7, 10, 45, 47) is significantly more involved in tasks requiring the manipulation of information, rather than tasks requiring only the maintenance of information (Rottschy et al., 2012; Wager & Smith, 2003). Moreover, several brain regions, such as the anterior insula, premotor area, pre-supplementary motor area, and precentral gyrus, have been shown to be activated depending on specific task contents. Specifically, visuospatial working memory has been shown to be related to activation in the right premotor areas, caudal superior frontal sulcus, and right inferior prefrontal gyrus (Nee et al., 2013; Rottschy et al., 2012).

Psychological studies have revealed that visuospatial working memory shows a linear increase in performance during early childhood in a battery of working memory tasks (Gathercole, Pickering, Ambridge, & Wearing, 2004). Tsujimoto et al. (2004) examined the neural correlates of working memory in 5- and 6-year-old children and adults in a visuospatial working memory task, and assessed activation in the lateral prefrontal region (BA 9/46) via NIRS. In this task, children and adult participants were instructed to memorize the location of a sample cue array, and were then asked to indicate whether a test cue location was the same as the sample cue location. Tsujimoto et al. found similarly activated brain regions and a comparable time course of activation between children and adults. That is, children exhibit sustained lateral prefrontal activation after onset of the sample cues.

Other researchers have examined the involvement of the fronto-parietal network in the development of working memory. In one such study, Buss, Fox, Boas, and Spencer (2014) showed participants a cue array and instructed them to keep the array in mind during a delay phase. Participants were shown a test array where either all items were the same as the cue array, or where some of the features were changed. Participants were then asked to report whether there were changes in the test array or not. Task difficulty depended on the number of items (1, 2, or 3) presented in the cue array. Buss et al. (2014) examined neural activation after presentation of the cue array with an NIRS system that covered the prefrontal regions (F3–5/4–6) (BA 9, 10) and the parietal regions (P3–5/4–6) (BA 7, 19) in 3- and 4-year-old children. Their results showed that children perform worse when shown three items, rather than just one. At the neural level, children appear to exhibit significant activation in the frontal and parietal regions after presentation of the cue array. Specifically, activation in the right PFC is significantly correlated with behavioral performance. In addition, activity in the left frontal areas and bilateral parietal areas appears significantly stronger when three items are presented than just one or two items.

Research in young children has consistently shown activation of the lateral prefrontal regions during visuospatial working memory tasks; however, the role of the parietal cortex is unclear. Research on older children has shown that the frontal-parietal network is significantly activated during visuospatial working memory tasks in school-aged children and adolescents, as well as in adults (Klingberg, 2006; Kwon, Reiss, & Menon, 2002). On the other hand, research has also shown that other regions, such as the caudate nucleus and anterior insula, are more involved during childhood, and that the frontal-parietal network is more activated during adolescence (Scherf, Sweeney, & Luna, 2006). Although the developmental period differs across studies, research consistently shows that developmental changes in functional connectivity between the prefrontal and parietal regions may be responsible for the development of visuospatial working memory (Edin, Macoveanu, Olesen, Tegnér, & Klingberg, 2007).

General Discussion and Future Directions

Here, I reviewed the relationship between EF and prefrontal area activation in infants, preschool children, school-aged children, and adults. Collectively, the results of these studies show that children exhibit significant activation of the PFC, and that this activation is strongly related to the development of cognitive shifting, inhibitory control, and working memory during infancy and preschool years. In contrast, school-aged children exhibit activation in the functional network underlying EF, which includes the DLPFC, VLPFC, and PPC, as well as other cortical and subcortical regions. The intensity of activation in such regions generally increases as a function of age. Nevertheless, the functional network in children is significantly different from adults. The development of the network would continue across adolescence and adulthood.

An important issue to address in future studies is how the fronto-parietal network contributes to EF development during infancy and preschool years. It has been shown that the functional connection between the prefrontal areas and the parietal areas would be stronger during middle childhood and adolescence (Fair et al., 2007; Richmond, Johnson, Seal, Allen, & Whittle, 2016). Moreover, existing evidence seems to suggest that behavioral performances in EF tasks are more correlated with activation in the PFC than in the parietal cortex during infancy and early childhood. However, the evidence does not mean that the parietal cortex is not related to the development of EF in infants and young children. Moreover, the network between subcortical regions and the fronto-parietal network may be important for the development of EF. Both fNIRS and fMRI studies, including analyses of resting states, can partly address the issue.

Another issue that remains unresolved is how the brain network underlying EF may change developmentally across different EF components (i.e., cognitive shifting, inhibitory control, and working memory). At the behavioral level, the factor structure of EF differentiates over the course of development. However, it is still unclear how the brain networks underlying different EF components develop. The available data suggests that lateral prefrontal regions may be generally activated across different tasks in younger children, but that such activation may localize to specific regions in older children. Johnson (2011) proposed that activation in some regions of the cerebral cortex may start with broad functionality, and consequently become partially activated in response to different stimuli and tasks. Thus, future research should assess developmental changes in brain activation in response to different EF components.

Finally, how and when activation in typically developed children and those in children with developmental disorders differs should be assessed during EF tasks. Several studies have been conducted to compare activation in the PFC in children with attention-deficit/hyperactivity disorder and autism spectrum disorders to those in typically developed children. The results consistently reveal that there are some qualitative and quantitative differences between the two groups (Yasumura et al., 2012). Nevertheless, most of these studies have been conducted on relatively older children. Thus, future studies should examine brain activation in younger children with developmental disorders or with risks for developmental disorders.

References

Adleman, N. E., Menon, V., Blasey, C. M., White, C. D., Warsofsky, I. S., Glover, G. H., et al. (2002). A developmental fMRI study of the Stroop color-word task. *NeuroImage, 16*, 61–75.

Anderson, P. (2002). Assessment and development of executive function (EF) during childhood. *Child Neuropsychology, 8*, 71–82.

Aron, A. R., Robbins, T. W., & Poldrack, R. A. (2014). Inhibition and the right inferior frontal cortex: One decade on. *Trends in Cognitive Sciences, 18*, 177–185.

Baddeley, A. D. (2000). The episodic buffer: A new component of working memory? *Trends in Cognitive Sciences, 4*, 417–423.

Baddeley, A. D., & Hitch, G. (1974). Working memory. In G. A. Bower (Ed.), *The psychology of learning and motivation* (pp. 48–79). New York: Academic Press.

Baird, A. A., Kagan, J., Gaudette, T., Walz, K. A., Hershlag, N., & Boas, D. A. (2002). Frontal lobe activation during object permanence: Data from near-infrared spectroscopy. *NeuroImage, 16*, 1120–1126.

Bell, M. A., & Fox, N. A. (1992). The relations between frontal brain electrical activity and cognitive development during infancy. *Child Development, 63*, 1142–1163.

Buchsbaum, B. R., Greer, S., Chang, W. L., & Berman, K. F. (2005). Meta-analysis of neuroimaging studies of the Wisconsin Card-Sorting task and component processes. *Human Brain Mapping, 25*, 35–45.

Bunge, S. A., Dudukovic, N. M., Thomason, M. E., Vaidya, C. J., & Gabrieli, J. D. E. (2002). Immature frontal lobe contributions to cognitive control in children: Evidence from fMRI. *Neuron, 33*, 301–311.

Buss, A. T., Fox, N., Boas, D. A., & Spencer, J. P. (2014). Probing the early development of visual working memory capacity with functional near-infrared spectroscopy. *NeuroImage, 85*, 314–325.

Cai, W., Ryali, S., Chen, T., Li, C.-S. R., & Menon, V. (2014). Dissociable roles of right inferior frontal cortex and anterior insula in inhibitory control: Evidence from intrinsic and task-related functional parcellation, connectivity, and response profile analyses across multiple datasets. *The Journal of Neuroscience, 34*, 14652–14667.

Crone, E. A., Donohue, S. E., Honomichl, R., Wendelken, C., & Bunge, S. A. (2006). Brain regions mediating flexible rule use during development. *Journal of Neuroscience, 26*, 11239–11247.

Dajani, D. R., & Uddin, L. Q. (2015). Demystifying cognitive flexibility: Implications for clinical and developmental neuroscience. *Trends in Neurosciences, 38*, 571–578.

Derrfuss, J., Brass, M., Neumann, J., & von Cramon, D. Y. (2005). Involvement of the inferior frontal junction in cognitive control: Meta-analyses of switching and Stroop studies. *Human Brain Mapping, 25*, 22–34.

Diamond, A., & Goldman-Rakic, P. S. (1989). Comparison of human infants and rhesus-monkeys on Piaget ABBAR task: Evidence for dependence on dorsolateral prefrontal cortex. *Experimental Brain Research, 74*, 24–40.

Doria, V., Beckmann, C. F., Arichi, T., Merchant, N., Groppo, M., Turkheimer, F. E., et al. (2010). Emergence of resting state networks in the preterm human brain. *Proceedings of the National Academy of Sciences, 107*, 20015–20020.

Durston, S., Davidson, M. C., Tottenham, N., Galvan, A., Spicer, J., Fossella, J. A., et al. (2006). A shift from diffuse to focal cortical activity with development. *Developmental Science, 9*, 1–8.

Edin, F., Macoveanu, J., Olesen, P., Tegnér, J., & Klingberg, T. (2007). Stronger synaptic connectivity as a mechanism behind development of working memory-related brain activity during childhood. *Journal of Cognitive Neuroscience, 19*, 750–760.

Espinet, S. D., Anderson, J. E., & Zelazo, P. D. (2012). N2 amplitude as a neural marker of executive function in young children: An ERP study of children who switch versus perseverate on the Dimensional Change Card Sort. *Developmental Cognitive Neuroscience, 2, Supplement 1*, S49–S58.

Espinet, S. D., Anderson, J. E., & Zelazo, P. D. (2013). Reflection training improves executive function in preschool-age children: Behavioral and neural effects. *Developmental Cognitive Neuroscience, 4*, 3–15.

Ezekiel, F., Bosma, R., & Morton, J. B. (2013). Dimensional change card sort performance associated with age-related differences in functional connectivity of lateral prefrontal cortex. *Developmental Cognitive Neuroscience, 5*, 40–50.

Fair, D. A., Dosenbach, N. U. F., Church, J. A., Cohen, A. L., Brahmbhatt, S., Miezin, F. M., et al. (2007). Development of distinct control networks through segregation and integration. *Proceedings of the National Academy of Sciences, 104*, 13507–13512.

Fuster, J. M. (1993). Frontal lobes. *Current Opinion in Neurobiology, 3*, 160–165.

Gao, W., Elton, A., Zhu, H., Alcauter, S., Smith, J. K., Gilmore, J. H., et al. (2014). Intersubject variability of and genetic effects on the brain's functional connectivity during infancy. *The Journal of Neuroscience, 34*, 11288–11296.

Gathercole, S. E., Pickering, S. J., Ambridge, B., & Wearing, H. (2004). The structure of working memory from 4 to 15 years of age. *Developmental Psychology, 40*, 177–190.

Giedd, J. N., Blumenthal, J., Jeffries, N. O., Castellanos, F. X., Liu, H., Zijdenbos, A., et al. (1999). Brain development during childhood and adolescence: A longitudinal MRI study. *Nature Neuroscience, 2*, 861–863.

Gogtay, N., Giedd, J. N., Lusk, L., Hayashi, K. M., Greenstein, D., Vaituzis, A. C., et al. (2004). Dynamic mapping of human cortical development during childhood through early adulthood. *Proceedings of the National Academy of Sciences of the United States of America, 101*, 8174–8179.

Grant, B. A., & Berg, E. A. (1948). A behavioral analysis of degree of reinforcement and ease of shifting to new responses in a Weigl-type card-sorting problem. *Journal of Experimental Psychology, 38*, 404–411.

Grossmann, T. (2013). Mapping prefrontal cortex functions in human infancy. *Infancy, 18*, 303–324.

Hampshire, A., & Sharp, D. (2015). Inferior PFC subregions have broad cognitive roles. *Trends in Cognitive Sciences, 19*, 712–713.

Harlow, J. M. (1868). Recovery from the passage of an iron bar through the head. *Publications of the Massachusetts Medical Society, 2*, 274–281.

Huizinga, M., Dolan, C. V., & van der Molen, M. W. (2006). Age-related change in executive function: Developmental trends and a latent variable analysis. *Neuropsychologia, 44*, 2017–2036.

Huttenlocher, P. R., & Dabholkar, A. S. (1997). Regional differences in synaptogenesis in human cerebral cortex. *Journal of Comparative Neurology, 387*, 167–178.

Inoue, Y., Sakihara, K., Gunji, A., Ozawa, H., Kimiya, S., Shinoda, H., et al. (2012). Reduced prefrontal hemodynamic response in children with ADHD during the Go/NoGo task: A NIRS study. *Neuroreport, 23*, 55–60.

Johnson, M. H. (2011). Interactive specialization: A domain-general framework for human functional brain development? *Developmental Cognitive Neuroscience, 1*, 7–21.

Kim, C., Cilles, S. E., Johnson, N. F., & Gold, B. T. (2012). Domain general and domain preferential brain regions associated with different types of task switching: A meta-analysis. *Human Brain Mapping, 33*, 130–142.

Kirkham, N. Z., Cruess, L., & Diamond, A. (2003). Helping children apply their knowledge to their behavior on a dimension-switching task. *Developmental Science, 6*, 449–467.

Klingberg, T. (2006). Development of a superior frontal–intraparietal network for visuospatial working memory. *Neuropsychologia, 44*, 2171–2177.

Konishi, S., Nakajima, K., Uchida, I., Kameyama, M., Nakahara, K., Sekihara, K., et al. (1998). Transient activation of inferior prefrontal cortex during cognitive set shifting. *Nature Neuroscience, 1*, 80–84.

Kwon, H., Reiss, A. L., & Menon, V. (2002). Neural basis of protracted developmental changes in visuo-spatial working memory. *Proceedings of the National Academy of Sciences, 99*, 13336–13341.

Laird, A. R., McMillan, K. M., Lancaster, J. L., Kochunov, P., Turkeltaub, P. E., Pardo, J. V., et al. (2005). A comparison of label-based review and ALE meta-analysis in the Stroop task. *Human Brain Mapping, 25*, 6–21.

Luria, A. R. (1966). *Higher cortical functions in man*. New York, NY: Basic Books.

Luria, A. R. (1973). *The working brain: An introduction to neuropsychology*. New York, NY: Basic Books.

Marsh, R., Zhu, H., Schultz, R. T., Quackenbush, G., Royal, J., Skudlarski, P., et al. (2006). A developmental fMRI study of self-regulatory control. *Human Brain Mapping, 27*, 848–863.

Marsh, R., Zhu, H., Wang, Z., Skudlarski, P., & Peterson, B. S. (2007). A developmental fMRI study of self-regulatory control in Tourette's syndrome. *American Journal of Psychiatry, 164*, 955–966.

Mehnert, J., Akhrif, A., Telkemeyer, S., Rossi, S., Schmitz, C. H., Steinbrink, J., et al. (2013). Developmental changes in brain activation and functional connectivity during response inhibition in the early childhood brain. *Brain and Development, 35*, 894–904.

Milner, B. (1963). Effects of different brain lesions on card sorting: The role of the frontal lobes. *Archives of Neurology, 9*, 90.

Miyake, A., & Friedman, N. P. (2012). The nature and organization of individual differences in executive functions: Four general conclusions. *Current Directions in Psychological Science, 21*, 8–14.

Miyake, A., Friedman, N. P., Emerson, M. J., Witzki, A. H., Howerter, A., & Wager, T. D. (2000). The unity and diversity of executive functions and their contributions to complex "frontal lobe" tasks: A latent variable analysis. *Cognitive Psychology, 41*, 49–100.

Monchi, O., Petrides, M., Petre, V., Worsley, K., & Dagher, A. (2001). Wisconsin card sorting revisited: Distinct neural circuits participating in different stages of the task identified by event-related functional magnetic resonance imaging. *The Journal of Neuroscience, 21*, 7733–7741.

Moriguchi, Y., & Hiraki, K. (2009). Neural origin of cognitive shifting in young children. *Proceedings of the National Academy of Sciences, 106*, 6017–6021.

Moriguchi, Y., & Hiraki, K. (2011). Longitudinal development of prefrontal function during early childhood. *Developmental Cognitive Neuroscience, 1*, 153–162.

Moriguchi, Y., & Hiraki, K. (2013). Prefrontal cortex and executive function in young children: A review of NIRS studies. *Frontiers in Human Neuroscience, 7*, 867.

Moriguchi, Y., & Hiraki, K. (2014). Neural basis of learning from television in young children. *Trends in Neuroscience and Education, 3*, 122–127.

Moriguchi, Y., Sakata, Y., Ishibashi, M., & Ishikawa, Y. (2015). Teaching others rule-use improves executive function and prefrontal activations in young children. *Frontiers in Psychology, 6*, 894.

Morton, J. B., Bosma, R., & Ansari, D. (2009). Age-related changes in brain activation associated with dimensional shifts of attention: An fMRI study. *NeuroImage, 46*, 249–256.

Nee, D. E., Brown, J. W., Askren, M. K., Berman, M. G., Demiralp, E., Krawitz, A., et al. (2013). A meta-analysis of executive components of working memory. *Cerebral Cortex, 23*, 264–282.

Peterson, B. S., Potenza, M. N., Wang, Z., Zhu, H., Martin, A., Marsh, R., et al. (2009). An fMRI study of the effects of psychostimulants on default-mode processing during Stroop task performance in youths with ADHD. *American Journal of Psychiatry, 166*, 1286–1294.

Richmond, S., Johnson, K. A., Seal, M. L., Allen, N. B., & Whittle, S. (2016). Development of brain networks and relevance of environmental and genetic factors: A systematic review. *Neuroscience & Biobehavioral Reviews, 71*, 215–239.

Roberts, K. L., & Hall, D. A. (2008). Examining a supramodal network for conflict processing: A systematic review and novel functional magnetic resonance imaging data for related visual and auditory Stroop tasks. *Journal of Cognitive Neuroscience, 20*, 1063–1078.

Rottschy, C., Langner, R., Dogan, I., Reetz, K., Laird, A. R., Schulz, J. B., et al. (2012). Modelling neural correlates of working memory: A coordinate-based meta-analysis. *NeuroImage, 60*, 830–846.

Rubia, K., Smith, A. B., Woolley, J., Nosarti, C., Heyman, I., Taylor, E., et al. (2006). Progressive increase of frontostriatal brain activation from childhood to adulthood during event-related tasks of cognitive control. *Human Brain Mapping, 27*, 973–993.

Scherf, K. S., Sweeney, J. A., & Luna, B. (2006). Brain basis of developmental change in visuospatial working memory. *Journal of Cognitive Neuroscience, 18*, 1045–1058.

Schroeter, M. L., Zysset, S., Wahl, M., & von Cramon, D. Y. (2004). Prefrontal activation due to Stroop interference increases during development: An event-related fNIRS study. *NeuroImage, 23*, 1317–1325.

Stroop, J. R. (1935). Studies of interference in serial verbal reactions. *Journal of Experimental Psychology, 18*, 643–662.

Tsujimoto, S., Yamamoto, T., Kawaguchi, H., Koizumi, H., & Sawaguchi, T. (2004). Prefrontal cortical activation associated with working memory in adults and preschool children: An event-related optical topography study. *Cerebral Cortex, 14*, 703–712.

Wager, T. D., & Smith, E. E. (2003). Neuroimaging studies of working memory. *Cognitive, Affective, & Behavioral Neuroscience, 3*, 255–274.

Wendelken, C., Munakata, Y., Baym, C., Souza, M., & Bunge, S. A. (2012). Flexible rule use: Common neural substrates in children and adults. *Developmental Cognitive Neuroscience, 2*, 329–339.

Wiebe, S. A., Espy, K. A., & Charak, D. (2008). Using confirmatory factor analysis to understand executive control in preschool children: I. Latent structure. *Developmental Psychology, 44*, 575–587.

Willoughby, M. T., Blair, C. B., Wirth, R., & Greenberg, M. (2012). The measurement of executive function at age 5: Psychometric properties and relationship to academic achievement. *Psychological Assessment, 24*, 226.

Wolfe, C. D., & Bell, M. A. (2004). Working memory and inhibitory control in early childhood: Contributions from physiology, temperament, and language. *Developmental Psychobiology, 44*, 68–83.

Xiao, T., Xiao, Z., Ke, X., Hong, S., Yang, H., Su, Y., et al. (2012). Response inhibition impairment in high functioning autism and attention deficit hyperactivity disorder: Evidence from near-infrared spectroscopy data. *PloS One, 7*, e46569.

Yakovlev, P. I., & Lecours, A.-R. (1967). The myelogenetic cycles of regional maturation of the brain. In A. Minkowski (Ed.), *Regional development of the brain in early life* (pp. 3–70). Oxford: Blackwell.

Yasumura, A., Kokubo, N., Yamamoto, H., Yasumura, Y., Moriguchi, Y., Nakagawa, E., et al. (2012). Neurobehavioral and hemodynamic evaluation of cognitive shifting in children with autism spectrum disorder. *Journal of Behavioral and Brain Science, 2*, 463–470.

Zelazo, P. D. (2004). The development of conscious control in childhood. *Trends in Cognitive Sciences, 8*(1), 12–17.

Zelazo, P. D., Anderson, J. E., Richler, J., Wallner-Allen, K., Beaumont, J. L., & Weintraub, S. (2013). NIH Toolbox Cognition Battery (NIHTB-CB): Measuring executive function and attention. *Monographs of the Society for Research in Child Development, 78*(4), 16–33.

Zelazo, P. D., Frye, D., & Rapus, T. (1996). An age-related dissociation between knowing rules and using them. *Cognitive Development, 11*, 37–63.

Zelazo, P. D., & Muller, U. (2002). Executive function in typical and atypical development. In U. Goswami (Ed.), *Blackwell handbook of childhood cognitive development* (pp. 445–469). Malden, MA: Blackwell Publishing.

6

AGING AND THE NEURAL CORRELATES OF EXECUTIVE FUNCTION

Robert West

The idea that aging is associated with a decline in the efficiency of executive function has a long history in the literature (for an extended review see Braver & West, 2008). For the last three decades, commentators have argued that those cognitive functions supported by the prefrontal cortex are more sensitive to the effects of aging than those supported by posterior brain structures (Moscovitch & Winocur, 1992; West, 1996). While this conceptualization has evolved over the years (Davis, Dennis, Daselaar, Fleck, & Cabeza, 2008; Greenwood, 2000; Verhaeghen, 2011), as a reader of this chapter will see, executive function and the prefrontal cortex remain central to our understanding of the cognitive neuroscience of aging, as well as development cognitive neuroscience more generally (Cuevas, Rajan, & Bryant, Chapter 1, this volume; Crone, Peters, & Steinbeis, Chapter 3, this volume). The centrality of executive function for cognitive aging becomes clear when one considers that as much as 30% of the aging population may experience impairment of executive function (Denburg et al., 2007). A poignant example of the societal implications of age-related decline in executive function is highlighted in a paper by Fisher, Franklin, and Post (2014), wherein the authors analyzed decision making in prominent world leaders, many of whom continued to have significant political influence well into their 70s or 80s, finding that some proportion of our older leaders were likely experiencing significant decline in executive functions while still in office.

Our understanding of the effects of aging on the neural basis of executive function has been enhanced by parallel developments in structural (e.g., diffusion tensor imaging) and functional magnetic resonance imaging (fMRI), and moderate to high density electroencephalogram (EEG). Functional MRI and EEG have been used to quantify alterations in the spatial and temporal distribution of neural activity associated with various cognitive processes (Grady, 2012;

West, 2016). Structural measures derived from MRI have also been used to examine the association between age-related variation in the integrity of gray and white matter and neuropsychological or psychometric measures of executive function (Bennett & Madden, 2014; Yuan & Raz, 2014). An interesting example of such a study is found in a paper by Head, Kennedy, Rodrigue, and Raz (2009). In this paper, the authors used path analysis to demonstrate that age-related increases in perseverative errors in the Wisconsin Card Sorting Task (WCST) – a classic measure of executive function – was related to age-related reduction in gray matter volume in the lateral prefrontal cortex, and that this effect was mediated through associations between lateral prefrontal cortex and inhibitory control, working memory, and temporal processing (Figure 6.1). In essence, these data demonstrate how age-related reduction in regional brain

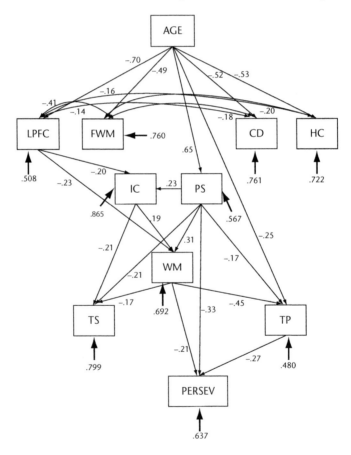

FIGURE 6.1 Path diagram demonstrating how age-related differences in volume within the left prefrontal cortex is related to perseveration errors in the WCST through basic cognitive processes related in inhibitory control, working memory and temporal processing (adapted with permission from figure 1 of Head et al., 2009).

volume may affect specific cognitive process, which may then be manifest as age-related differences in complex behavior.

Knowledge of the neural basis of executive function has greatly increased over the last two decades owing to complementary work within cognitive and social neuroscience, computational neuroscience, and neuropsychology (Asp & Tranel, 2012; Koechlin, 2012; Stuss & Knight, 2012). Paralleling the general development of these fields, we have also come to have a much greater appreciation for alterations in executive function associated with both typical and pathological aging (Lindenberger, Burzynska, & Nagel, 2012; Verhaeghen, 2011). Perhaps one of the most significant advances in the study of executive function has been the movement from a rather general description of executive or frontal lobe processes to a more nuanced account of how specific neurocognitive processes are affected by the aging process (e.g., Paxton, Barch, Racine, & Braver, 2008; West & Travers, 2008). While this clearly represents a welcome advancement for the field, it also presents a challenge in providing a concise overview of the state of the literature related to the effects of aging on the neural bases of executive function. Therefore, in considering the structure of this chapter I have adopted an integrative framework of anterior attentional functions proposed by Stuss, Shallice, Alexander, and Picton (1995) that has its foundation in the Supervisory System model of Norman and Shallice (1986).

Supervisory System Framework

The framework proposed by Stuss et al. (1995) incorporates five supervisory processes that include energizing schemata, inhibiting schemata, adjusting contention-scheduling, monitoring the level of activity in schemata, and the control of "if–then" logical processes that act upon schemata to facilitate goal-directed information processing and action (Figure 6.2). The five processes described in the Supervisory System framework overlap to some degree with those described in a model of executive control introduced by Miyake et al. (2000) wherein control is realized through processes that support shifting, updating, and inhibition. The highly influential tripartite model of executive control has served as the foundation for a number of studies examining the development of executive function across the life span as considered in Chapters 5 and 6 of this volume.

In the Supervisory System framework, schemata can be considered acquired knowledge structures that guide established perception–action cycles and facilitate routine information processing. As an example, in the Stroop task (Stroop, 1935) – which is commonly considered a measure of executive function – an individual might possess two relevant schemata: one that supports word reading and another that supports color naming. These schemata may have developed through long-term interactions with the environment (i.e., learning to read and the names of colors during childhood) or emerge through direct instruction and

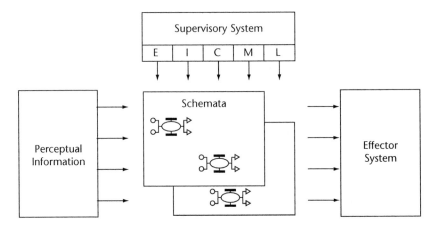

FIGURE 6.2 Conceptual representation of the five supervisory processes (E – energizing, I – inhibiting, C – adjustment of contention scheduling, M – monitoring, L – if–then logical processes) as described within the Supervisory System framework (adapted with permission from figure 1 of Stuss et al., 1995).

practice (i.e., learning that a set of colors is associated with a specific set of response keys used for a laboratory task). Contention scheduling is thought to operate through the lateral inhibition of associated or competing schemata. For instance, in the Stroop task naming the color of a stimulus over several trials would be expected to increase lateral inhibition of the word reading schemata.

The contribution of each of the supervisory processes to the control of cognition and action can also be appreciated within the context of the Stroop task. Energizing the color naming schema and inhibiting the word reading schema are perhaps the most obvious supervisory processes utilized during performance of the Stroop task, and there is empirical (Lindsay & Jacoby, 1994; West & Alain, 1999) and computational (Cohen, Dunbar, & McCelland, 1990) evidence demonstrating the contribution of both processes to task performance. Monitoring processes may be sensitive to both the level of interference that is experienced on incongruent trials (West & Bailey, 2012), and to the presence of intrusion errors where the task irrelevant dimension guides the response (West, 2004). The experience of greater interference and intrusion errors is associated with the recruitment of the anterior cingulate cortex (ACC) when either color naming or word reading is required, demonstrating the sensitivity of monitoring processes to dynamic task demands (Kerns et al., 2004; West, 2004). The recruitment of if–then logical analysis in the Stroop task may be realized in two or more ways. Within the typical Stroop task, if–then logical analysis might be used to restore goal-directed action following an intrusion error, being activated by monitoring processes, and serving to tune processes that support energizing or inhibiting, or adjustment of contention scheduling. In contrast, within the switching Stroop task, if–then logical analysis would use information from a cue consistently associated with an increase in mixing costs (i.e., the slowing of response time resulting from interleaving two or more tasks over time), while age-related differences in switching costs are less robust across studies (Wasylyshyn, Verhaeghen, & Sliwinski, 2011). In the neuroimaging literature, aging can be associated with either increases or decreases in task-related neural recruitment in older adults relative to younger adults (e.g., Kunimi, Kiyama, & Nakai, 2016; Madden et al., 2010).

A study by West and Travers (2008) provides an example of how ERPs have been used to examine the effects of aging on processes related to mixing cost, cue encoding, and task switching. This study revealed that the age-related increase in mixing costs is associated with age-related differences in slow wave ERP activity between younger and older adults. In younger adults this slow wave activity was greatest over the frontal-polar and parietal regions, while in older adults this activity extended from the frontal to the parietal regions. In contrast, cue encoding was associated with an age-related decrease in ERP activity, while switching from word to color processing was related to an increase in the amplitude of ERP activity in older adults relative to younger adults. Together these data reveal that aging may have different effects on subprocesses that underlie the implementation of if–then processes that may interact with task demands or prior experience.

Adjustment of Contention Scheduling and Aging

Processes underlying the adjustment of contention scheduling are probably the least clearly understood within the original description of the Supervisory System framework (Stuss et al., 1995); this makes it difficult to evaluate the potential effects of aging on this component of executive function. If the tuning of information processing related to the conflict adaptation effect is taken as a marker of such adjustments, then the available behavioral evidence indicates that these processes may be relatively immune to the aging process (Larson et al., 2016). As described in a previous section of this chapter, the one study that has examined neural activity related to conflict adaptation reveals that the age-related invariance seen at the behavioral level is associated with differential neural recruitment in younger and older adults (West & Moore, 2005). Based upon these limited findings, one fruitful line of future research may be to both identify consistent neural correlates of adjustments of contention scheduling across relevant tasks, and to then examine the effects of aging on neural recruitment in these paradigms.

Conclusions

The last two decades have witnessed tremendous advances in our understanding of the effects of aging on the neural correlates of executive function supported by conceptual developments regarding what executive function represents (Miyake et al., 2000) and methodological developments in our ability to measure the structural and functional correlates of neurocognitive processes (Grady, 2012). Within the context of the Supervisory System framework, aging is clearly associated with alterations in neural recruitment related to some aspects of energizing, inhibiting, monitoring, and if–then logical processing, while other aspects of executive processing may be preserved in later adulthood. One avenue for future research to pursue is the development of a tighter coupling of the linkage between neural and behavioral markers of executive function that should emerge from the creative efforts of current and new investigators within the field.

References

Asp, E., & Tranel, D. (2012). False tagging theory: Toward a unitary account of prefrontal cortex function. In D. T. Stuss & R. T. Knight (Eds.), *Principles of frontal lobe function* (2nd ed., pp. 383–416). Oxford: Oxford University Press.

Bennett, I. J., & Madden, D. J. (2014). Disconnected aging: Cerebral white matter integrity and age-related difference in cognition. *Neuroscience, 12*, 187–205.

Botvinick, M. M., Braver, T. S., Barch, D. M., Carter, C. S., & Cohen, J. D. (2001). Conflict monitoring and cognitive control. *Psychological Review, 108*, 624–652.

Braver, T. S. (2012). The variable nature of cognitive control: A dual mechanisms framework. *Trends in Cognitive Sciences, 16*, 106–113.

Braver, T. S., Satpute, A. B., Rush, B. K., Racine, C. A., & Barch, D. M. (2005). Context processing and context maintenance in healthy aging and early stage dementia of the Alzheimer's type. *Psychology and Aging, 20*, 33–46.

Braver, T. S., & West, R. (2008). Working memory, executive control and aging. In F. I. M. Craik & T. A. Salthouse (Eds.), *The handbook of aging and cognition* (3rd ed., pp. 311–372). New York, NY: Psychology Press.

Bugg, J. M., Jacoby, L. L., & Toth, J. P. (2008). Multiple levels of control in the Stroop task. *Memory & Cognition, 36*, 1484–1494.

Carlson, M. C., Hasher, L., Connelly, C., & Zacks, R. T. (1995). Aging, distraction, and the benefits of predictable location. *Psychology and Aging, 10*, 427–436.

Carter, C. S., Braver, T. S., Barch, D. M., Botvinick, M. W., Noll, D., & Cohen, J. D. (1998). Anterior cingulate cortex, error detection, and the online monitoring of performance. *Science, 280*, 747–749.

Chadick, J. Z., Zanto, T. P., & Gazzaley, A. (2014). Structural and functional differences in medial prefrontal cortex underlie distractibility and suppression deficits in ageing. *Nature Communications, 5*, 4223.

Cohen, J. D., Dunbar, K., & McCelland, J. L. (1990). On the control of automatic processes: A parallel distributed processing account of the Stroop effect. *Psychological Review, 97*, 332–361.

Davis, S. W., Dennis, N. A., Daselaar, S. M., Fleck, M. S., & Cabeza, R. (2008). Que PASA? The posterior-anterior shift in aging. *Cerebral Cortex, 18*, 1201–1209.

Dehaene, S., Posner, M. I., & Tucker, D. M. (1994). Localization of a neural system for error detection and compensation. *Psychological Science, 5*, 303–305.

Denburg, N. L., Cole, C. A., Hernandez, M., Yamada, T. H., Tranel, D., Bechara, A., & Wallice, R. B. (2007). The orbitofrontal cortex, real-world decision making, and normal aging. *Annals of the New York Academy of Sciences, 1121*, 480–498.

Eppinger, B., Kray, J., Mock, B., & Mecklinger, A. (2008). Better or worse than expected? Aging, learning, and the ERN. *Neuropsychologia, 46*, 521–539.

Fisher, M., Franklin, D. L., & Post, J. M. (2014). Executive dysfunction, brain aging, and political leadership. *Politics and the Life Sciences, 33*, 93–102.

Gazzaley, A., Clapp, W., Kelley, J., McEvoy, K., Knight, R. T., & D'Esposito, M. (2008). Age-related top-down suppression deficit in the early stages of cortical visual memory processing. *Proceeding of the National Academy of Sciences, 105*, 13122–13126.

Gazzaley, A., Cooney, J. W., Rissman, J., & D'Esposito, M. (2005). Top-down suppression deficit underlies working memory impairment in normal aging. *Nature Neuroscience, 8*, 1298–1300.

Gehring, W. J., Goss, B., Coles, M. G. H., Meyer, D. E., & Donchin, E. (1993). A neural system for error detection and compensation. *Psychological Science, 4*, 385–390.

Gehring, W. J., & Willoughby, A. R. (2002). The medial frontal cortex and the rapid processing of monetary gains and losses. *Science, 295*, 2279–2282.

Grady, C. (2012). The cognitive neuroscience of aging. *Nature Reviews Neuroscience, 13*, 491–505.

Greenwood, P. M. (2000). The frontal aging hypothesis evaluated. *Journal of the International Neuropsychological Society, 6*, 705–726.

Hämmerer, D., Li, S., Müller, V., & Lindenberger, U. (2010). Life span differences in electrophysiological correlates of monitoring gains and losses during probabilistic reinforcement learning. *Journal of Cognitive Neuroscience, 23*, 579–592.

Hasher, L., & Zacks, R. T. (1988). Working memory, comprehension, and aging: A review and a new view. In G. G. Bower (Ed.), *The psychology of learning and motivation* (Vol. 22, pp. 193–225). San Diego, CA: Academic Press.

Head, D., Kennedy, K. M., Rodrigue, K. M., & Raz, N. (2009). Age differences in perseveration: Cognitive and neuroanatomical mediators of performance on the Wisconsin Card Sorting Test. *Neuropsychologia, 47*, 1200–1203.

Jacoby, L. L., Lindsay, D. S., & Hessels, S. (2003). Item-specific control of automatic processes: Stroop process dissociations. *Psychonomic Bulletin & Review, 10*, 638–644.

Kerns, J. G., Cohen, J. D., MacDonald, A. W., III, Cho, R. Y., Stenger, V. A., & Carter, C. S. (2004). Anterior cingulate conflict monitoring and adjustments in control. *Science, 303*, 1023–1026.

Koechlin, E. (2012). Motivation, control, and human prefrontal executive function. In D. T. Stuss & R. T. Knight (Eds.), *Principles of frontal lobe function* (2nd ed., pp. 279–291). Oxford: Oxford University Press.

Kunimi, M., Kiyama, S., & Nakai, T. (2016). Investigation of age-related changes in brain activity during the divalent task-switching paradigm using functional MRI. *Neuroscience Research, 103*, 18–26.

Larson, M. J., Clayson, P. E., Keith, C. M., Hunt, I. J., Hedges, D. W., Nielson, B. L., & Call, V. R. A. (2016). Cognitive control adjustments in healthy older and younger adults: Conflict adaption, the error-related negativity (ERN), and evidence of generalized decline with age. *Biological Psychology, 115*, 50–63.

Lindenberger, U., Burzynska, A. Z., & Nagel, I. E. (2012). Heterogeneity in frontal lobe aging. In D. T. Stuss & R. T. Knight (Eds.), *Principles of frontal lobe function* (2nd ed., pp. 609–627). Oxford: Oxford University Press.

Lindsay, D. S., & Jacoby, L. L. (1994). Stroop process dissociations: The relationship between facilitation and interference. *Journal of Experimental Psychology: Human Performance and Perception, 20*, 219–234.

Lustig, C., Hasher, L., & Zacks, R. T. (2007). Inhibitory deficit theory: Recent developments in a "New View". In D. S. Gorfein & C. M. MacLeod (Eds.), *Inhibition in cognition* (pp. 145–162). Washington, DC: American Psychological Association.

Lustig, C., & Jantz, T. (2014). Questions of age difference in interference control: When and how, not if? *Brain Research, 1612*, 59–69.

Madden, D. J., Costello, M. C., Dennis, N. A., Davis, S. W., Shepler, A. M., Spaniol, J., et al. (2010). Adult age differences in functional connectivity during executive control. *NeuroImage, 15*, 643–657.

Mathalon, D. H., Bennett, A., Askari, N., Gray, E. M., Rosenbloom, M. J., & Ford, J. M. (2003). Response-monitoring dysfunction in aging and Alzheimer's disease: An event-related potential study. *Neurobiology of Aging, 24*, 675–685.

Mathewson, K. J., Dywan, J., & Segalowitz, S. J. (2005). Brain bases of error-related ERPs as influenced by age and task. *Biological Psychology, 70*, 88–104.

Miyake, A., Friedman, N. P., Emerson, M. J., Witzki, A. H., Howerter, A., & Wager, T. D. (2000). The unity and diversity of executive functions and their contributions to complex "frontal lobe" tasks: A latent variable analysis. *Cognitive Psychology, 41*, 49–100.

Moscovitch, M., & Winocur, G. (1992). The neuropsychology of memory and aging. In F. I. M. Craik & T. A. Salthouse (Eds.), *The handbook of aging and cognition* (pp. 315–372). Hillsdale, NJ: Lawrence Erlbaum Associates.

Norman, D. A., & Shallice, T. (1986). Attention to action: Willed and automatic control of behaviour. In R. J. Davidson, G. E. Schwartz, & D. Shapiro (Eds.), *Consciousness and self-regulation: Advances in research and theory* (Vol. 4, pp. 1–18). New York, NY: Plenum.

Paxton, J. L., Barch, D. M., Racine, C. A., & Braver, T. S. (2008). Cognitive control, goal maintenance, and prefrontal function in healthy aging. *Cerebral Cortex, 18*, 1010–1028.

Pratte, M. S., Rouder, J. N., Morey, R. D., & Feng, C. (2010). Exploring the differences in distributional properties between Stroop and Simon effects using delta plots. *Attention, Perception, & Psychophysics, 7*, 2013–2025.

Rowe, G., Valderrama, S., Hasher, L., & Lenartowicz, A. (2006). Attentional disregulation: A benefit for implicit memory. *Psychology and Aging, 21*, 826–830.

Stroop, J. R. (1935). Studies of interference in serial verbal reactions. *Journal of Experimental Psychology, 18*, 643–661.

Stuss, D. T., & Knight, R. T. (2012). Introduction: Past and future. In D. T. Stuss & R. T. Knight (Eds.), *Principles of frontal lobe function* (2nd ed., pp. 1–11). Oxford: Oxford University Press.

Stuss, D. T., Shallice, T., Alexander, M. P., & Picton, T. W. (1995). A multidisciplinary approach to anterior attentional functions. *Annals of the New York Academy of Sciences, 769*, 191–211.

Travers, S., & West, R. (2008). Neural correlates of cue retrieval, task set configuration, and rule mapping in the explicit cue task switching paradigm. *Psychophysiology, 45*, 588–601.

van Veen, V., & Carter, C. S. (2002). The timing of action-monitoring processes in the anterior cingulate cortex. *Journal of Cognitive Neuroscience, 14*, 593–602.

Verhaeghen, P. (2011). Aging and executive control: Reports of a demise greatly exaggerated. *Current Directions in Psychological Sciences, 20*, 174–180.

Walsh, M. M., & Anderson, J. R. (2012). Learning from experience: Event-related potential correlates of reward processing, neural adaptation, and behavioral choice. *Neuroscience and Biobehavioral Reviews, 36*, 1870–1884.

Wasylyshyn, C., Verhaeghen, P., & Sliwinski, M. J. (2011). Aging and task switching: A meta-analysis. *Psychology and Aging, 26*, 15–20.

West, R. (2003). Neural correlates of cognitive control and conflict detection in the Stroop and digit-location tasks. *Neuropsychologia, 41*, 1122–1135.

West, R. (2004). The effects of aging on controlled attention and conflict processing in the Stroop task. *Journal of Cognitive Neuroscience, 16*, 103–113.

West, R. (2016). Event-related potentials. In N. A. Pachana (Ed.), *Encyclopedia of gerontology*. Singapore: Springer Science.

West, R., & Alain, C. (1999). Event-related neural activity associated with the Stroop task. *Cognitive Brain Research, 8*, 157–164.

West, R., & Alain, C. (2000). Age-related decline in inhibitory control contributes to the increased Stroop effect observed in older adults. *Psychophysiology, 37*, 179–189.

West, R., & Bailey, K. (2012). ERP correlates of dual mechanisms of control in the counting Stroop task. *Psychophysiology, 49*, 1309–1318.

West, R., & Baylis, G. C. (1998). Effects of increased response dominance and contextual disintegration on the Stroop interference effect in older adults. *Psychology and Aging, 13*, 206–217.

West, R., & Moore, K. (2005). Adjustments of cognitive control in younger and older adults. *Cortex, 41*, 570–581.

West, R., & Schwarb, H. (2006). The influence of aging and frontal status on the neural correlates of regulative and evaluative aspects of cognitive control. *Neuropsychology, 20*, 468–481.

West, R., Tiernan, B. N., Kieffaber, P. D., Bailey, K., & Anderson, S. (2014). The effect of age on the neural correlates of feedback processing in a naturalistic gambling game. *Psychophysiology, 51*, 734–745.

West, R., & Travers, S. (2008). Differential effects of aging on processes underlying task switching. *Brain and Cognition, 68*, 67–80.

West, R., & Travers, S. (2008). Tracking the temporal dynamics of updating cognitive control: An examination of error processing. *Cerebral Cortex, 18*, 1112–1124.

West, R. L. (1996). An application of prefrontal cortex function theory to cognitive aging. *Psychological Bulletin, 120*, 272–292.

Wild-Wall, N., Falkenstein, M., & Hohnsbein, J. (2008). Flanker interference in young and old participants as reflected in event-related potentials. *Brain Research, 1211*, 72–84.

Yuan, P., & Raz, N. (2014). Prefrontal cortex and executive functions in healthy adults: A meta-analysis of structural neuroimaging studies. *Neuroscience and Biobehavioral Reviews, 42*, 180–192.

Zanto, T. P., Hennigan, K., Ostberg, M., Clapp, W. C., & Gazzaley, A. (2010). Predictive knowledge of stimulus relevance does not influence top-down suppression of irrelevant information in older adults. *Cortex, 46*, 564–574.

7
GENETIC INFLUENCES ON EXECUTIVE FUNCTIONS ACROSS THE LIFE SPAN

James J. Li & Delanie K. Roberts

There is consensus that genetic influences play an important role in just about every psychological trait, including executive functions (EF) (Friedman et al., 2008; Miyake, Friedman, Emerson, Witzki, & Howerter, 2000). This consensus emerged out of several decades of behavioral genetic studies – primarily those involving twins and families – which consistently indicate robust genetic influences contributing to individual differences with respect to cognition and behavior (see meta-analysis by Bergen, Gardner, & Kendler, 2007). However, the field of genetics has undergone tremendous shifts over the last 20 years. The purpose of this chapter is to provide a general overview of how genetic factors contribute to variation in EF over the life span, summarizing studies from across the genetics landscape, including twin studies, candidate gene studies, and, more recently, genome-wide association studies (GWAS). We will answer three key questions in this chapter: (1) how heritable (i.e., genetically influenced) are EF? (2) Is there any evidence that genetic influences affect the stability (or malleability) of EF across development? And (3) what genes contribute to EF?

How Heritable Are EFs?

One way of discerning the heritability of a trait is through the classic twin study. The twin study methodology is described elsewhere (see Neale & Cardon, 1992), but the basic premise is summarized here. The twin study involves trait comparisons between monozygotic (MZ) and dizygotic (DZ) twins. Given that MZ twins share 100% of their genes (and their family environment as well), trait differences within MZ twin-pairs are assumed to be due to environmental factors (i.e., non-shared environmental influences). DZ twins share roughly 50% of their genes, thus allowing researchers to estimate heritability (h^2) by comparing how correlated MZ and DZ twins are for a trait of interest (Dick, 2011). Generally speaking, higher trait correlations among MZ twins than DZ twins suggest the prominence of a genetic effect whereas equal correlations between MZ and DZ twins suggest the likelihood of a strong environmental effect. The strength of the twin design is that it permits the study of genetic influences through a natural experiment of sorts, allowing researchers to understand the "impact of genetic variation all at once" without the need to collect DNA from each individual (Kendler, 2011, p. 7).

Heritability estimates assessed from twin studies of EF are presented in Table 7.1 (for child and adolescent twin studies) and Table 7.2 (for adult twin studies). Generally, studies indicate that among children and adolescents, heritability for EF during this developmental epoch are small to moderate, with h^2 estimates ranging from 0.09 to 0.56 across the EF dimensions. During adulthood, where one would expect a higher heritability in EF (i.e., Bergen et al., 2007), heritability estimates of EF were about in line with estimates reported in child and adolescent samples, with h^2 estimates ranging from 0.05 to 0.73 (see Table 7.2). On the surface, these studies suggest that genes, on average, matter about as much as environments in contributing to individual differences in EF. Perhaps one reason why heritability estimates tend to be so modest is due to the genetic and phenotypic heterogeneity among the "lower-level processes" of EF (Friedman & Miyake, 2004). For instance, a substantial proportion of the variance underlying inhibition may be driven by other EF dimensions, such as working memory and set-shifting (Young et al., 2009). As EF dimensions tend to be highly correlated with each other, they may constitute a "collection of related but separable abilities" that can be accounted for by a single common factor (Friedman et al., 2008, p. 201). Young and colleagues (2009) showed that when inhibition was characterized as a common latent factor that included multiple laboratory measures of EF (i.e., antisaccade task, Stop-Signal task, and Stroop task), individual differences on EF were almost *entirely* driven by genetic influences ($h^2 = \sim 0.99$). Similarly, a twin analysis was conducted using a sample of 272 school-aged twins to examine genetic and environmental influences of inhibition, switching, working memory, and updating and on a higher-order general factor of EF (Engelhardt, Briley, Mann, Harden, & Tucker-Drob, 2015). Consistent with Young et al. (2009), individual tasks of EF were only moderately heritable ($h^2 = 0.17$–0.47), but a higher-order common factor was *entirely* heritable ($h^2 = 1.00$).

Thus, genetic factors potentially operate on multiple levels as they contribute to individual differences in EF. A common set of highly influential genes may have top-down effects on lower-level EF processes, governing how one can "actively maintain goals and use them to bias lower-level processing" (Friedman et al., 2016, p. 336). At the same time, there may be a set of genes that uniquely contribute to each dimension of EF, which are also influenced by environmental factors. Twin studies using a hierarchical factor framework for EF,

TABLE 7.1 EF Heritability Estimates in Children and Adolescents

EF Dimension	Study	Sample Ages	Sample Size	Heritability
Inhibition				
Stop-Signal	Schachar et al., 2011	8	131 twin pairs (55 MZ, 76 DZ)	0.27–0.50
	Young et al., 2009	11–20	293 twin pairs (159 MZ, 134 DZ)	0.17–0.18
	Friedman et al., 2016	16–28	401 twin pairs (214 MZ, 187 DZ)	0.12–0.44
Stroop	Young et al., 2009	11–20	293 twin pairs (159 MZ, 134 DZ)	0.18–0.20
	Stins et al., 2004	12	145 twin pairs (78 MZ, 67 DZ)	0.50
	Friedman et al., 2016	16–28	401 twin pairs (214 MZ, 187 DZ)	0.38–0.49
Go/No-Go (other)	Groot et al., 2004	6	237 twin pairs (125 MZ, 112 DZ)	0.30–0.54
	Kuntsi et al., 2006	7–9	400 twin pairs (156 MZ, 244 DZ)	0.10–0.54
Other	Wang et al., 2012	6–8	202 twin pairs (90 MZ, 112 DZ)	0.64
	Friedman et al., 2016	16–28	401 twin pairs (214 MZ, 187 DZ)	0.42–0.54
Working Memory				
Digit Span	Kuntsi et al., 2006	7–9	400 twin pairs (156 MZ, 244 DZ)	0.36
	Polderman et al., 2006	12	177 twin pairs (97 MZ, 80 DZ)	0.56
	Ando et al., 2001	16–29	236 twin pairs (143 MZ, 93 DZ)	0.43–0.49

N-back	Friedman et al., 2016	16–28	401 twin pairs (214 MZ, 187 DZ)	0.25–0.53
Other	Wang & Saudino, 2013	3	304 twin pairs (140 MZ, 164 DZ)	0.29
	Wang et al., 2012	6–8	202 twin pairs (90 MZ, 112 DZ)	0.30
	Polderman et al., 2006	12	177 twin pairs (97 MZ, 80 DZ)	0.43–0.54
	Luciano et al., 2001	15–18	390 twin pairs (184 MZ, 206 DZ)	0.36
	Friedman et al., 2016	16–28	401 twin pairs (214 MZ, 187 DZ)	0.53–0.69
Cognitive Flexibility				
Wisconsin Card Sort	Anokhin et al., 2010	12–14	367 twin pairs (166 MZ, 201 DZ)	0.19–0.49
	Chou et al., 2010	12–16	350 twin pairs (257 MZ, 93 DZ) and 47 same-sex siblings	0.09
	Godinez et al., 2012	15–20	356 twin pairs (191 MZ, 165 DZ) and 40 individuals	0.10–0.42
Other	Friedman et al., 2016	16–28	401 twin pairs (214 MZ, 187 DZ)	0.28–0.53
Common EF Factor	Coolidge et al., 2000	8	112 twin pairs (70 MZ, 42 DZ)	0.77
	Engelhardt et al., 2015	7–15	272 twin pairs (84 MZ, 188 DZ)	1.00
	Friedman et al., 2008	15–20	291 twin pairs (158 MZ, 133 DZ)	0.99

TABLE 7.2 EF Heritability Estimates in Adults

EF Dimension	Study	Sample Ages	Sample Size	Heritability
Inhibition				
Stroop	Lee et al., 2012	65–88	215 twin pairs (117 MZ, 98 DZ) and 42 individuals	0.57
Go/No-Go (other)	Anokhin et al., 2004	18–28	194 female twin pairs (52 MZ, 45 DZ)	0.37–0.46
Working Memory				
Digit Span	Finkel et al., 1995	27–64	124 twin pairs (62 MZ, 62 DZ): Swedish	0.33
	Finkel et al., 1995	27–88	191 twin pairs (119 MZ, 72 DZ): MTSADA	0.39
	Kremen et al., 2007	41–58	345 twin pairs (176 MZ, 169 DZ)	0.27
	Pedersen et al., 1992	50–84	302 twin pairs (113 MZ, 189 DZ)	0.44
	Plomin et al., 1994	64.1 (SD = 7.5)	223 twin pairs (82 MZ, 141 DZ)	0.49–0.52
	Finkel et al., 1995	65–85	85 twin pairs (31 MZ, 51 DZ): Swedish	0.28
	Lee et al., 2012	65–88	215 twin pairs (117 MZ, 98 DZ) and 42 individuals	0.62
N-back	Blokland et al., 2008	21–27	60 twin pairs (29 MZ, 31 DZ)	0.57–0.73
Other	Pedersen et al., 1992	50–84	302 twin pairs (113 MZ, 189 DZ)	0.32–0.42
	Plomin et al., 1994	64.1 (SD = 7.5)	223 twin pairs (82 MZ, 141 DZ)	0.36–0.51
Cognitive Flexibility				
Trail-Making	Vasilopoulos et al., 2012	51–60	615 twin pairs (347 MZ, 268 DZ) and 7 unpaired twins	0.34–0.65

accounting for the "unity and diversity" of EF (Miyake et al., 2000), have only recently emerged in the literature to provide circumstantial support for this hypothesis (see Friedman et al., 2008, 2016; Engelhardt et al., 2015; Vasilopoulos et al., 2012). Moving forward, it will be important to identify the specific genes and environments that account for this variation (Friedman et al., 2016).

In sum, heritability estimates of the individual dimensions of EF are fairly modest. However, a higher-order common factor of EF, which accounts for the shared variation among the dimensions of EF, is almost entirely heritable across the life span. It is possible that shared genetic influences contribute the common factor of EF, whereas unique genetic and environmental influences contribute to variation in the individual dimensions of EF. However, this hypothesis has yet to be explicitly tested.

How Much Do Genes Affect Individual Variation in EF?

As described in previous chapters, improvements in EF are expected as a function of age, particularly during the transition from adolescence into adulthood, as the brain undergoes a rapid synaptic pruning and myelination process during this developmental period that should lead to greater efficiency in brain functions (De Luca et al., 2003; Chevalier & Clark, Chapter 2, this volume; Crone, Peters, & Steinbeis, Chapter 3, this volume). However, there are also vast individual differences in EF maturation during this period, where some individuals improve over time while others may decline. Given that the transition into adulthood is a critical period for EF development because it also coincides with drastic life changes, including new social roles, increase in responsibilities, and exposure to new peer groups (Crone et al., Chapter 3, this volume), understanding the contribution of genes on individual variability in EF maturation may provide important insights into the emerging field of EF intervention (Zelazo & Carlson, 2012).

Friedman and her colleagues (2016) tested EF abilities in 749 MZ and DZ same-sex twins from the Colorado Longitudinal Twin Sample at Wave 1 (adolescence; age 17) and Wave 2 (young adulthood; age 23). The longitudinal design permitted them to examine the phenotypic stability of a latent common factor of EF and separate latent dimensions for updating (i.e., working memory) and shifting (i.e., cognitive flexibility) from Wave 1 to Wave 2. Not surprisingly, individual differences in EF were highly (although not entirely) stable across Waves (common $r = 0.86$, updating $r = 1.00$, shifting $r = 0.91$). The stability of EF was almost entirely driven by genetic influences, but the change in EF from adolescence into adulthood was due to unique environmental factors that came online at Wave 2 (during early adulthood), rather than to any new genetic or shared environmental factors.

These findings reinforce the notion that EF are highly stable and heritable across development, but also that environments may meaningfully impact individual variation in EF maturation during the period of transition into early adulthood. Moving to a new residence, starting a new job, or creating new social networks may have small but significant effects on EF abilities, suggesting that EF may in fact be malleable even during adulthood. It is worth noting here that high heritability estimates can also be confounded by aspects of the environment that are also under genetic control (i.e., gene–environment correlation; Jaffee & Price, 2007). Gene–environment correlations exist when the environment is influenced in part by the person's own genotype. Certain individuals may seek environments that stimulate their EF development by virtue of their genetic constitution (i.e., active gene–environment correlation). In addition, parents (who each pass on 50% of their genetic material to their offspring) may create an environment that nurtures the development of EF that are similar to their own abilities (i.e., passive gene–environment correlation). Discerning the role of gene–environment correlation in twin studies of EF has traditionally been a challenge, but promising methods have been developed as an important refinement to the traditional twin study design (Narusyte et al., 2008).

What Specific Genes Influence EF?

Twin studies have provided valuable information regarding *if* and *how much* genes and environments matter for EF, but they do not tell us anything about the biological substrates or genetic variants that contribute to individual differences in EF. Candidate gene studies test for the association between a trait of interest and specific genetic variants (e.g., single nucleotide polymorphism, SNP; variable number of tandem repeat, VNTR). These studies typically involve genetic variants that regulate functioning in neurobiological systems with relevance to the trait (Barnes, Dean, Nandam, O'Connell, & Bellgrove, 2011). For example, the medial prefrontal cortex (mPFC) and orbital prefrontal cortex (OFC) play central roles in EF based on animal and human studies (Kesner & Churchwell, 2011; Logue & Gould, 2014). Given that the frontal regions of the brain receive input from dopamine, norepinephrine, and serotonin (Logue & Gould, 2014) and that the morphology and functioning of these brain regions are known to be under tight genetic control (see reviews by Winterer & Goldman, 2003; Barnes et al., 2011), polymorphisms that regulate monoamine function in these brain regions are attractive candidates for study in EF. In this section, we will highlight two well-characterized candidate genes, including polymorphisms in the catechol-o-methyltransferase (*COMT*) and serotonin transporter (*5-HTTLPR*) genes, and summarize their role in EF development.

COMT

Among the most-studied genetic variants for EF is the *val158met* single nucleotide polymorphism (rs4680) in the *COMT* gene, located on 22q11.2. This polymorphism transcribes an enzyme that breaks down catecholamine and modulates the transmission of serotonin, norepinephrine, and dopamine in the PFC (Cumming, Brown, Damsma, & Fibiger, 1992). The valine (val) to methionine (met) amino acid substitution at this locus influences the activity level of the enzyme, where the met variant has 3 to 4 times less enzyme activity (and consequently greater dopamine availability in the PFC) compared to the val variant (Zubieta et al., 2003). Given that stimulant medications (such as methylphenidate) increase dopamine availability and are associated with therapeutic effects (i.e., better EF performance), the met/met genotype has been hypothesized to be associated with better EF performance.

However, evidence of this association is equivocal. One study showed impaired performance across measures of EF among children with the met/met genotype (Bellgrove et al., 2005), whereas other child studies found opposite associations of this genotype on EF performance (i.e., leading to superior EF performance) (Diamond, Briand, Fossella, & Gehlbach, 2004) or no association at all (Taerk et al., 2004). In adult samples, the met/met polymorphism was found to be associated with superior performance across the dimensions of EF, including cognitive flexibility (Egan et al., 2001) and working memory (Goldberg et al., 2003), but non-associations have also been reported (Glaser et al., 2006). Barnett, Scoriels, and Munafò (2008) conducted a meta-analysis of EF outcomes in both healthy and clinical samples of children and adults, including Trail-Making, verbal fluency, verbal recall, N-back accuracy (2-back), and the Wisconsin Card Sorting test. Across 46 independent studies, the meta-analysis showed that associations between *COMT* and EF were not significant. For instance, on the Wisconsin Card Sort (Crone et al., Chapter 3, this volume), the mean Cohen's d effect size of the *COMT* met/met polymorphism across 25 independent samples was -0.04 ($p=0.36$). Similarly, across 10 independent studies on *COMT* and Trail-Making, meta-analysis yielded no significant effect of the met/met polymorphism on performance ($d=0.04$, $p=0.66$). Despite strong biological plausibility for its association, meta-analytic evidence suggests that the *COMT val158met* polymorphism is not strongly associated with EF performance.

5-HTTLPR

Another polymorphism with an association to EF is a functional polymorphism in the promoter region of the *5-HTTLPR* gene. This polymorphism plays a role in regulating the expression of serotonin in the PFC (Puig & Gulledge, 2011), and studies have shown that variation in serotonergic activity in the PFC has been associated with EF performance in humans (Cools, Roberts, & Robbins, 2008) and nonhuman primates (Walker, Mikheenko, Argyle, Robbins, & Roberts, 2006). *5-HTTLPR* contains two allelic variants – the short (S) allele, which is associated with reduced serotonergic reuptake (hence, greater serotonin availability in the PFC), and the long (L) allele, which is associated with relatively greater transcriptional activity (i.e., less serotonin availability in the PFC) (Hu et al., 2006). The S allele has been shown to increase vulnerability for a variety of clinical phenotypes, including depression (Karg, Burmeister, Shedden, & Sen, 2011), aggression (Reif et al., 2007), and attention-deficit/hyperactivity disorder (Retz et al., 2008). Supporting this hypothesis, Weiss and colleagues (2014) recruited a sample of healthy adult women and found that women with the S/S genotype performed worse on a working memory task than women carrying at least one L allele. Similar associations between the S/S genotype and impaired EF performance have also been shown in older adult samples (O'Hara et al., 2007). However, divergent results have also been reported. Meta-analysis of studies of youth with attention-deficit/hyperactivity disorder found an association between the L/L genotype and worse performance on measures of impulsivity, inattention, and working memory (Gizer, Ficks, & Waldman, 2009). Similarly, adults carrying the L/L genotype performed worse on tasks of risky decision making and visual planning (Roiser, Rogers, Cook, & Sahakian, 2006), set-shifting (Borg et al., 2009), and inhibition (Roiser et al., 2006) compared to individuals without this genotype.

One reason why genetic associations have not been consistently replicated across studies is due to the possibility of a gene–environment interaction, where the magnitude and/or direction of allelic effects on an outcome may depend on exposure to certain types of environments (Caspi et al., 2002; Jaffee et al., 2005). A gene–environment interaction was detected in our recent study of EF conducted in a sample of typically developing adolescents (Li et al., 2015). We were interested in the role of adolescent parental supervision as a potential moderator of *5-HTTLPR* genotype on EF abilities, given that positive parental influences have been known to protect against a wide range of maladaptive outcomes pertaining to socioemotional (Li, Berk, & Lee, 2013), behavioral (Clark, Thatcher, & Maisto, 2005), and academic (Li, Walker, & Armstrong, 2014) development. We predicted that adolescents carrying the "susceptibility" (i.e., L/L) genotype would be more sensitive to parental influences during adolescence, such that these individuals may perform worse on EF under conditions of low parental supervision relative to those without the susceptibility genotype. First, we tested a hierarchical factor model of EF, where we found that the covariation between EF dimensions (including conceptual flexibility, fluency, and inhibition) was largely accounted for by a single common factor of EF, which is in line with previous studies (Friedman et al., 2008). We also found evidence of gene–environment interaction that was specific to conceptual flexibility (Figure 7.1). Adolescents who carried the L/L genotype had significantly lower scores on conceptual flexibility compared to adolescents carrying the S/L or S/S genotypes, but this effect was only significant at low levels of parental supervision. We did not detect any evidence of *5-HTTLPR*–parental supervision interaction for the common EF factor, inhibition, or fluency, suggesting specificity of this genotype–environment configuration with respect to conceptual flexibility.

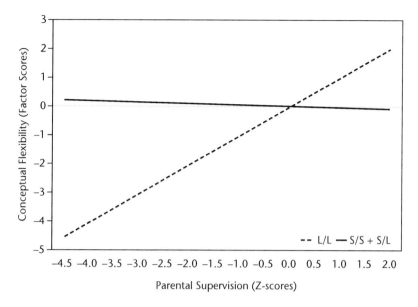

FIGURE 7.1 5-HTTLPR genotype × parental supervision on EF.

Note
Lines represent *5-HTTLPR* genotypes, L/L and S/S + S/L. Figure from Li et al. (2015).

The findings also provide evidence that an individual's genetic constitution does not necessarily determine their EF ability, but rather, it may serve to attenuate or amplify their abilities in the context of their environment (Belsky & Pluess, 2009). This helps to explain why genetic main effects are so difficult to detect across the literature – significant genetic effects on EF may be masked in the absence of incorporating some measurement of a plausible environmental influence. Yet, there are almost no studies of EF that have rigorously examined gene–environment interaction. This will be an important area of scientific inquiry moving forward, particularly once more genes of significance for EF are identified.

Genome-Wide Association Studies

There are limitations to the candidate gene approach, however. For one, it focuses on only the smallest possible portion of human genetic variation (traditionally a single polymorphism), despite evidence that complex traits are most likely influenced by a multitude of genetic variants from across the genome (Plomin, Haworth, & Davis, 2009). Second, while it makes sense to focus on functional variants with biological plausibility, the utility of this approach is also predicated on having full knowledge about gene function and other biological pathways that are associated with a trait (Neale & Sham, 2004). These constraints may actually preclude the study of many other plausible genes in relation to EF, which have simply not yet been identified.

Hence, there has been an increasing interest in recent years to find novel genetic variants from across the entire genome that may be associated with EF (genome-wide association studies, GWAS; Bush & Moore, 2012; Hirschhorn & Daly, 2005). Several GWAS of EF have emerged in recent years (see Koiliari et al., 2014; Vaidyanathan et al., 2014; Luciano et al., 2011, Seshadri et al., 2007; Cirulli et al., 2010; Need et al., 2009), but given the nascent state of the field, most have been underpowered by modern-day GWAS-standards ($N < 5{,}000$), leading to inconclusive and un-replicated results thus far.

One notable exception is the recent (and, to date, largest) GWAS on EF using data from the Cohorts for Heart and Aging Research in Genomic Epidemiology (CHARGE) consortium (Ibrahim-Verbaas et al., 2016), which consists of over 20 community- and European-based cohort studies with healthy and dementia-free adult participants (age 45 years and older; $N > 13{,}000$ with a comparably sized independent replication sample). Genome-wide association analyses were conducted for several EF dimensions, including tasks of cognitive flexibility and processing speed (i.e., Trail-Making Test, Stroop, and Digit-Symbol Tests). One significant association was observed at an intronic variant (rs17518584) in the gene encoding cell adhesion molecule 2 (*CADMA2*), explaining about 0.05% of the variance in scores for tests of processing speed for one of the largest cohorts of the sample (i.e., Rotterdam Study). This variant is functionally associated with the encoding of proteins involved in neuronal adhesion in the developing brain (Thomas, Akins, & Biederer, 2008), and has previously been found to be associated with body mass index (Speliotes et al., 2010), autism spectrum disorder (Casey et al. 2012), and dimensions of personality (Service et al., 2012). However, this variant was not significant in the replication sample. Importantly, non-replication for a genome-wide significant variant in EF does not necessarily imply that the variant is not relevant for EF. Rather, this may be a matter of sample size, where having a larger sample size may produce a stronger signal or yield additional genome-wide significant variants (Wood et al., 2014). Fortunately, international cross-collaborative efforts (such as CHARGE) have continued to increase sample sizes across multiple cohorts, so the prospect of identifying new genetic associations for EF are hopefully just on the horizon.

Molecular genetic studies of EF have shed light on the specific genetic mechanisms that may contribute to individual variation in EF abilities. The ascendency of GWAS in recent years has led to the promise of unraveling the precise genetic architecture for EF, and the success of these efforts are predicated on having a large enough sample size (Wood et al., 2014). The possibility of gene–environment interaction remains understudied in the post-genomic era.

Conclusions and Future Directions

Despite the tremendous progress in the field of genetics over the past several decades, gaps in our knowledge regarding the genetic influences underlying EF remain immensely wide. Twin studies have largely confirmed that genetic influences play a crucial role in contributing to individual differences in EF across development, but molecular genetic studies have yet to uncover the specific genes that make up this variation. Furthermore, we know that the environment plays a role in the development of EF, but there are few studies that have interrogated gene–environment interplay. Clearly, there is still much research to be conducted in terms of understanding how genes influence individual variation in EF across development.

As sample sizes in genetic studies of EF continue to increase, GWAS-based methods will continue being the primary tool used to identify genetic associations for EF. What is exciting about GWAS is that it has applications well beyond identifying genes for EF. For example, Mukherjee and colleagues (2014) performed gene-based tests of association for EF and found evidence of several genetic pathways that were significantly associated with EF resilience (defined as having better than expected EF performance following brain damage), including those that regulate the dendritic/neuronal spine, presynaptic membrane, postsynaptic density, and others. This type of GWAS provides an understanding about how a set of genes within a biological system (rather than a single variant) influence a particular behavior. Down the line, it will be possible for researchers to focus on genes within these associated genetic pathways, providing a far greater scope for the study of gene–environment interplay.

GWAS have also led to the development of promising methods to *predict* variation for a given trait, through a "polygenic score" (Purcell et al., 2009). As there are potentially thousands of common genetic variants that exert very small effects on a trait, genome-wide methods to index the aggregate effects of all associated genetic variants are now possible. Polygenic scores reflect a weighted sum of the associated alleles at each locus, such that an individual who carries more copies of associated alleles will have a higher polygenic score and hence, a higher genetic predisposition for the trait. The predictive power of polygenic scores has been documented for a variety of medical conditions, including breast cancer, prostate cancer, heart disease, and schizophrenia. The strength of this approach is that it may also be useful for gene–environment interaction studies, providing yet another alternative to the long-standing tradition of employing single-gene methods to study trait variation that ignore the multi-factorial genomic architecture of complex phenotypes.

Kendler (2011, p. 5) discussed the once-held notion that genetic and environmental influences were "separable, passive components" of an etiological model that could simply be added together to account for the entirety of trait variation. He notes that this perspective has become an increasingly untenable way to develop etiologic models for psychological traits, as we know that genes and environments are far from "fixed and static" entities (Kendler, 2011, p. 6). Rather, they are dynamic and interrelated influences on each other over time. Yet, there has been very little research on EF using integrative models that incorporate measured genes, environments, and development. Longitudinal models of EF that incorporate these influences may inform efforts relevant to EF intervention, helping us identify genetically sensitive populations who may stand to benefit from early interventions (Brody, Beach, Philibert, Chen, & Murry, 2009).

Author Note

This work was supported by a core grant to the Waisman Center from the National Institute on Child Health and Human Development (P30-HD03352).

References

Ando, J., Ono, Y., & Wright, M. J. (2001). Genetic structure of spatial and verbal working memory. *Behavior Genetics, 31*, 615–624.

Anokhin, A. P., Golosheykin, S., Grant, J. D., & Heath, A. C. (2010). Developmental and genetic influences on prefrontal function in adolescents: A longitudinal twin study of WCST performance. *Neuroscience Letters, 472*, 119–122.

Anokhin, A. P., Heath, A. C., & Ralano, A. (2003). Genetic influences on frontal brain function: WCST performance in twins. *NeuroReport, 14*, 1975–1978.

Barnes, J. J. M., Dean, A. J., Nandam, L. S., O'Connell, R. G., & Bellgrove, M. A. (2011). The molecular genetics of executive function: Role of monoamine system genes. *Biological Psychiatry, 69*, e127–e143.

Barnett, J. H., Scoriels, L., & Munafò, M. R. (2008). Meta-analysis of the cognitive effects of the catechol-O-methyltransferase gene Val158/108Met polymorphism. *Biological Psychiatry, 64*, 137–144.

Bellgrove, M. A., Domschke, K., Hawi, Z., Kirley, A., Mullins, C., Robertson, I. H., & Gill, M. (2005). The methionine allele of the COMT polymorphism impairs prefrontal cognition in children and adolescents with ADHD. *Experimental Brain Research, 163*, 352–360.

Belsky, J., & Pluess, M. (2009). Beyond diathesis stress: Differential susceptibility to environmental influences. *Psychological Bulletin, 135*, 885–908.

Bergen, S. E., Gardner, C. O., & Kendler, K. S. (2007). Age-related changes in heritability of behavioral phenotypes over adolescence and young adulthood: A meta-analysis. *Twin Research and Human Genetics, 10*, 423–433.

Blokland, G. A. M., McMahon, K. L., Hoffman, J., Zhu, G., Meredith, M., Martin, N. G., et al. (2008). Quantifying the heritability of task-related brain activation and performance during the N-back working memory task: A twin fMRI study. *Biological Psychology, 79*, 70–79.

Borg, J., Henningsson, S., Saijo, T., Inoue, M., Bah, J., Westberg, L., et al. (2009). Serotonin transporter genotype is associated with cognitive performance but not regional 5-HT1A receptor binding in humans. *International Journal of Neuropsychopharmacology, 12*, 783–792.

Brody, G. H., Beach, S. R., Philibert, R. A., Chen, Y. F., & Murry, V. M. (2009). Prevention effects moderate the association of 5-HTTLPR and youth risk behavior initiation: Gene × environment hypotheses tested via a randomized prevention design. *Child Development, 80,* 645–661.

Bush, W. S., & Moore, J. H. (2012). Genome-wide association studies. *PLoS Computational Biology, 8,* e1002822.

Casey, J. P., Magalhaes, T., Conroy, J. M., Regan, R., Shah, N., Anney, R., et al. (2012). A novel approach of homozygous haplotype sharing identifies candidate genes in autism spectrum disorder. *Human Genetics, 131,* 565–579.

Caspi, A., McClay, J., Moffitt, T. E., Mill, J., Martin, J., Craig, I. W., et al. (2002). Role of genotype in the cycle of violence in maltreated children. *Science, 297,* 851–854.

Chou, L., Kuo, P., Lin, C. C. H., & Chen, W. J. (2010). Genetic and environmental influences on the Wisconsin Card Sorting test performance in healthy adolescents: A twin/sibling study. *Behavior Genetics, 40,* 22–30.

Cirulli, E. T., Kasperaviciute, D., Attix, D. K., Need, A. C., Ge, D., Gibson, G., & Goldstein, D. B. (2010). Common genetic variation and performance on standardized cognitive tests. *European Journal of Human Genetics, 18,* 815–820.

Clark, D. B., Thatcher, D. L., & Maisto, S. A. (2005). Supervisory neglect and adolescent alcohol use disorders: Effects on AUD onset and treatment outcome. *Addictive Behaviors, 30,* 1737–1750.

Coolidge, F. L., Thede, L. L., & Young, S. E. (2000). Heritability and the comorbidity of attention deficit hyperactivity disorder with behavioral disorders and executive function deficits: A preliminary investigation. *Developmental Neuropsychology, 17,* 273–287.

Cools, R., Roberts, A. C., & Robbins, T. W. (2008). Serotoninergic regulation of emotional and behavioural control processes. *Trends in Cognitive Sciences, 12,* 31–40.

Cumming, P., Brown, E., Damsma, G., & Fibiger, H. (1992). Formation and clearance of interstitial metabolites of dopamine and serotonin in the rat striatum: An in vivo microdialysis study. *Journal of Neurochemistry, 59,* 1905–1914.

De Luca, C. R., Wood, S. J., Anderson, V., Buchanan, J., Proffitt, T. M., Mahony, K., & Pantelis, C. (2003). Normative data from the CANTAB. I: Development of executive function over the lifespan. *Journal of Clinical and Experimental Neuropsychology, 25,* 242–254.

Diamond, A., Briand, L., Fossella, J., & Gehlbach, L. (2004). Genetic and neurochemical modulation of prefrontal cognitive functions in children. *The American Journal of Psychiatry, 161,* 125–132.

Dick, D. M. (2011). Gene–environment interaction in psychological traits and disorders. *Annual Review of Clinical Psychology, 7,* 383–409.

Egan, M. F., Goldberg, T. E., Kolachana, B. S., Callicott, J. H., Mazzanti, C. M., Straub, R. E., et al. (2001). Effect of COMT Val108/158 Met genotype on frontal lobe function and risk for schizophrenia. *Proceedings of the National Academy of Sciences, 98,* 6917–6922.

Engelhardt, L. E., Briley, D. A., Mann, F. D., Harden, K. P., & Tucker-Drob, E. (2015). Genes unite executive functions in childhood. *Psychological Science, 26,* 1151–1163.

Finkel, D., Pedersen, N., & McGue, M. (1995). Genetic influences on memory performance in adulthood: Comparison of Minnesota and Swedish twin data. *Psychology and Aging, 10,* 437–446.

Friedman, N. P., & Miyake, A. (2004). The relations among inhibition and interference control functions: A latent-variable analysis. *Journal of Experimental Psychology: General, 133*, 101–135.

Friedman, N. P., Miyake, A., Altamirano, L. J., Corley, R. P., Young, S. E., Rhea, S. A., & Hewitt, J. K. (2016). Stability and change in executive function abilities from late adolescence to early adulthood: A longitudinal twin study. *Developmental Psychology, 52*, 326–340.

Friedman, N. P., Miyake, A., Young, S. E., DeFries, J. C., Corley, R. P., & Hewitt, J. K. (2008). Individual differences in executive functions are almost entirely genetic in origin. *Journal of Experimental Psychology: General, 137*, 201–225.

Gizer, I. R., Ficks, C., & Waldman, I. D. (2009). Candidate gene studies of ADHD: A meta-analytic review. *Human Genetics, 126*, 51–90.

Glaser, B., Debbane, M., Hinard, C., Morris, M. A., Dahoun, S. P., Antonarakis, S. E., & Eliez, S. (2006). No evidence for an effect of COMT Val158Met genotype on executive function in patients with 22q11 deletion syndrome. *American Journal of Psychiatry, 163*, 537–539.

Godinez, D. A., Friedman, N. P., Rhee, S. H., Miyake, A., & Hewitt, J. K. (2012). Phenotypic and genetic analyses of the Wisconsin card sort. *Behavior Genetics, 42*, 209–220.

Goldberg, T. E., Egan, M. F., Gscheidle, T., Coppola, R., Weickert, T., Kolachana, B. S., et al. (2003). Executive subprocesses in working memory: Relationship to catechol-O-methyltransferase Val158Met genotype and schizophrenia. *Archives of General Psychiatry, 60*, 889–896.

Groot, A. S., De Sonneville, L. M., Stins, J. F., & Boomsma, D. I. (2004). Familial influences on sustained attention and inhibition in preschoolers. *Journal of Child Psychology and Psychiatry, 45*, 306–314.

Hirschhorn, J. N., & Daly, M. J. (2005). Genome-wide association studies for common diseases and complex traits. *Nature Reviews Genetics, 6*, 95–108.

Hu, X. Z., Lipsky, R. H., Zhu, G., Akhtar, L. A., Taubman, J., Greenberg, B. D., et al. (2006). Serotonin transporter promoter gain-of-function genotypes are linked to obsessive-compulsive disorder. *The American Journal of Human Genetics, 78*, 815–826.

Ibrahim-Verbaas, C. A., Bressler, J., Debette, S., Schuur, M., Smith, A. V., Bis, J. C., et al. (2016). GWAS for executive function and processing speed suggests involvement of the CADM2 gene. *Molecular Psychiatry, 21*, 189–197.

Jaffee, S. R., Caspi, A., Moffitt, T. E., Dodge, K. A., Rutter, M., Taylor, A., & Tully, L. A. (2005). Nature × nurture: Genetic vulnerabilities interact with physical maltreatment to promote conduct problems. *Development and Psychopathology, 17*, 67–84.

Jaffee, S. R., & Price, T. S. (2007). Gene–environment correlations: A review of the evidence and implications for prevention of mental illness. *Molecular Psychiatry, 12*, 432–442.

Karg, K., Burmeister, M., Shedden, K., & Sen, S. (2011). The serotonin transporter promoter variant (5-HTTLPR), stress, and depression meta-analysis revisited: Evidence of genetic moderation. *Archives of General Psychiatry, 68*, 444–454.

Kendler, K. S. (2011). A conceptual overview of gene–environment interaction and correlation in a developmental context. In K. S. Kendler, S. R. Jaffee, & D. Romer (Eds.), *The dynamic genome and mental health: The role of genes and environments in youth development* (pp. 5–29). New York, NY: Oxford University Press.

Kesner, R. P., & Churchwell, J. C. (2011). An analysis of rat prefrontal cortex in mediating executive function. *Neurobiology of Learning and Memory, 96*, 417–431.

Koiliari, E., Roussos, P., Pasparakis, E., Lencz, T., Malhotra, A., Siever, L. J., et al. (2014). The CSMD1 genome-wide associated schizophrenia risk variant rs10503253 affects general cognitive ability and executive function in healthy males. *Schizophrenia Research, 154,* 42–47.

Kremen, W. S., Jacobsen, K. C., Xian, H., Eisen, S. A., Eaves, L. J., Tsuang, M. T., & Lyons, M. J. (2007). Genetics of verbal working memory processes: A twin study of middle-aged men. *Neuropsychology, 21,* 569–580.

Kuntsi, J., Rogers, H., Swinard, G., Börger, N., van der Meere, J., Rijsdijk, F., & Asherson, P. (2006). Reaction time, inhibition, working memory and "delay aversion" performance: Genetic influences and their interpretation. *Psychological Medicine, 36,* 1613–1624.

Lee, T., Mosing, M. A., Henry, J. D., Trollor, J. N., Lammel, A., Ames, D., et al. (2012). Genetic influences on five measures of processing speed and their covariation with general cognitive ability in the elderly: The older Australian twins study. *Behavior Genetics, 42,* 96–106.

Li, H., Walker, R., & Armstrong, D. (2014). International note: Parenting, academic achievement and problem behaviour among Chinese adolescents. *Journal of Adolescence, 37,* 387–389.

Li, J. J., Berk, M. S., & Lee, S. S. (2013). Differential susceptibility in longitudinal models of gene–environment interaction for adolescent depression. *Development and Psychopathology, 25,* 991–1003.

Li, J. J., Chung, T. A., Vanyukov, M. M., Wood, D. S., Ferrell, R., & Clark, D. B. (2015). A hierarchical factor model of executive functions in adolescents: Evidence of gene–environment interplay. *Journal of the International Neuropsychological Society, 21,* 62–73.

Logue, S. F., & Gould, T. J. (2014). The neural and genetic basis of executive function: Attention, cognitive flexibility, and response inhibition. *Pharmacology, Biochemistry and Behavior, 123,* 45–54.

Luciano, M., Hansell, N. K., Lahti, J., Davies, G., Medland, S. E., Räikkönen, K., et al. (2011). Whole genome association scan for genetic polymorphisms influencing information processing speed. *Biological Psychology, 86,* 193–202.

Luciano, M., Smith, G. A., Wright, M. J., Geffen, G. M., Geffen, L. B., & Martin, N. G. (2001). On the heritability of inspection time and its covariance with IQ: A twin study. *Intelligence, 29,* 443–457.

Miyake, A., Friedman, N. P., Emerson, M. J., Witzki, A. H., Howerter, A., & Wager, T. D. (2000). The unity and diversity of executive functions and their contributions to complex "frontal lobe" tasks: A latent variable analysis. *Cognitive Psychology, 41,* 49–100.

Mukherjee, S., Kim, S., Ramanan, V. K., Gibbons, L. E., Nho, K., Glymour, M. M., et al. (2014). Gene-based GWAS and biological pathway analysis of the resilience of executive functioning. *Brain Imaging and Behavior, 8,* 110–118.

Narusyte, J., Neiderhiser, J. M., D'Onofrio, B. M., Reiss, D., Spotts, E. L., Ganiban, J., & Lichtenstein, P. (2008). Testing different types of genotype-environment correlation: An extended children-of-twins model. *Developmental Psychology, 44,* 1591–1603.

Neale, B. M., & Sham, P. C. (2004). The future of association studies: Gene-based analysis and replication. *The American Journal of Human Genetics, 75,* 353–362.

Neale, M., & Cardon, L. (1992). *Methodology for genetic studies of twins and families.* Dordrecht: Springer Netherlands.

Need, A. C., Attix, D. K., McEvoy, J. M., Cirulli, E. T., Linney, K. L., Hunt, P., et al. (2009). A genome-wide study of common SNPs and CNVs in cognitive performance in the CANTAB. *Human Molecular Genetics, 18*, 4650–4661.

O'Hara, R., Schröder, C. M., Mahadevan, R., Schatzberg, A. F., Lindley, S., Fox, S., et al. (2007). Serotonin transporter polymorphism, memory and hippocampal volume in the elderly: Association and interaction with cortisol. *Molecular Psychiatry, 12*, 544–555.

Pedersen, N. L., Plomin, R., Nesselroade, J. R., & McClearn, G. E. (1992). A quantitative genetic analysis of cognitive abilities during the second half of the life span. *Psychological Science, 3*, 346–353.

Plomin, R., Haworth, C. M., & Davis, O. S. (2009). Common disorders are quantitative traits. *Nature Reviews Genetics, 10*, 872–878.

Plomin, R., Pedersen, N. L., Lichtenstein, P., & McClearn, G. E. (1994). Variability and stability in cognitive abilities are largely genetic later in life. *Behavior Genetics, 24*, 207–215.

Polderman, T. J. C., Stins, J. F., Posthuma, D., Gosso, M. F., Verhulst, F. C., & Boomsma, D. I. (2006). The phenotypic and genotypic relation between working memory speed and capacity. *Intelligence, 34*, 549–560.

Puig, M. V., & Gulledge, A. T. (2011). Serotonin and prefrontal cortex function: Neurons, networks, and circuits. *Molecular Neurobiology, 44*, 449–464.

Purcell, S. M., Wray, N. R., Stone, J. L., Visscher, P. M., O'Donovan, M. C., Sullivan, P. F., et al. (2009). Common polygenic variation contributes to risk of schizophrenia and bipolar disorder. *Nature, 460*, 748–752.

Reif, A., Rösler, M., Freitag, C. M., Schneider, M., Eujen, A., Kissling, C., et al. (2007). Nature and nurture predispose to violent behavior: Serotonergic genes and adverse childhood environment. *Neuropsychopharmacology, 32*, 2375–2383.

Retz, W., Freitag, C. M., Retz-Junginger, P., Wenzler, D., Schneider, M., Kissling, C., et al. (2008). A functional serotonin transporter promoter gene polymorphism increases ADHD symptoms in delinquents: Interaction with adverse childhood environment. *Psychiatry Research, 158*, 123–131.

Roiser, J. P., Rogers, R. D., Cook, L. J., & Sahakian, B. J. (2006). The effect of polymorphism at the serotonin transporter gene on decision-making, memory and executive function in ecstasy users and controls. *Psychopharmacology, 188*, 213–227.

Schachar, R. J., Forget-Dubois, N., Dionne, G., Boivin, M., & Robaey, P. (2011). Heritability of response inhibition in children. *Journal of the International Neuropsychological Society, 17*, 238–247.

Service, S. K., Verweij, K. J. H., Lahti, J., Congdon, E., Ekelund, J., Hintsanen, M., et al. (2012). A genome-wide meta-analysis of association studies of Cloninger's Temperament Scales. *Translational Psychiatry, 2*, e116.

Seshadri, S., DeStefano, A. L., Au, R., Massaro, J. M., Beiser, A. S., Kelly-Hayes, M., et al. (2007). Genetic correlates of brain aging on MRI and cognitive test measures: A genome-wide association and linkage analysis in the Framingham Study. *BMC Medical Genetics, 8*.

Speliotes, E. K., Willer, C. J., Berndt, S. I., Monda, K. L., Thorleifsson, G., Jackson, A. U., et al. (2010). Association analyses of 249,796 individuals reveal 18 new loci associated with body mass index. *Nature Genetics, 42*, 937–948.

Stins, J. F., Van Baal, G. C. M., Polderman, T. J., Verhulst, F. C., & Boomsma, D. I. (2004). Heritability of Stroop and flanker performance in 12-year old children. *BMC Neuroscience, 5*, 8.

Taerk, E., Grizenko, N., Amor, L. B., Lageix, P., Mbekou, V., Deguzman, R., et al. (2004). Catechol-O-Methyltransferase (COMT) Val 108/158 Met polymorphism does not modulate executive function in children with ADHD. *BMC Medical Genetics, 5*, 1.

Thomas, L. A., Akins, M. R., & Biederer, T. (2008). Expression and adhesion profiles of SynCAM molecules indicate distinct neuronal functions. *Journal of Comparative Neurology, 510*, 47–67.

Vaidyanathan, U., Malone, S. M., Donnelly, J. M., Hammer, M. A., Miller, M. B., McGue, M., & Iacono, W. G. (2014). Heritability and molecular genetic basis of antisaccade eye tracking error rate: A genome-wide association study. *Psychophysiology, 51*, 1272–1284.

Vasilopoulos, T., Franz, C. E., Panizzon, M. S., Xian, H., Grant, M. D., Lyons, M. J., et al. (2012). Genetic architecture of the Delis-Kaplan executive function system trail making test: Evidence for distinct genetic influences on executive function. *Neuropsychology, 26*, 238–250.

Walker, S. C., Mikheenko, Y. P., Argyle, L. D., Robbins, T. W., & Roberts, A. C. (2006). Selective prefrontal serotonin depletion impairs acquisition of a detour-reaching task. *European Journal of Neuroscience, 23*, 3119–3123.

Wang, M., & Saudino, K. J. (2013). Genetic and environmental influences on individual differences in emotion regulation and its relation to working memory in toddlerhood. *Emotion, 13*, 1055–1067.

Wang, Z., Deater-Deckard, K., Cutting, L., Thompson, L. A., & Petrill, S. A. (2012). Working memory and parent-rated components of attention in middle childhood: A behavioral genetic study. *Behavior Genetics, 42*, 199–208.

Weiss, E. M., Schulter, G., Fink, A., Reiser, E. M., Mittenecker, E., Niederstätter, H., et al. (2014). Influences of COMT and 5-HTTLPR polymorphisms on cognitive flexibility in healthy women: Inhibition of prepotent responses and memory updating. *PloS One, 9*, e85506.

Winterer, G., & Goldman, D. (2003). Genetics of human prefrontal function. *Brain Research Reviews, 43*, 134–163.

Wood, A. R., Esko, T., Yang, J., Vedantam, S., Pers, T. H., Gustafsson, S., et al. (2014). Defining the role of common variation in the genomic and biological architecture of adult human height. *Nature Genetics, 46*, 1173–1186.

Young, S. E., Friedman, N. P., Miyake, A., Willcutt, E. G., Corley, R. P., Haberstick, B. C., & Hewitt, J. K. (2009). Behavioral disinhibition: Liability for externalizing spectrum disorders and its genetic and environmental relation to response inhibition across adolescence. *Journal of Abnormal Psychology, 118*, 117–130.

Zelazo, P. D., & Carlson, S. M. (2012). Hot and cool executive function in childhood and adolescence: Development and plasticity. *Child Development Perspectives, 6*, 354–360.

Zubieta, J., Heitzeg, M. M., Smith, Y. R., Bueller, J. A., Xu, K., Xu, Y., et al. (2003). COMT val158met genotype affects μ-opioid neurotransmitter responses to a pain stressor. *Science, 299*, 1240–1243.

8

COMPUTATIONAL MODELS OF EXECUTIVE FUNCTION DEVELOPMENT

Aaron T. Buss

Computational models (CMs) have increased in popularity in the psychological sciences over the previous three decades since the parallel distributed processing (PDP) revolution of the mid-1980s. As the use of CMs has increased in popularity, diverse frameworks have been developed to implement a range of theoretical concepts. The broad application of CMs reflects the utility of mathematically formulated descriptions of a neurocognitive system and how it learns or changes over development (Schlesinger & McMurray, 2012; Turner, Forstmann, Love, Palmeri, & Van Maanen, 2017; Yermolayeva & Rakison, 2014). To the extent that a particular model reproduces a pattern of data, we can gain access to the inner mechanisms driving cognition. The real value of CMs, however, goes beyond simply reproducing a dataset. As I will discuss in this chapter, CMs provide an engine for research and theory development.

Why Model?

The domain of executive function (EF) development presents unique challenges for theoretical explanations. The first challenge concerns definitions and terminology. EF is commonly characterized as a multi-faceted construct. Central aspects of EF such as working memory, inhibition, and attention have been the focus of theoretical debate. For example, it is not clear what role inhibition plays in the development of cognitive flexibility (Happaney & Zelazo, 2003; Kirkham & Diamond, 2003; Munakata, Morton, & Yerys, 2003) or how best to characterize the processes underlying EF over development (Brydges, Fox, Reid, & Anderson, 2014; Howard, Okely, & Ellis, 2015). By quantifying the processes that give rise to EF, CMs provide means of studying development from the inside out – models can implement hypothesized mechanisms to determine whether those mechanisms can explain existing data or predict data for future experiments.

EF is also challenging to study because of the transformational nature of changes over development. As discussed in Chapters 1 and 2 (Cuevas, Rajan, & Bryant; Chevalier & Clark), the period from birth to 5 years reveals various qualitative transitions in behavior (Carlson, 2005; Smith, Thelen, Titzer, & McLin, 1999; Zelazo, Muller, Frye, & Marcovitch, 2003). In this regard, CMs can provide tools for linking timescales to understand how transformations over developmental timescales can emerge from continuous accumulation of changes from moment to moment (Spencer & Perone, 2008). Finally, challenges arise when trying to link changes between the behavioral and neural levels. Despite the fact that the neural basis of EF has been a central topic since the first identification of EF as a concept (e.g., Milner, 1963), integration across levels remains challenging due to the qualitatively distinct concepts comprising cognitive science and neuroscience (Ashby & Waldschmidt, 2008; Turner et al., 2017). In this context, CMs can provide a language to translate and integrate between different levels of analysis.

Tasks

Computational models of EF development have focused on the same touchstone paradigms: the A-not-B task and the Dimensional Change Card Sort (DCCS) task. These tasks have garnered so much attention because they provide rich information about development. Both tasks reveal a qualitative shift in performance that is accompanied by change in neural activation (Baird et al., 2002; Moriguchi & Hiraki, 2009). Additionally, these tasks are thought to tap into multiple aspects of EF, including inhibition, working memory, and switching (see Garon et al., 2008), making these paradigms powerful arenas for the development of theories and CMs. However, there are also drawbacks to this narrow focus on these tasks because it is often difficult to connect models of these tasks with the broader literature on EF development (see Conclusion section).

The A-not-B Task

Although the A-not-B task was initially developed by Piaget as a probe of object permanence (Piaget, 1954), it is now thought of as an EF task that measures inhibitory control and working memory (Baird et al., 2002; Clearfield, Dineva, Smith, Diedrich, & Thelen, 2009; Cuevas et al., Chapter 1, this volume; Diedrich, Highlands, Spahr, Thelen, & Smith, 2001; Smith et al., 1999; Spencer, Smith, & Thelen, 2001; Thelen, Schöner, Scheier, & Smith, 2001). In this task, infants between the ages of 8 and 12 months are presented with a toy and two hiding locations. The toy is first hidden at one location (the 'A' location) and after a delay the infant is given an opportunity to search for and retrieve the toy. After a series of trials of retrieving the object from the 'A' location, the object is then hidden at the 'B' location. Typically, children under 10 months will perseverate on 'B' trials and continue to search at the 'A' location, whereas older infants will have little difficulty updating their behavior to search at the 'B' location. Importantly, a toy is not required and infants will make the same pattern of errors even when only the hiding lids are used, suggesting that object permanence does not need to be invoked to explain development in this task (Smith et al., 1999). Performance in this task requires multiple components of EF: inhibitory control is needed initially on 'A' trials to suppress reaching to the alternative 'B' location, inhibitory control needs to be updated on 'B' trials in order to suppress habits associated with the history of reaching to the 'A' location, and working memory is required to maintain a representation of the most recently cued location. Finally, neuroimaging data using functional near-infrared spectroscopy (fNIRS) has revealed that increases in lateral frontal cortex activation is associated with the development of successful performance on 'B' trials (Baird et al., 2002).

The Dimensional Change Card Sort (DCCS) Task

As discussed in Chapter 2 (Chevalier & Clark), the DCCS task is a flexible rule-use task that requires children to first sort cards by shape or color and then to switch to sort by the other dimension (see Figure 8.1). When instructed to switch rules, 3-year-olds typically fail to do so and will continue to sort by the pre-switch rules. By age 5, the majority of children have little difficulty switching rules (Zelazo et al., 2003). Like the A-not-B task, performance in this task also requires multiple components of EF: inhibitory control is needed to initially suppress attention to the irrelevant dimension, working memory is required to maintain a representation of the current rules, and switching is required to update these processes after the rules change. Additionally, neuroimaging data using fNIRS has revealed that increases in activation across frontal, temporal, and parietal cortices is associated with the development of successful switching (Buss & Spencer, submitted; Moriguchi & Hiraki, 2009).

Computational Approaches

Connectionism

Connectionist, or parallel distributed processing (PDP), models are the most common type of computational model used in the cognitive and developmental sciences (for a review, see Rumelhart & McClelland, 1986; Yermolayeva & Rakison, 2014). These models are composed of neuron-like units that have levels of activation and synaptic connections to other units. Information can be represented across many units. Connection weights specify how activation

Computational Models of EF Development   **127**

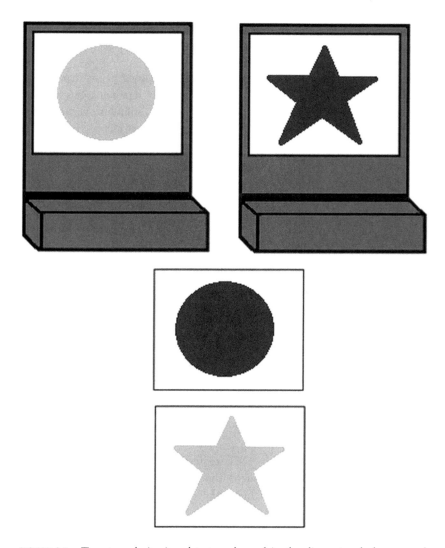

FIGURE 8.1 Target cards (top) and test cards used in the dimensional change card sort (DCCS).

propagates through layers in the network. Positive connection weights indicate excitatory connection and negative connection weights indicate inhibitory connections. Between input and output, intermediate or hidden layers can be used. Neurally realistic activation functions are used to simulate activity in the model. Activation of a unit can, for example, indicate the presence of a feature or object to which that unit is tuned. Connectionist models also implement neurally realistic learning mechanisms. Learning in a connectionist model refers to the modification of connection weights based on supervised or unsupervised learning algorithms. Through the concepts of activation and weight modification, the model is able to speak to the timescales of performance, learning, and development.

Performance of these networks in the EF tasks discussed below depends on two types of representations: latent and active. Latent representations refer to the weight changes that occur from the processing of a stimulus. As stimuli are processed, weights between units that are activated together increase in strength. Thus, latent representations enable the formation of the memory traces that can make future information processing faster and more efficient. Active representations, on the other hand, reflect an activation state that is driven by recurrent connectivity and maintains beyond the presentation of an input similar to a working memory representation. Both active and latent representations can vary in their strength and performance in EF tasks over development is a function of the competition between these representations.

Dynamic Field Theory

Dynamic field theory (DFT) is a computational modeling framework that simulates the dynamical systems that unfold in real time in a neural system (Schoner, Spencer, & the DFT Research Group, 2015). The view from DFT is not fundamentally different from connectionist approaches (Spencer, Thomas, & McClelland, 2009). Like connectionism, DFT simulates cognition using neuron-like units that interact through excitation and inhibition, information is distributed across units, and representations can be graded. However, DFT also emphasizes new principles: stability of activation and grounding cognition in perceptual-motor representations.

In the framework provided by DFT, representations arise from stable attractor states that emerge in real time within populations of neurons tuned to dimensions of perception and action. The fundamental unit of cognition in DFT is a 'peak' of activation, which refers to the stabilized activation of a unit or group of units. Peaks are built through a dynamic competition that involves local-excitation and lateral-inhibition type interactions and correspond to an active representational state. It is important to distinguish this activation state from the active representations used in the connectionist framework. Although active representations in DFT can have a similar role as active representations in connectionist models by maintaining activation over time, the formation of a peak of activation in DFT represents a qualitative shift in the attractor state of the system. In this way, the formation of a peak can be viewed as a selection process whereby a subset of neural units is activated and other neural units are deactivated through lateral inhibition. Through the active selection of neural units, the model can directly map onto behavioral measures such as reaction time or accuracy of the type of response activated (Buss & Spencer, 2014; Buss, Wifall, Hazeltine, & Spencer, 2014; Simmering, Schutte, & Spencer, 2008).

In the context of embodied representations of perceptual and motor dimensions, DF models can be situated in a task similar to how infants or children are given a task by using inputs within continuous and metrically defined dimensions. Additional constraints can then be imposed based on the temporal and perceptual structure of a task.

DFT is able to provide a functional account of the underlying neural states involved in cognitive development. The equations implemented by DFT grew out of foundational work in the field of neuroscience (Grossberg, 1970, 1980; Wilson & Cowan, 1973; Amari, 1977) and implement neural processes at the population level (Bastian, Schöner, & Riehle, 2003; Jancke et al., 1999; Markounikau, Igel, Grinvald, & Jancke, 2010). Recent applications have extended the methods of integrate-and-fire models (Deco, Rolls, & Horwitz, 2004; Edin, Macoveanu, Olesen, Tegnér, & Klingberg, 2007) to demonstrate that DF models can also be used to concurrently simulate real-time behavioral and hemodynamic data (Buss & Spencer, submitted; Buss et al., 2014). Thus, DF models are able to maintain close ties to behavioral and neural data.

Simulating the A-not-B Task

Munakata (1998) presented a connectionist model of the A-not-B task that is constructed with an input layer, a hidden layer, and a response layer (see Figure 8.2). The input units represent the locations (A and B), the covers for those locations, and the toys involved in the task. These units converge on a single hidden layer. Activation in the hidden layer propagates to an output layer for looking and reaching behaviors. That is, the model can look or reach to the 'A' or 'B' locations. The hidden layer and output layer have recurrent connectivity that allows them to form active representations. The output units for the two modalities have different coupling properties to the hidden layer reflecting the structure imposed by the task. Specifically, the looking units are able to respond to every stimulus event, whereas the reaching units do not get to respond or update its activation from the hidden layer until after the delay when the model is allowed to reach. This corresponds to the tray being pushed into the reaching space of infants.

As the model completes trials, the connection weights between units that are simultaneously activated are strengthened, creating latent memory traces that bias the flow of activation in the model toward previously activated units. Thus, the model will fail to generate a response at the 'B' location on 'B' trials because of the extra activity created by the previous responses at the 'A' location. Development is implemented in the model through increases in the strength of recurrent connectivity on the hidden units and output units. With stronger recurrent connections, the model is able to maintain activation associated with the 'B' location in order to generate a reaching response to that location.

Munakata (1998) used this model to unify a diverse set of findings with this task. For example, the model offers an explanation for the dissociation in

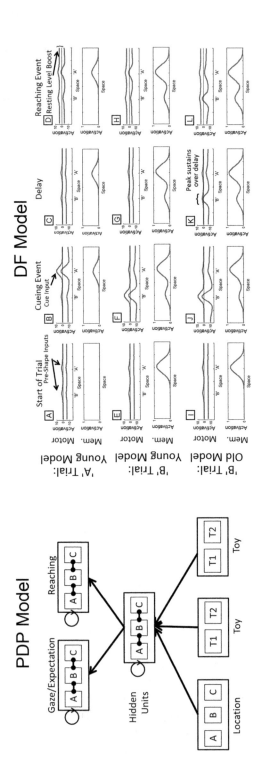

FIGURE 8.2 Connectionist network (left; adapted from Munakata, 1998) and DF model (right; adapted from Thelen et al., 2001) used to simulate A-not-B performance.

looking and reaching performance. Specifically, young infants will look at the 'B' location on 'B' trials despite failing to correctly reach to the 'B' location (Hofstadter & Reznick, 1996). As explained above, the modalities in the response layer have distinct coupling properties to the hidden layer. The looking system responds to every event over the course of a trial and has more opportunity to update its activation. Thus, the looking system is driven by the cueing event to look at the 'B' location. The reaching system, on the other hand, is unable to update its activation until the reaching event. At that point in time, activation in the hidden layer associated with the cue has dissipated, leading the unit for the 'A' location to gain stronger activation during the reaching event. This illustrates how models can be used to reverse-engineer a cognitive system that can produce an initially perplexing pattern of behavior.

Munakata (1998) simulated additional behavioral effects that arise from the active/latent memory distinction. First, young infants are able to reach to the 'B' location if the delay between the hiding and search is short (Diamond, 1985). In the model, this is because the reaching units are able to update its activation before the active representations have dissipated in the hidden layer. Further, young infants are less likely to make errors if the 'B' location has a distinctive cover (Wellman, Cross, Bartsch, & Harris, 1987). In the model, this result stems from weaker learning on 'A' trials as weight changes are normalized across additional units that are included to represent a different cover. In this case, weaker levels of recurrence are sufficient to drive a response at 'B' on 'B' trials.

The explanation offered by DFT bears many similarities to that offered by the connectionist model. The A-not-B model (Thelen et al., 2001) is implemented with a spatial motor planning field (field labeled Motor in Figure 8.2) which has neural units that are tuned to the spatial dimension relevant for a task. Each unit has a Gaussian tuning curve specifying its sensitivity to task inputs, excitatory neural interactions, and inhibitory neural interactions. To simulate performance in the task, task inputs are presented at the two reaching locations throughout the entire trial. These inputs are subthreshold and do not induce neural output. Rather, these inputs structure the activation of the field based on the structure of the task. These inputs are equal strength for the two reaching locations and, thus, do not provide any information regarding which location might be relevant. At the cueing or hiding event, a stronger input is presented at the 'A' location. This input brings units in the field above the activation threshold and neural interactions are engaged to create a stabilized peak of activation. During this event, the model forms an active representation of the 'A' location (note the peak formed in Figure 8.2B). Along with this peak of activation, memories accumulate for the activated neurons (see fields labeled Mem. in Figure 8.2). The memory system accumulates activation while associated neural units are activated and decays while those units are inactive. These memories feed onto the motor planning field and increase the excitability of associated units. To simulate the test event when infants are allowed to reach, the motor planning field is given a global boost in activation to bring the field closer to its activation threshold and induce a 'peak' of activity. At this point, the model generates a response based on the spatial location that becomes activated in the field at this point in time (see Figure 8.2D). That is, the excitatory and inhibitory interactions give rise to a selection process that allows the model to generate an active response on each trial. Variations in the performance across trials, such as response latency and which response is selected, can be generated by noisy fluctuations in activation.

After multiple 'A' trials, the model will have accumulated memory at the 'A' location. Thus, the model will tend to make an error and form a motor plan to reach to the 'A' location on 'B' trials (see Figure 8.2H). To simulate development in this task, Thelen et al. (2001) increased the strength of excitatory and inhibitory neural interactions in the motor planning field. These parameter changes lead to a qualitative shift in the dynamics of the model. Specifically, the 'old' model is able to maintain a working memory when a location is cued or a toy is hidden, indicating the formation of an active working memory representation of the relevant spatial location (see Figure 8.2K–L). During the test phase of the trial, the model will have maintained activation at the 'B' location that serves as the basis for generating a motor plan to reach to that location. Notably, similar parameter changes have been used to simulate changes in spatial working memory and visual working memory from infancy to adulthood to adulthood (Perone, Simmering, & Spencer, 2011; Schutte, Spencer, & Schoner, 2003; Simmering & Patterson, 2012; Simmering et al., 2008).

Like the PDP model, this DF model is able to explain the delay effect. Similar to the explanation offered by the connectionist model, the activation associated with the cue at the 'B' location at short delays does not fully dissipate by the time of the reaching event and the model is able to successfully reach to 'B'. Unlike the PDP model, this model was able to shed light on the role of spontaneous errors on 'A' trials. That is, young infants will sometimes make errors on 'A' trials and reach to 'B'. Infants who make spontaneous errors are more likely to reach to 'B' on 'A' trials (Smith et al., 1999). Spontaneous errors can occur in the model because it generates active responses on every trial. During the reaching phase on 'A' trials, as the motor planning field is boosted to its threshold, noisy fluctuations can raise activation at the 'B' location above threshold sooner than the influence of memory formation can raise activation to threshold at the 'A' location. Spontaneous errors build up memories at the 'B' location that make responding at the 'B' location more likely during 'B' trials. The PDP model does not generate active responses; thus, it does not have a mechanism for producing spontaneous errors.

These spontaneous errors also serve to explain the pattern of performance when the 'B' lid is visually distinct from the 'A' lid. Specifically, young infants can successfully reach to 'B' on 'B' trials if the hiding lid for the 'B' location is visually distinct from that for the 'A' location or if a toy is not used on 'A' trials but is used on 'B' trials. In the model, these effects were explained through increased spontaneous errors on 'A' trials. A visually distinct 'B' lid was implemented through a stronger task input at the 'B' location which serves to produce more spontaneous errors in the model and infants (Clearfield et al., 2009). In this case, the PDP model and the DF model have unique explanations for the influence of a distinctive cover at the 'B' location. Finally, this model has also been extended beyond infancy to explain and predict patterns of performance in tasks similar to the A-not-B in later childhood and adulthood (Simmering et al., 2008). The DF model implements representations along a continuous spatial dimension, allowing this model to generalize to task variations that do not involve discrete spatial locations.

Summary

Both models demonstrate how the ability to maintain a representation of the 'B' location in the A-not-B task can arise from continuous changes in the underlying system. Both models implemented neurally plausible systems to explain the emergence of a new skill. These models focused on some similar effects, but also diverged in their applications. Whereas the connectionist model addressed differences between looking and reaching behavior, the DF model shed light on the role of spontaneous errors on performance during 'B' trials and the role of spatial representations. In this way, both modeling approaches have been used to gain different types of insight into the dynamics underlying behavior and development.

Simulating the DCCS Task

Using a similar distinction between active and latent representations, the connectionist approach has been extended beyond infancy to explain development in the DCCS task. Morton and Munakata (2002) developed a model of the DCCS task (see Figure 8.3) composed of input units for shape features, color features, and instructions for sorting by shape or color. The hidden layer is composed of units for each of the features involved in the task (two shapes and two colors) that have specific connections to the response location where those features are located based on the target cards. For example, the hidden units for 'blue' and 'star' are connected to the response location with the 'blue-star' target card, and 'red' and 'circle' are connected to the response location with the 'red-circle' target card. Propagation between the hidden layer and the output layer is modulated by a prefrontal cortex (PFC) layer that serves to represent the currently relevant rule. When the shape unit is active, the connections between the hidden layer units for shape features and their corresponding output units are boosted. Conversely, if the color unit is active, the connections between the

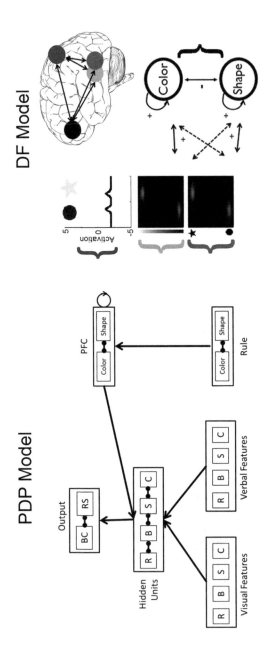

FIGURE 8.3 Connectionist network (left; adapted from Morton & Munakata, 2002) and DF model (right; adapted from Buss & Spencer, 2014) used to simulate DCCS performance.

hidden layer units for color features and their corresponding output units are increased.

As with the model of the A-not-B task, latent representations form as the model sorts cards. Going into the post-switch phase, the model has latent representations that bias processing toward sorting by the pre-switch dimension. With weak recurrent activation on the PFC layer, the model perseverates and continues to sort by the pre-switch dimension. The developmental transition to flexibly switching between rules comes about as recurrent connectivity of the PFC layer increases in strength. Using the concept of competition between active and latent representations, various qualitative predictions have been generated to test this explanation (see also Yerys & Munakata, 2006). For example, Brace, Morton, and Munakata (2006) administered an intermediate phase between the pre- and post-switch phases that was aimed at building latent traces that would support post-switch performance. The sequence of cards that children sorted in the intermediate phase started with only post-switch dimension information. For example, when the post-switch dimension is shape, these cards contained black and white outlines of the shapes. As children progressed through this phase, though, the pre-switch dimension was gradually morphed into the image on the test cards. Continuing with this example, color would gradually shade into the card as children progressed through the intermediate phase. These cards, then, reduce interference from the pre-switch dimension and scaffold children's performance by building latent traces for the post-switch features at their correct sorting locations. Three-year-olds who were given this intermediary phase performed significantly better during the post-switch phase compared to children who had intermediary cards with features that were irrelevant to the task (e.g., oriented lines).

One of the most important advantages of these connectionist models is that they provide neurally plausible explanations of behavior. For instance, neuroimaging data from both the A-not-B and DCCS tasks show that increases in frontal cortex activation are associated with improvements in behavior (Baird et al., 2002; Moriguchi & Hiraki, 2009), which is consistent with the increases in recurrent connectivity implemented in the two models. Additionally, the connectionist modeling framework has been used to conceptualize how the brain changes over development from the level of neurotransmitters to neural populations, and how these changes in brain function are related to changes in behavior (Munakata, Chatham, & Snyder, 2013). The model of the DCCS task, for example, was recently expanded to include a PFC updating mechanism that is grounded in frontostriatal function based on observations that the original model was unable to explain the effects of feedback on children's sorting behavior (Chatham, Yerys, & Munakata, 2012).

DFT has also been applied to the development of cognitive flexibility in the DCCS task. This model implements two interactive systems: an object representation system and a label system (see Figure 8.3). The object representation system has a spatial field that corresponds to representations in parietal cortex. This field is coupled to two feature-space fields that correspond to neural populations that bind spatial information to object feature information (e.g., shape or color) in temporal cortex. This system is able to form object representations through stabilized patterns of activation that link shapes with colors at a spatial location. This object representation system is coupled to a label system corresponding to lateral frontal cortex. In the case of the DCCS, these are labels for visual dimensions (i.e., 'shape' and 'color'). Labels come to have meaning by forming long-range connections with visual feature representations in temporal cortex. These systems are reciprocally connected such that the activation of labels can project onto feature representations in the temporal component, and the activation of features project to label representations in the frontal component. In the context of the DCCS, this creates a distributed rule-representation system: the frontal component is concerned with which dimension is relevant, whereas the temporal component is concerned with which features go where.

Buss and Spencer (2014) demonstrated that these coupled systems can explain the developmental transition seen in the DCCS. There are two central properties of the model that enables it to simulate a wide range of behavioral findings: the formation of feature-space memories in the pre-switch phase and the changes in the strength of frontal-temporal connectivity. First, when sorting cards during the pre-switch phase, the model builds memory traces for the conjunctions of spatial locations and features which then boost activation at the location of the target card inputs in this field. For the post-switch features, however, the memories are at the opposite location of the target card inputs for those features. Thus, memory representations and the target inputs cooperate through local excitation within the pre-switch field because they overlap at the same spatial locations, but compete through lateral inhibition in the post-switch field as a result of the conflicting spatial locations. Second, changes in the strength of coupling between the frontal and temporal components also affect sorting behavior. As mentioned above, development is implemented through stronger connectivity between the frontal and temporal model components. Along with this, changes in connectivity within the frontal cortex were also implemented to create stability in the activation of this region in the face of stronger coupling to the temporal system. Thus, an 'old' model has label representations with strong connectivity to feature representations has the ability to stably activate frontal representations and modulate visual feature representations more robustly. With stronger boosting of the relevant feature dimension, the model is able to overcome the influence of memory traces and use the target inputs in the post-switch feature field to make decisions during the post-switch phase. The 'young' model with weak connectivity, however, perseverates and makes decisions based on the memory traces during this phase.

This DF model explains a complex pattern of results across 14 different conditions in quantitative detail. For example, if the features of only the pre- or the post-switch dimensions are changed before the post-switch phase, then 3-year-olds will continue to perseverate. However, if both features are changed, 3-year-olds switch rules at a significantly higher rate (Zelazo et al., 2003). In the model, this pattern of data is explained by the configuration of memories and target inputs. When only one dimension is changed to new features, the model still has a source of interference, either cooperation in the pre-switch field or competition in the post-switch field. Buss and Spencer (2014) demonstrated that either source of interference is sufficient to induce perseveration in the 'young' model. In another condition, it was demonstrated that conflict during the pre-switch phase is necessary for 3-year-olds' failures. For example, if the pre-switch test cards match the target cards along both dimensions, then 3-year-olds have little difficulty switching rules during the post-switch phase with standard conflict cards (Zelazo et al., 2003). In the model, the no-conflict cards establish a pattern of memories that support post-switch sorting. In this case, the post-switch field now has cooperation between memories and target card inputs that facilitates correct switching in the 'young' model. Importantly, the model provides a means to directly probe the mechanisms proposed to explain children's behavior through empirically testable manipulations. Buss and Spencer (2014) presented data supporting quantitative predictions regarding the role of spatially specific memories for features by manipulating the spatial configuration of target cards between the pre- and post-switch phases.

DFT also creates new insights about the functional aspects of neural activation as measured in traditional cognitive neuroscience methods, such as the hemodynamic response. For example, recent modeling work using DFT developed applications of a linking hypothesis generated from biophysical work to simulate hemodynamic data (Deco et al., 2004; Edin et al., 2007; Lee et al., 2010; Logothetis, Pauls, Augath, Trinath, & Oeltermann, 2001). This method was implemented in the DCCS model to explain and predict neural changes over development. The old model that switches rules produces significantly stronger hemodynamics from the frontal component compared to the young model that perseverates, replicating the fNIRS results from the literature (Moriguchi & Hiraki, 2009).

The model, however, goes beyond this initial observation and makes a series of quantitative hemodynamic predictions. First, the model predicts that children who switch rules should also show stronger activation across parietal and temporal cortex relative to children that perseverate. Second, the model makes the prediction that children who perseverate in the standard version of the task should show strong frontal cortex activation in a no-conflict version. As discussed above, when no-conflict cards are used during the pre-switch phase, the feature-space fields become more strongly activated due to the memory traces which cooperate with the target card inputs. This increased activation sends stronger activation to the frontal component which increases the simulated hemodynamic response from this region relative to the standard condition. These predictions were supported by data from a fNIRS study with 3- and 4-year-olds (Buss & Spencer, submitted) providing converging data to support the mechanisms implemented by the model. In this case, the model provides a cognitively functional account that explains how neural activation is linked to cognitive functioning over development.

Summary

Both models provide an explanation of how children develop the ability to switch between different sets of rules. These models explain patterns of behavior in terms of competition between active and latent representations. However, the nature of these representations differs between models. Whereas the PDP model implements an abstract representational system, the strength of the DF model comes from embodied representations of visual dimensions. The DF model explained a wide array of effects that stem from altering the features or combinations of features involved in the task. Further, the DF model explains and predicts hemodynamic data. Whereas the PDP model focuses on changes in the connectivity and activity of the frontal cortex model component, the DF model is marked by a move away from centering on frontal cortex and focuses on understanding how EF arises from the interactions between cortical regions. Finally, the DF model is unique because it generates active responses in real-time whereas the PDP model explains aggregate rates of performance.

Conclusions

CMs provide the tools for paving a path forward in the study of EF development. Although hurdles lie in the way of computational approaches becoming more prevalent in the field (see Simmering, Triesch, Deák, & Spencer, 2010), the benefits of CMs are very clear. For the sake of clarity and brevity, I focused on CMs that have been applied most extensively (for other approaches, see Marcovitch & Zelazo, 2009; Ramscar, Dye, Witten, & Klein, 2013). The models discussed here have generated new understanding and new insights into the developmental mechanisms that can explain EF development.

First, each model implemented an interactive computational system. There were no explicit modules corresponding to concepts of EF such as inhibition, working memory, or switching; rather, these processes emerged from network dynamics as the models reproduced patterns of behavior. Thus, these models provide explanations that can be grounded in system interactions. A challenge for future work is to go beyond single-task models to shed light on the structure of EF observed in performance across many tasks. Structural equation modeling suggests that EF in adults can be explained through three components corresponding to inhibition, working memory, and updating (Miyake et al., 2000). Developmental studies suggest that this structure becomes differentiated from a single component, but the exact nature of differentiation is currently under debate (Brydges et al., 2014; Howard et al., 2015; Wiebe et al., 2011). In this regard, one task for future modeling work can be to administer batteries of tasks to models in order to determine if a similar structure can emerge from the developmental processes instantiated by particular models. This kind of work can shed light on the boundaries between different functions, establish a common mechanistic basis that gives rise to associations between particular measures of EF, and can provide a means of testing CMs in a broader context beyond the specific tasks that were the focus of this chapter.

Second, these models demonstrate how quantitative changes in the parameters governing interactions in the models can give rise to qualitative shifts in performance. However, development in these models was implemented 'by hand' through changing parameters. It is unclear from these simulations how children's experiences or learning plays a role into these developmental changes. Thus, a second challenge for future modeling work is to determine how experience and learning can develop EF. Specifically, models can be used as a tool for testing how specific experiences can give rise to changes in EF. All of the models have been used to generate different ways of experimentally improving performance (Brace et al., 2006; Buss & Spencer, 2014; Perone, Molitor, Buss, Spencer, & Samuelson, 2015; Yerys & Munakata, 2006). However, no studies have yet tested whether changes can be induced that persist over development. DFT provides an intriguing avenue for this work. The model proposes that dimensional label learning is involved in developmental changes in the ability to flexibly allocate attention across visual dimensions. Future work can test the role of dimensional label comprehension and production in the development of dimensional attention.

Finally, the models discussed above explain behavior using neural activation dynamics and neurally plausible architectures. A third challenge for future modeling work is to begin to build principled bridges between brain and behavior. Recent approaches using integrate-and-fire models and DFT (Buss & Spencer, submitted) have established linking hypotheses to quantitatively simulate aspects of neural data. In one example not discussed above, Edin et al. (2007) implemented an integrate-and-fire network to test different mechanisms for changes in hemodynamic activation measured using fMRI in the context of a working memory task. As working memory develops, frontal and parietal cortex show larger amplitudes in the hemodynamic response and greater correlations in activity. The same linking hypothesis discussed in the context of DFT (Buss et al., 2014; see also Deco et al., 2004) was used to simulate hemodynamic data from populations of neurons that are tuned to the features relevant for the working memory task used in the fMRI study. These authors tested a series of mechanisms to demonstrate that greater connection strength between cortical regions and higher contrast in tuning within regions were the only mechanisms that could reproduce the observed pattern of fMRI data. Computational work using these methods can establish a basis for testing hypotheses about brain development that can be directly and quantitatively compared to experimental observations.

Ultimately, there is agreement on very basic aspects of EF development across the different CMs presented here. All models implemented some form of competition between habits/memories and active representational processes. All models linked changes in EF to changes in frontal cortex. As models continue to be used to predict data and test competing explanations, we can move closer to a principled understanding of the nature of EF development at the behavioral and neural levels. Different computational frameworks provide different tools that are best suited for different types of questions or problems. Thus, which model is more correct does not matter as much as using them to gain new insight into developmental processes and mechanisms.

References

Anderson, P. (2002). Assessment and development of executive function (EF) during childhood. *Child Neuropsychology, 8*, 71–82.

Amari, S. (1977). Dynamics of pattern formation in lateral-inhibition type neural fields. *Biological Cybernetics, 27*, 77–87.

Ashby, F. G., & Waldschmidt, J. G. (2008). Fitting computational models to fMRI data. *Behavior Research Methods, 40*, 713–721.

Baird, A. A., Kagan, J., Gaudette, T., Walz, K. A., Hershlag, N., & Boas, D. A. (2002). Frontal lobe activation during object permanence: Data from near-infrared spectroscopy. *NeuroImage, 16*, 1120–1126.

Bastian, A., Schöner, G., & Riehle, A. (2003). Preshaping and continuous evolution of motor cortical representations during movement preparation. *European Journal of Neuroscience, 18*, 2047–2058.

Brace, J. J., Morton, J. B., & Munakata, Y. (2006). When actions speak louder than words improving children's flexibility in a card-sorting task. *Psychological Science, 17*, 665–669.

Brydges, C. R., Fox, A. M., Reid, C. L., & Anderson, M. (2014). The differentiation of executive functions in middle and late childhood: A longitudinal latent-variable analysis. *Intelligence, 47*, 34–43.

Buss, A. T., & Spencer, J. P. (2014). The emergent executive: A dynamic neural field theory of the development of executive function. *Monographs of the Society for Research in Child Development, 79*(2), 1–104.

Buss, A. T., & Spencer, J. P. (submitted). Changes in frontal-posterior connectivity underlie the early emergence of executive function.

Buss, A. T., Wifall, T., Hazeltine, E., & Spencer, J. P. (2014). Integrating the behavioral and neural dynamics of reponse selection in a dual-task paradigm: A dynamic neural field model of Dux et al. (2009). *Journal of Cognitive Neuroscience, 26*, 334–351.

Carlson, S. M. (2005). Developmentally sensitive measures of executive function in preschool children. *Developmental Neuropsychology, 28*, 595–616.

Chatham, C. H., Yerys, B. E., & Munakata, Y. (2012). Why won't you do what I want? The informative failures of children and models. *Cognitive Development, 27*, 349–366.

Clearfield, M. W., Dineva, E., Smith, L. B., Diedrich, F. J., & Thelen, E. (2009). Cue salience and infant perseverative reaching: Tests of the dynamic field theory. *Developmental Science, 12*, 26–40.

Deco, G., Rolls, E. T., & Horwitz, B. (2004). 'What' and 'Where' in visual working memory: A computational neurodynamical perspective for integrating fMRI and single-neuron data. *Journal of Cognitive Neuroscience, 16*, 683–701.

Diamond, A. (1985). Development of the ability to use recall to guide action, as indicated by infants' performance on AB. *Child Development, 56*, 868–883.

Diedrich, F. J., Highlands, T. M., Spahr, K. A., Thelen, E., & Smith, L. B. (2001). The role of target distinctiveness in infant perseverative reaching. *Journal of Experimental Child Psychology, 78*, 263–290.

Edin, F., Macoveanu, J., Olesen, P., Tegnér, J., & Klingberg, T. (2007). Stronger synaptic connectivity as a mechanism behind development of working memory-related brain activity during childhood. *Journal of Cognitive Neuroscience, 19*, 750–760.

Garon, N., Bryson, S. E., & Smith, I. M. (2008). Executive function in preschoolers: A review using an integrative framework. *Psychological Bulletin, 134*, 31–60.

Grossberg, S. (1970). Some networks that can learn, remember, and reproduce any number of complicated space-time patterns, II. *Studies in Applied Mathematics, 49*, 135–166.

Grossberg, S. (1980). Biological competition: Decision rules, pattern formation, and oscillations. *Proceedings of the National Academy of Sciences USA, 77*, 2338–2342.

Happaney, K., & Zelazo, P. D. (2003). Inhibition as a problem in the psychology of behavior. *Developmental Science, 6*, 468–470.

Hofstadter, M., & Reznick, J. S. (1996). Response modality affects human infant delayed-response performance. *Child Development, 67*, 646–658.

Howard, S. J., Okely, A. D., & Ellis, Y. G. (2015). Evaluation of a differentiation model of preschoolers' executive functions. *Frontiers in Psychology, 6*, 285.

Jancke, D., Erlhagen, W., Dinse, H. R., Akhavan, A. C., Giese, M., Steinhage, A., & Schöner, G. (1999). Parametric population representation of retinal location: Neuronal interaction dynamics in cat primary visual cortex. *The Journal of Neuroscience, 19*, 9016–9028.

Kirkham, N. Z., & Diamond, A. (2003). Sorting between theories of perseveration: Performance in conflict tasks requires memory, attention and inhibition. *Developmental Science, 6*, 474–476.

Lee, J. H., Durand, R., Gradinaru, V., Zhang, F., Goshen, I., Kim, D.-S., et al. (2010). Global and local fMRI signals driven by neurons defined optogenetically by type and wiring. *Nature, 465*, 788–792.

Logothetis, N. K., Pauls, J., Augath, M., Trinath, T., & Oeltermann, A. (2001). Neurophysiological investigation of the basis of the fMRI signal. *Nature, 412*, 150–157.

Marcovitch, S., & Zelazo, P. D. (2009). A hierarchical competing systems model of the emergence and early development of executive function. *Developmental Science, 12*, 1–18.

Markounikau, V., Igel, C., Grinvald, A., & Jancke, D. (2010). A dynamic neural field model of mesoscopic cortical activity captured with voltage-sensitive dye imaging. *PLOS Computational Biology, 6*, e1000919.

Milner, B. (1963). Effects of different brain lesions on card sorting: The role of the frontal lobes. *Archives of Neurology, 9*, 90–100.

Moriguchi, Y., & Hiraki, K. (2009). Neural origin of cognitive shifting in young children. *Proceedings of the National Academy of Sciences of the United States of America, 106*, 6017–6021.

Morton, J. B., & Munakata, Y. (2002). Active versus latent representations: A neural network model of perseveration, dissociation, and decalage. *Developmental Psychobiology, 40*, 255–265.

Munakata, Y. (1998). Infant perseveration and implications for object permanence theories: A PDP model of the A B task. *Developmental Science, 1*, 161–184.

Munakata, Y., Chatham, C. H., & Snyder, H. R. (2013). Mechanistic accounts of frontal lobe development. In D. T. Stuss & R. T. Knight (Eds.), *Principles of frontal lobe function* (2nd ed., pp. 185–206). Oxford: Oxford University Press.

Munakata, Y., Morton, J. B., & Yerys, B. E. (2003). Children's perseveration: Attentional inertia and alternative accounts. *Developmental Science, 6*, 471–473.

Perone, S., Simmering, V. R., & Spencer, J. P. (2011). Stronger neural dynamics capture changes in infants' visual working memory capacity over development. *Developmental Science, 14*, 1379–1392.

Piaget, J. (1954). *The construction of reality in the child*. New York: Basic Books.

Ramscar, M., Dye, M., Witten, J., & Klein, J. (2013). Two routes to cognitive flexibility: Learning and response conflict resolution in the dimensional change card sort task. *Child Development, 84*, 1308–1323.

Rumelhart, D. E., & McClelland, J. L. (1986). *Parallel distributed processing: Explorations in the microstructure of cognition. No. 1: Foundations*. Cambridge, MA: MIT Press.

Schlesinger, M., & McMurray, B. (2012). The past, present, and future of computational models of cognitive development. *Cognitive Development, 27*, 326–348.

Schoner, G., Spencer, J. P., & the DFT Research Group. (2015). *Dynamic thinking: A primer on dynamic field theory*. New York: Oxford University Press.

Schutte, A. R., Spencer, J. P., & Schoner, G. (2003). Testing the dynamic field theory: Working memory for locations becomes more spatially precise over development. *Child Development, 74*, 1393–1417.

Simmering, V. R., & Patterson, A. R. (2012). Models provide specificity: Testing a proposed mechanism of visual working memory capacity development. *Cognitive Development, 27*, 419–439.

Simmering, V. R., Schutte, A. R., & Spencer, J. P. (2008). Generalizing the dynamic field theory of spatial cognition across real and developmental time scales. *Brain Research, 1202*, 68–86.

Simmering, V. R., Triesch, J., Deák, G. O., & Spencer, J. P. (2010). A dialogue on the role of computational modeling in developmental science. *Child Development Perspectives, 4*, 152–158.

Smith, L. B., Thelen, E., Titzer, R., & McLin, D. (1999). Knowing in the context of acting : The task dynamics of the A-not-B error. *Psychological Review, 106*, 235–260.

Spencer, J. P., & Perone, S. (2008). Defending qualitative change: The view from dynamical systems theory. *Child Development, 79*, 1639–1647.

Spencer, J. P., Smith, L. B., & Thelen, E. (2001). Tests of a dynamic systems account of the A-not-B error: The influence of prior experience on the spatial memory abilities of 2-year-olds. *Child Development, 72*, 1327–1346.

Spencer, J. P., Thomas, M. S. C., & McClelland, J. L. (Eds.). (2009). *Toward a unified theory of development: Connectionism and dynamic systems theory re-considered*. Oxford: Oxford University Press.

Thelen, E., Schöner, G., Scheier, C., & Smith, L. B. (2001). The dynamics of embodiment: A field theory of infant perseverative reaching. *The Behavioral and Brain Sciences, 24*, 1–86.

Turner, B. M., Forstmann, B. U., Love, B. C., Palmeri, T. J., & Van Maanen, L. (2017). Approaches to analysis in model-based cognitive neuroscience. *Journal of Mathematical Psychology, 76* Part B, 65–79.

Wellman, H. M., Cross, D., Bartsch, K., & Harris, P. L. (1987). Infant search and object permanence: A meta-analysis of the A-not-B error. *Monographs of the Society for Research in Child Development, 51*(3), 1–67.

Wiebe, S. A., Sheffield, T., Nelson, J. M., Clark, C. A. C., Chevalier, N., & Espy, K. A. (2011). The structure of executive function in 3-year-olds. *Journal of Experimental Child Psychology, 108*, 436–452.

Yermolayeva, Y., & Rakison, D. H. (2014). Connectionist modeling of developmental changes in infancy: Approaches, challenges, and contributions. *Psychological Bulletin, 140*, 224–255.

Yerys, B. E., & Munakata, Y. (2006). When labels hurt but novelty helps: Children's perseveration and flexibility in a card-sorting task. *Child Development, 77*, 1589–1607.

Zelazo, P. D., Muller, U., Frye, D., & Marcovitch, S. (2003). The development of executive function in early childhood. *Monographs of the Society for Research in Child Development, 68*(3), 1–137.

Part III

Environmental, Cultural, and Lifestyle Factors That Shape Executive Function Development Across the Life Span

9

ADVERSITY AND STRESS

Implications for the Development of Executive Functions

Jenna E. Finch & Jelena Obradović

Executive functions (EFs) play a key role in children's abilities to learn, as well as engage in positive social interactions. The development of EFs, which support the self-regulation of behavior and attention, are influenced by both biology and experiences, with transactional influences among the environment, behavior, neural activity, and genetic activity over time (Gottlieb, 1991; Chevalier & Clark, Chapter 2, this volume). The prefrontal cortex, which supports EFs, is malleable to environmental influences into young adulthood, and thus children and adolescents' experiences have significant and long-lasting effects on the development of their EFs. Due to the importance of environmental influences for EF development, adverse experiences have been consistently linked to EF deficits. We first provide an overview of the literature linking economic adversity, namely poverty, and associated risk factors to EFs. Second, we focus on quality of caregiving as an important mediator of socioeconomic risk on children's EFs. Third, we discuss how adversity works through and interacts with neurobiological mechanisms, including genetics, stress physiology, and brain structure and functionality, to impact EFs. We conclude by discussing the need for causal research to better understand how income and experiences of adversity undermine the development of EFs. We also describe new interventions that may benefit EF development, at-scale.

Adversity and Executive Functions

While many types of adverse experiences have negative consequences for EF development, our chapter primarily focuses on how income disparities and associated risks relate to the development of EFs. In particular, we explore links between family-level poverty and EF development. In the United States, poverty refers to a family income below a certain threshold, and is used to determine access for means-based social services. For a family of four, the 2016 poverty threshold was set at US$24,300 per year (U.S. Department of Health and Human Services, 2016). In the global context, poverty has a more multi-dimensional definition and refers to deprivation in the immediate aspects of life including nutrition, health, water, education, protection, and shelter. The pernicious effects of poverty impact 43.1 million people in the United States. Over 702 million people globally live in "extreme poverty", on less than US$1.90 per day (Cruz, Foster, Quillin, & Schellekens, 2015). Given that early experiences of poverty are most strongly linked to EF deficits, it is particularly concerning that one out of five American children is poor (DeNavas-Walt & Proctor, 2015) and more than one-third of the poor, worldwide, are children (Olinto, Beegle, Sobrado, & Uematsu, 2013).

In the United States, family income during the first 3 years of the child's life has been associated with better working memory and planning skills in elementary school students, and these effects were consistent through middle childhood (Hackman, Gallop, Evans, & Farah, 2015). Chronic exposure to poverty has been linked to increased EF deficits (Raver, Blair, & Willoughby, 2013), suggesting that the amount of time a child spends in poverty is predictive of their EF difficulties. In a sample of children living in Madagascar, Fernald and colleagues found socioeconomic gradients in children's EFs within this context of extreme poverty (Fernald, Weber, Galasso, & Ratsifandrihamanana, 2011). The effects of family-level poverty are exacerbated by the frequent co-occurrence of poor neighborhood and school quality. Researchers have found unique contributions of residential stressors (housing and neighborhood problems) over and above socio-demographic stressors (low caregiver education and low-income status) for children's EFs (Li-Grining, 2007). Moves out of low-poverty and into high-poverty neighborhoods were detrimental for elementary school children's EFs, even after controlling for household income and caregiver education, along with other socio-demographic variables (Roy, McCoy, & Raver, 2014). Further, enrollment in unsafe elementary schools, as rated by students, was significantly predictive of teacher-rated EF difficulties in the classroom, for children who began with elevated levels of EF difficulties (Raver, McCoy, Lowenstein, & Pess, 2013).

Further, researchers have examined how more proximal measures of adverse experiences are related to EF development. Low income is associated with increased exposure to risk factors, such as residential instability, family conflict, violence, and marital conflict. Measures of family instability and household chaos, as indexed by changes in residence and changes in household members, were detrimental to children's EFs (Sturge-Apple, Davies, Cicchetti, Hentges, & Coe, 2016; Vernon-Feagans, Willoughby, Garrett-Peters, & The Family Life Project Key Investigators, 2016). An aggregate measure of cumulative risk, including residential instability, household density, and negative life events, accounted for a significant proportion of the association between income and children's delay of gratification skills (Lengua et al., 2015). Future research should explore how specific poverty-related risk factors uniquely mediate links between income and EFs, and whether these relations differ depending on a child's age.

Caregiving Mediates the Effects of Adversity on Executive Functions

The effects of low-income and related risk factors, such as high household density and multiple residential moves, on children's EFs were partially mediated by the quality of children's home environments. High-quality caregiving, including responsive and sensitive parent–child interactions, consistency, and cognitively stimulating activities, has had positive effects on children's EFs (Fay-Stammbach, Hawes, & Meredith, 2014; Hughes & Devine, Chapter 10, this volume). In impoverished contexts, parents tend to experience more stressors that lower the quality of home environments, which has a cascading effect on children's EFs (Blair & Raver, 2012). In a cross-sectional study of elementary school children, the quality of the home environment mediated associations between family socioeconomic status and children's inhibitory control and working memory (Sarsour et al., 2011).

Further, specific parenting behaviors, including maternal sensitivity and scaffolding behaviors, have also been linked to EF development in early childhood. Blair and colleagues (2011) found that income-to-needs and maternal education worked through parenting behaviors, as coded during parent–child interactions at 7 and 15 months, to influence children's EFs at 3 years. Both negative parenting behaviors, such as harshness, and positive parenting behaviors, such as sensitivity and scaffolding, mediated associations between socioeconomic status and EFs at 3 years in expected directions (Blair, Granger, et al., 2011). A short-term longitudinal study showed that limit setting and scaffolding behaviors mediated the link between cumulative risk and EFs (Lengua, Honorado, & Bush, 2007). Preschoolers with more demographic and psychosocial risk factors had decreased EFs, due to the effects of risk factors on mothers' parenting quality. Scaffolding has also been shown to mediate the effects of both income and cumulative risk on *growth* in EFs from 3 to 5.5 years (Lengua et al., 2014). Hackman and colleagues found that both a global measure of home enrichment and maternal sensitivity during the preschool years fully mediated the effects of early income-to-needs on working memory at school entry (Hackman et al., 2015). Since both mediators were included in the same model, this suggests that more global measures of the children's home learning environments and specific parenting behaviors uniquely impact the development of EFs.

These findings also extend to recent research on child development in low- and middle-income countries. For example, in Zambia, the effects of wealth and caregiver education on 6-year-olds' EFs were mediated by home-based cognitive stimulation, as indexed by number of books in the home and time spent reading (McCoy, Zuilkowski, & Fink, 2015). Finally, both the general quality of the home environment and maternal scaffolding behaviors have been found to mediate the longitudinal effect of an early responsive parenting intervention designed on preschoolers' EFs in rural Pakistan (Obradović, Yousafzai, Finch, & Rasheed, 2016). This provided evidence that mothers could adapt messages provided by the intervention in the first 2 years of the child's life to the cognitive scaffolding of preschoolers.

Separately, high-quality preschool settings have been associated with the development of young children's EFs (Finch, Johnson, & Phillips, 2015; Weiland, Ulvestad, Sachs, & Yoshikawa, 2013). Given the increasing number of children attending early childhood education settings, improving the quality of preschool may help reduce educational inequities between disadvantaged children and their more advantaged peers (Raver, 2012). Preschool interventions for low-income children benefit their EFs and school-readiness skills. A public prekindergarten program in the Boston Public Schools, with research-based curricula and coaching, had significant effects on EFs (Weiland & Yoshikawa, 2013). Effects of the program on EFs were limited to children who were eligible for free or reduced-price lunch. In Head Start classrooms, two interventions focused on self-regulation and social-emotional skills had significant effects on children's EF skills (Bierman, Nix, Greenberg, Blair, & Domitrovich, 2008; Raver et al., 2011). These studies also showed that EFs partially mediated associations between the intervention and children's academic and social-emotional outcomes.

Stress and the Biological Underpinnings of Executive Functions

According to Blair and Raver's experiential canalization model (2012), stress physiology acts as a primary mechanism through which experiences shape children's self-regulation. In turn, changes in stress physiology influence the structure and connectivity in the brain regions that support EFs. Indeed, research shows that measures of socioeconomic status have associations with the structure and function of brain regions known to support EFs (Hackman & Farah, 2009). For example, parental education was linked to cortical thickness in regions of the prefrontal cortex (Lawson, Duda, Avants, Wu, & Farah, 2013). Noble and colleagues (2015) found that income was logarithmically associated with brain surface area, such that, among children from lower-income families, small differences in income were associated with relatively large differences in surface area for the brain regions supporting EFs and spatial and language skills. This suggests that differences in income may be most salient for the brain development of low-income children. Studies also found socioeconomic disparities in functional brain activation patterns during selective attention tasks, even when children's behavioral performance on the tasks did not differ (D'Angiulli, Herdman, Stapells, & Hertzman, 2008; Stevens, Lauinger, & Neville, 2009). More recent research has shown differential brain activation during a demanding working memory task, such that higher-income middle school students showed more activation of the fronto-parietal executive network compared to their lower-income peers (Finn et al., 2016). There is also evidence that the relation between brain structure and EFs skills varies by children's socioeconomic status. A recent paper found that lower white matter volume was associated with cognitive flexibility deficits only for low-income children. High-income children's cognitive flexibility skills did not differ by the volume of white matter (Ursache, Noble, & the Pediatric Imaging, Neurocognition and Genetics Study, 2016).

Research shows that experiences, such as low socioeconomic status and poor housing quality, can be linked to deregulated stress hormones in children (Blair, Raver, et al., 2011; Chen, Cohen, & Miller, 2010). The over- and underproduction of stress hormones, specifically cortisol, have been shown to adversely impact the development and activation of brain regions implicated in EFs (Lupien, McEwen, Gunnar, & Heim, 2009). Stress physiology acts as a key mediator of the effects of adversity on neurobiology (Blair & Raver, 2012). One study, by Kim and colleagues, found that chronic stressors experienced throughout middle childhood and adolescence fully explained the effects of poverty at 9 years on prefrontal cortex activity during an emotion regulation task at 24 years (Kim et al., 2013). Poverty was linked to stressful experiences, such as low housing quality, family turmoil and violence, which then predicted reduced activity in the ventrolateral and dorsolateral prefrontal cortex. This study provides preliminary evidence supporting theorized pathways (Blair & Raver, 2012), but more research is needed to test the mediation pathway from adversity to brain development through measures of stress physiology, such as cortisol.

Increased literature has found that stress physiology partially explains the pathway between adverse experiences and EF development. Chronic stress across ages 9 to 13, as measured by mean allostatic load, mediated the effects of childhood poverty on young adults' working memory (Evans & Fuller-Rowell, 2013). After controlling for maternal education and child ethnicity, basal cortisol levels partially mediated associations between family instability and changes in hot EFs during the preschool years (Sturge-Apple et al., 2016). Our recent research showed that there is an interplay between stress physiology and EFs, such that individual differences in children's EFs were related to dynamic measures of physiological reactivity and recovery during a stressful task (Obradović & Finch, 2016). This suggests that it is important to understand the role of EFs in regulating physiological arousal.

Further, differential susceptibility theory posits that some individuals are more sensitive to both negative and positive contexts (Ellis, Boyce, Belsky,

Bakermans-Kranenburg, & van Ijzendoorn, 2011). Specifically, individual difference in temperamental reactivity, physiological reactivity, and genetic polymorphisms predispose children to be more or less susceptible to contextual influences (Obradović & Boyce, 2009; Li & Roberts, Chapter 7, this volume). Different markers of sensitivity, such as temperamental emotional reactivity and stress physiology, have been shown to interact with income to predict children's EFs, such that income effects are restricted to children with high sensitivity (Obradović, Portilla, & Ballard, 2015; Raver, Blair, et al., 2013). Researchers have shown that allelic variation moderated the effects of parenting on EFs in preschoolers (Kochanska, Philibert, & Barry, 2009; Smith, Kryski, Sheikh, Singh, & Hayden, 2013), parenting on self-regulation skills in adolescent males (Belsky & Beaver, 2011), and decline in EF skills over a 10-year period in elderly adults (Erickson et al., 2008).

At the same time, children's behaviors have effects on their environments and experiences (Portilla, Ballard, Adler, Boyce, & Obradović, 2014; Yates, Obradović, & Egeland, 2010), which may explain why EFs can serve as a protective factor in contexts of risk and adversity (Wenzel & Gunnar, 2014). Self-regulation skills also enable people to use more efficient coping strategies, such as better directed attentional skills. Within a sample of very-low-income youth, assessor-rated self-regulation skills were linked to achievement, social competency, and mental health symptomology (Buckner, Mezzacappa, & Beardslee, 2009). For homeless children, EFs were associated with adaptive functioning in school settings, even after controlling for child IQ, parenting quality, and sociodemographic risks (Masten et al., 2012; Obradović, 2010). Young children with low EFs had significantly higher levels of problem behaviors when experiencing socioeconomic risks (Lengua, Bush, Long, Kovacs, & Trancik, 2008). A longitudinal study found that hot EFs at 9 years of age moderated associations between childhood poverty and EFs in young adulthood (Evans & Fuller-Rowell, 2013). For both of these studies, higher EFs acted as a protective factor, such that poverty and associated risks had a less harmful influence on later skills when children had greater EF abilities.

Causal Links between Adversity and EFs: New Directions and Interventions

While there has been a multitude of research citing socioeconomic disparities in children's cognitive development, the majority of this work has been correlational (Duncan, Magnuson, & Votruba-Drzal, 2017). Experimental lab work from the cognitive psychology and behavioral economics fields have provided some causal evidence for why subjective feelings of low socioeconomic status may have a negative effect on cognitive functioning. Experimental manipulations show that priming adults to feel "powerless" affected their ability to focus on the goal at hand (Smith, Jostmann, Galinsky, & van Dijk, 2008). Further,

Mani and colleagues (2013) found that an experimental manipulation forcing poor adults to think about a financially difficult situation (e.g. paying $1,500 for car repairs) caused decreased performance on fluid intelligence and EF tasks. The experimental manipulation had no effect on high-income adults. The authors hypothesized that the additional psychological demands of poverty limited the ability of poor adults to make decisions and engage in cognitively demanding tasks. This type of work needs to be extended from subjective measures of socioeconomic status to experimental manipulations of family income. Duncan and colleagues have proposed an experiment where a significant amount of money is provided to families with young children to elucidate how income boosts causally impact neurocognitive development (Duncan et al., 2017).

Intervention research provides insights into how policymakers can improve EF development for children. In the past, small-scale interventions that targeted EFs, such as computerized training, mindfulness, and aerobic exercise interventions have succeeded at increasing EFs in the short term (Bierman & Torres, 2015; Diamond & Lee, 2011; Kliegel, Hering, Ihle, & Zuber, Chapter 13, this volume). Further, studies of preschool programs that included specific supports for the development of self-regulation and social-emotional skills benefited young children's EF development (Barnett et al., 2008; Raver et al., 2011). Currently, researchers are focused on scaling up early childhood programs, by providing more general interventions during important developmental periods that encourage positive interactions between caregivers and children. These interventions aimed to benefit disadvantaged children are not specifically targeting EFs, but more broad measures of positive adaptation and development. EFs are viewed as an intermediary mechanism for interventions, as they are implicated in academic achievement, social-emotional development, and long-term outcomes.

High-quality preschool programs have been shown to improve academic outcomes as well as EF skills. In Boston, a large-scale public prekindergarten program with strong literacy and math curricula, along with coaching support, improved children's EF and academic skills (Weiland & Yoshikawa, 2013). Moreover, a recent intervention targeting family child-care providers benefited young preschoolers' EFs (Merz, Landry, Johnson, Williams, & Jung, 2016). The intervention group took an online course including weekly mentoring focused on responsive and contingent interactions. These more "general" programs have the potential to reach a large percentage of children and promote EFs along with other important social-emotional and academic skills. There is a strong need to do more research in school-age children to see how their educational settings could be adjusted to promote the development of self-regulation skills. Most research on EFs focuses on the early childhood and adolescent periods, failing to explore the environmental influences on development of EFs during middle childhood.

Finally, it will be particularly important to explore how caregivers' EFs can impact the developmental trajectories of children's EFs. Direct measures of maternal EFs, as well as maternal report of EF skills, have been linked to increased positive parenting behaviors and decreased negative parenting behaviors (Chico, Gonzalez, Ali, Steiner, & Fleming, 2014; Deater-Deckard, Wang, Chen, & Bell, 2012). In a highly disadvantaged context, we found significant associations between directly assessed maternal EFs and maternal cognitive scaffolding for mothers living in rural Pakistan (Obradović et al., 2017). Building on an evidence base that maternal EFs promote parenting practices known to foster EFs (Crandall, Deater-Deckard, & Riley, 2015), a growing number of studies examine how maternal EFs are associated with children's EF development. Cuevas and colleagues (2014) provided preliminary evidence that there are unique effects of maternal EF on child EF through caregiving practices, for an advantaged sample. There is a need to extend this work to children experiencing a wider range of risks, as maternal EF may matter more for children who experience high levels of adversity. This knowledge should be used to inform two-generational intervention programs, as EF abilities enable parents to care for themselves and their children (Shonkoff & Fisher, 2013).

For infants and toddlers, two-generational home visiting programs have shown great promise for promoting the EF skills of young children, at-scale. A meta-analysis of responsive stimulation interventions in low- and middle-income contexts during the first 2 years showed medium effect sizes on children's cognitive and language development (Aboud & Yousafzai, 2015). Our recent work in Pakistan demonstrated long-term effects of a responsive stimulation parenting intervention during the first 2 years of life on children's EFs skills (Yousafzai et al., 2016). The intervention worked through improved parenting skills to improve children's EFs and IQ skills (Obradović et al., 2016). Similar efforts are underway in the United States. Chang and colleagues (2015) evaluated a small-scale program that seeks to improve parenting practices for low-income families. The intervention had significant impacts on toddler's EFs, through improvements in parenting practices, and specifically scaffolding. In 2010, the Maternal, Infant, and Early Childhood Home Visiting (MIECHV) program was created as part of the Affordable Care Act, to provide at-risk parents with the resources and skills needed to promote positive development for their children (Michalopoulos et al., 2015). The program is now being evaluated by MDRC to test how four different kinds of home visiting programs impact parental and child well-being, as well as parenting, family economic self-sufficiency, and rates of violence in children's homes.

Conclusion

Evidence from multiple disciplinary backgrounds, including developmental psychology, neuroscience, and education, cites income-based disparities in children's EFs, which have lasting consequences for their academic and social success. Experiences of adversity in the early years, as children's biological systems are rapidly developing, appear to have the strongest effects. EF development is complex and dynamic, with interplay between experiences and biology across the life span. Although we have learned much about specific predictors of EF development, there is a need for interdisciplinary, longitudinal research to examine how adverse experiences across the life span interact with individual characteristics to affect EFs. Studies that identify causal links between adversity and EFs will be essential, as researchers work with policymakers to promote positive development for a new generation of children.

References

Aboud, F. E., & Yousafzai, A. K. (2015). Global health and development in early childhood. *Annual Review of Psychology, 66*, 433–457.

Barnett, W. S., Jung, K., Yarosz, D. J., Thomas, J., Hornbeck, A., Stechuk, R., & Burns, S. (2008). Educational effects of the Tools of the Mind curriculum: A randomized trial. *Early Childhood Research Quarterly, 23*, 299–313.

Belsky, J., & Beaver, K. M. (2011). Cumulative-genetic plasticity, parenting and adolescent self-regulation: Gene × environment interaction, parenting and adolescent self-regulation. *Journal of Child Psychology and Psychiatry, 52*, 619–626.

Bierman, K. L., Nix, R. L., Greenberg, M. T., Blair, C., & Domitrovich, C. E. (2008). Executive functions and school readiness intervention: Impact, moderation, and mediation in the Head Start REDI program. *Development and Psychopathology, 20*, 821–843.

Bierman, K. L., & Torres, M. (2015). Promoting the development of executive functions through early education and prevention programs. In J. A. Griffin, L. S. Freund, & P. McCardle (Eds.), *Executive function in preschool age children: Integrating measurement neurodevelopment, and translational research*. Washington, DC: American Psychological Association.

Blair, C., Granger, D. A., Willoughby, M., Mills-Koonce, R., Cox, M., Greenberg, M. T., et al. (2011). Salivary cortisol mediates effects of poverty and parenting on executive functions in early childhood. *Child Development, 82*, 1970–1984.

Blair, C., & Raver, C. C. (2012). Child development in the context of adversity: Experiential canalization of brain and behavior. *American Psychologist, 67*, 309–318.

Blair, C., Raver, C. C., Granger, D., Mills-Koonce, R., Hibel, L., & The Family Life Project Key Investigators. (2011). Allostasis and allostatic load in the context of poverty in early childhood. *Development and Psychopathology, 23*, 845–857.

Buckner, J. C., Mezzacappa, E., & Beardslee, W. R. (2009). Self-regulation and its relations to adaptive functioning in low income youths. *American Journal of Orthopsychiatry, 79*, 19–30.

Chang, H., Shaw, D. S., Dishion, T. J., Gardner, F., & Wilson, M. N. (2015). Proactive parenting and children's effortful control: Mediating role of language and indirect intervention effects. *Social Development, 24*, 206–223.

Chen, E., Cohen, S., & Miller, G. E. (2010). How low socioeconomic status affects 2-year hormonal trajectories in children. *Psychological Science, 21*, 31–37.

Chico, E., Gonzalez, A., Ali, N., Steiner, M., & Fleming, A. S. (2014). Executive function and mothering: Challenges faced by teenage mothers – Executive function and parenting. *Developmental Psychobiology, 56*, 1027–1035.

Crandall, A., Deater-Deckard, K., & Riley, A. W. (2015). Maternal emotion and cognitive control capacities and parenting: A conceptual framework. *Developmental Review, 36*, 105–126.

Cruz, M., Foster, J., Quillin, B., & Schellekens, P. (2015). *Ending extreme poverty and sharing prosperity: Progress and policies* (Policy Research Note). Washington, DC: World Bank Group.

Cuevas, K., Deater-Deckard, K., Kim-Spoon, J., Watson, A. J., Morasch, K. C., & Bell, M. A. (2014). What's mom got to do with it? Contributions of maternal executive function and caregiving to the development of executive function across early childhood. *Developmental Science, 17*, 224–238.

D'Angiulli, A., Herdman, A., Stapells, D., & Hertzman, C. (2008). Children's event-related potentials of auditory selective attention vary with their socioeconomic status. *Neuropsychology, 22*, 293–300.

Deater-Deckard, K., Wang, Z., Chen, N., & Bell, M. A. (2012). Maternal executive function, harsh parenting, and child conduct problems. *Journal of Child Psychology and Psychiatry and Allied Disciplines, 53*, 1084–1091.

DeNavas-Walt, C., & Proctor, B. D. (2015). *Income and Poverty in the United States* (Current Population Reports, U.S. Census Bureau No. P60-252). Washington, DC: U.S. Census Bureau.

Diamond, A., & Lee, K. (2011). Interventions shown to aid executive function development in children 4 to 12 years old. *Science, 333*, 959–964.

Duncan, G. J., Magnuson, K., & Votruba-Drzal, E. (2017). Moving beyond correlations in assessing the consequences of poverty. *Annual Review of Psychology, 68*, 413–434.

Ellis, B. J., Boyce, W. T., Belsky, J., Bakermans-Kranenburg, M. J., & van Ijzendoorn, M. H. (2011). Differential susceptibility to the environment: An evolutionary-neurodevelopmental theory. *Development and Psychopathology, 23*, 7–28.

Erickson, K., Kim, J., Suever, B. L., Voss, M., Francis, B. M., & Kramer, A. F. (2008). Genetic contributions to age-related decline in executive function: A 10-year longitudinal study of COMT and BDNF polymorphisms. *Frontiers in Human Neuroscience, 2*, 11.

Evans, G. W., & Fuller-Rowell, T. E. (2013). Childhood poverty, chronic stress, and young adult working memory: The protective role of self-regulatory capacity. *Developmental Science, 16*, 688–696.

Fay-Stammbach, T., Hawes, D. J., & Meredith, P. (2014). Parenting influences on executive function in early childhood: A review. *Child Development Perspectives, 8*, 258–264.

Fernald, L. C. H., Weber, A., Galasso, E., & Ratsifandrihamanana, L. (2011). Socioeconomic gradients and child development in a very low income population: Evidence from Madagascar. *Developmental Science, 14*, 832–847.

Finch, J. E., Johnson, A. D., & Phillips, D. A. (2015). Is sensitive caregiving in child care associated with children's effortful control skills? An exploration of linear and threshold effects. *Early Childhood Research Quarterly, 31*, 125–134.

Finn, A. S., Minas, J. E., Leonard, J. A., Mackey, A. P., Salvatore, J., Goetz, C., et al. (2016). Functional brain organization of working memory in adolescents varies in relation to family income and academic achievement. Advance online publication. *Developmental Science.* doi: 10.1111/desc.12450.

Gottlieb, G. (1991). Experiential canalization of behavioral development: Theory. *Developmental Psychology, 27*, 4–13.

Hackman, D. A., & Farah, M. J. (2009). Socioeconomic status and the developing brain. *Trends in Cognitive Sciences, 13*, 65–73.

Hackman, D. A., Gallop, R., Evans, G. W., & Farah, M. J. (2015). Socioeconomic status and executive function: Developmental trajectories and mediation. *Developmental Science, 18*, 686–702.

Kim, P., Evans, G. W., Angstadt, M., Ho, S. S., Sripada, C. S., Swain, J. E., et al. (2013). Effects of childhood poverty and chronic stress on emotion regulatory brain function in adulthood. *Proceedings of the National Academy of Sciences, 110*, 18442–18447.

Kochanska, G., Philibert, R. A., & Barry, R. A. (2009). Interplay of genes and early mother–child relationship in the development of self-regulation from toddler to preschool age. *Journal of Child Psychology and Psychiatry, 50*, 1331–1338.

Lawson, G. M., Duda, J. T., Avants, B. B., Wu, J., & Farah, M. J. (2013). Associations between children's socioeconomic status and prefrontal cortical thickness. *Developmental Science, 16*, 641–652.

Lengua, L. J., Bush, N. R., Long, A. C., Kovacs, E. A., & Trancik, A. M. (2008). Effortful control as a moderator of the relation between contextual risk factors and growth in adjustment problems. *Development and Psychopathology, 20*, 509–528.

Lengua, L. J., Honorado, E., & Bush, N. R. (2007). Contextual risk and parenting as predictors of effortful control and social competence in preschool children. *Journal of Applied Developmental Psychology, 28*, 40–55.

Lengua, L. J., Kiff, C., Moran, L., Zalewski, M., Thompson, S., Cortes, R., & Ruberry, E. (2014). Parenting mediates the effects of income and cumulative risk on the development of effortful control. *Social Development, 23*, 631–649.

Lengua, L. J., Moran, L., Zalewski, M., Ruberry, E., Kiff, C., & Thompson, S. (2015). Relations of growth in effortful control to family income, cumulative risk, and adjustment in preschool-age children. *Journal of Abnormal Child Psychology, 43*, 705–720.

Li-Grining, C. P. (2007). Effortful control among low-income preschoolers in three cities: Stability, change, and individual differences. *Developmental Psychology, 43*, 208–221.

Lupien, S. J., McEwen, B. S., Gunnar, M. R., & Heim, C. (2009). Effects of stress throughout the lifespan on the brain, behaviour and cognition. *Nature Reviews Neuroscience, 10*, 434–445.

Mani, A., Mullainathan, S., Shafir, E., & Zhao, J. (2013). Poverty impedes cognitive function. *Science, 341*, 976–980.

Masten, A. S., Herbers, J. E., Desjardins, C. D., Cutuli, J. J., McCormick, C. M., Sapienza, J. K., et al. (2012). Executive function skills and school success in young children experiencing homelessness. *Educational Researcher, 41*, 375–384.

McCoy, D. C., Zuilkowski, S. S., & Fink, G. (2015). Poverty, physical stature, and cognitive skills: Mechanisms underlying children's school enrollment in Zambia. *Developmental Psychology, 51*, 600–614.

Merz, E. C., Landry, S. H., Johnson, U. Y., Williams, J. M., & Jung, K. (2016). Effects of a responsiveness-focused intervention in family child care homes on children's executive function. *Early Childhood Research Quarterly, 34*, 128–139.

Michalopoulos, C., Lee, H., Duggan, A., Lundquist, E., Tso, A., Crowne, S., et al. (2015). *The Mother and Infant Home Visiting Program evaluation: Early findings on the Maternal, Infant, and Early Childhood Home Visiting Program* (OPRE Report No. 2015-11). Washington, DC: Office of Planning, Research and Evaluation, Administration for Children and Families, U.S. Department of Health and Human Services.

Noble, K. G., Houston, S. M., Brito, N. H., Bartsch, H., Kan, E., Kuperman, J. M., et al. (2015). Family income, parental education and brain structure in children and adolescents. *Nature Neuroscience, 18*, 773–778.

Obradović, J. (2010). Effortful control and adaptive functioning of homeless children: Variable-focused and person-focused analyses. *Journal of Applied Developmental Psychology, 31,* 109–117.

Obradović, J., & Boyce, W. T. (2009). Individual differences in behavioral, physiological, and genetic sensitivities to contexts: Implications for development and adaptation. *Developmental Neuroscience, 31,* 300–308.

Obradović, J., & Finch, J. E. (2016). Linking executive function skills and physiological challenge response: Piecewise latent growth curve modeling. Advance Online Publication. *Developmental Science.* doi: 10.1111/desc.12476.

Obradović, J., Portilla, X. A., & Ballard, P. J. (2015). Biological sensitivity to family income: Differential effects on early executive functioning. *Child Development, 87,* 374–384.

Obradović, J., Portilla, X., Tirado-Strayer, N., Siyal, S., Rasheed, M. A., & Yousafzai, A. K. (2017). Maternal scaffolding in a disadvantaged global context: The role of maternal cognitive capacities. *Journal of Family Psychology, 31,* 139–149

Obradović, J., Yousafzai, A. K., Finch, J. E., & Rasheed, M. A. (2016). Maternal scaffolding and home stimulation: Key mediators of early intervention effects on children's cognitive development. *Developmental Psychology, 52,* 1409–1421.

Olinto, P., Beegle, K., Sobrado, C., & Uematsu, H. (2013). *The state of the poor: Where are the poor, where is extreme poverty harder to end, and what is the current profile of the world's poor?* (Economic Premise No. 125). Washington, DC: World Bank.

Portilla, X. A., Ballard, P. J., Adler, N. E., Boyce, W. T., & Obradović, J. (2014). An integrative view of school functioning: Transactions between self-regulation, school engagement, and teacher–child relationship quality. *Child Development, 85,* 1915–1931.

Raver, C. C. (2012). Low-income children's self-regulation in the classroom: Scientific inquiry for social change. *American Psychologist, 67,* 681–689.

Raver, C. C., Blair, C., & Willoughby, M. (2013). Poverty as a predictor of 4-year-olds' executive function: New perspectives on models of differential susceptibility. *Developmental Psychology, 49,* 292–304.

Raver, C. C., Jones, S. M., Li-Grining, C., Zhai, F., Bub, K., & Pressler, E. (2011). CSRP's impact on low-income preschoolers' preacademic skills: Self-regulation as a mediating mechanism. *Child Development, 82,* 362–378.

Raver, C. C., McCoy, D. C., Lowenstein, A. E., & Pess, R. (2013). Predicting individual differences in low-income children's executive control from early to middle childhood. *Developmental Science, 16,* 394–408.

Roy, A. L., McCoy, D. C., & Raver, C. C. (2014). Instability versus quality: Residential mobility, neighborhood poverty, and children's self-regulation. *Developmental Psychology, 50,* 1891–1896.

Sarsour, K., Sheridan, M., Jutte, D., Nuru-Jeter, A., Hinshaw, S., & Boyce, W. T. (2011). Family socioeconomic status and child executive functions: The roles of language, home environment, and single parenthood. *Journal of the International Neuropsychological Society, 17,* 120–132.

Shonkoff, J. P., & Fisher, P. A. (2013). Rethinking evidence-based practice and two-generation programs to create the future of early childhood policy. *Development and Psychopathology, 25,* 1635–1653.

Smith, H. J., Kryski, K. R., Sheikh, H. I., Singh, S. M., & Hayden, E. P. (2013). The role of parenting and dopamine D4 receptor gene polymorphisms in children's inhibitory control. *Developmental Science, 16,* 515–530.

Smith, P. K., Jostmann, N. B., Galinsky, A. D., & van Dijk, W. W. (2008). Lacking power impairs executive functions. *Psychological Science, 19,* 441–447.

Stevens, C., Lauinger, B., & Neville, H. (2009). Differences in the neural mechanisms of selective attention in children from different socioeconomic backgrounds: An event-related brain potential study. *Developmental Science, 12*, 634–646.

Sturge-Apple, M. L., Davies, P. T., Cicchetti, D., Hentges, R. F., & Coe, J. L. (2016). Family instability and children's effortful control in the context of poverty: Sometimes a bird in the hand is worth two in the bush. Advance online publication. *Development and Psychopathology*. doi: 10.1017/S0954579416000407.

Ursache, A., Noble, K. G., & the Pediatric Imaging, Neurocognition and Genetics Study. (2016). Socioeconomic status, white matter, and executive function in children. *Brain and Behavior, 6*, e00531.

U.S. Department of Health and Human Services. (2016). *U.S. federal poverty guidelines used to determine financial eligibility for certain federal programs*. Office of the Assistant Secretary for Planning and Evaluation. Retrieved from https://aspe.hhs.gov/poverty-guidelines.

Vernon-Feagans, L., Willoughby, M., Garrett-Peters, P., & The Family Life Project Key Investigators. (2016). Predictors of behavioral regulation in kindergarten: Household chaos, parenting, and early executive functions. *Developmental Psychology, 52*, 430–441.

Weiland, C., Ulvestad, K., Sachs, J., & Yoshikawa, H. (2013). Associations between classroom quality and children's vocabulary and executive function skills in an urban public prekindergarten program. *Early Childhood Research Quarterly, 28*, 199–209.

Weiland, C., & Yoshikawa, H. (2013). Impacts of a prekindergarten program on children's mathematics, language, literacy, executive function, and emotional skills. *Child Development, 84*, 2112–2130.

Wenzel, A. J., & Gunnar, M. R. (2014). Protective role of executive function skills in high-risk environments. *Encyclopedia on early childhood development: Executive functions*. Center for Excellence in Early Childhood Development. Retrieved from www.child-encyclopedia.com/sites/default/files/dossiers-complets/en/executive-functions.pdf. 25.

Yates, T. M., Obradović, J., & Egeland, B. (2010). Transactional relations across contextual strain, parenting quality, and early childhood regulation and adaptation in a high-risk sample. *Development and Psychopathology, 22*, 539–555.

Yousafzai, A. K., Obradović, J., Rasheed, M. A., Rizvi, A., Portilla, X. A., Tirado-Strayer, N., et al. (2016). Effects of responsive stimulation and nutrition interventions on children's development and growth at age 4 years in a disadvantaged population in Pakistan: A longitudinal follow-up of a cluster-randomised factorial effectiveness trial. *The Lancet Global Health, 4*, e548–e558.

10

PARENTAL INFLUENCES ON CHILDREN'S EXECUTIVE FUNCTION

A Differentiated Approach

Claire Hughes & Rory T. Devine

Introduction

In a review of the biological, psychological, and social factors that contribute to the onset and maintenance of disruptive behaviors in childhood and adolescence, Hill (2002) noted the interplay between environmental disadvantages, such as hostile or intrusive parenting, and individual factors, such as impairments in language and in executive functions (EF). Over the intervening 15 years, research interest in family influences on EF has grown dramatically. One catalyst for this research interest is the recognition that the prefrontal cortex (the core neural substrate for EF) shows a very protracted maturation (Kolb et al., 2012) and is, as a result, particularly susceptible to environmental influence and robustly associated with factors such as socioeconomic status (SES; Noble, Norman, & Farah, 2005). Recent work has highlighted just how early in life the adverse effects of poverty on brain development become manifest. For example, in an imaging study of normal brain development, Hanson et al. (2015) showed that from the early postnatal period to the age of 4, children from low-income families were nearly half a standard deviation below their more affluent peers in their mean volume of total gray matter. The impact of stress and poverty on EF development is discussed in detail by Finch and Obradović (Chapter 9, this volume). In this chapter, we focus on more proximal aspects of home environment and parenting as influences on children's EF. This topic complements the material presented in two chapters from the life-span section of this volume: Cuevas, Rajan, and Bryant (Chapter 1) refer to parental influences on EF in infancy, while Crone, Peters, and Steinbeis (Chapter 3) report on findings from EF training studies with adolescents.

In the first section of this chapter we focus on studies that use rating scales to measure either general aspects of the home learning environment or more specific risk factors, such as poor parental well-being and family chaos. In the second section we review the studies that have adopted direct observations of parent–child interactions in order to examine how families can help or hinder young children's developing EF skills. Next we consider whether parenting behaviors mediate the intergenerational association in EF that has been reported in a few studies. Wherever possible, we provide information on the effect sizes. The categorization of effect sizes as small, medium, or large correspond to values of 0.10, 0.30, and 0.50 for correlation (r) and regression coefficients (β) and to values of 0.20, 0.50, and 0.80 for mean differences (Cohen's d) (Ellis, 2010). We conclude our chapter with a brief commentary on limitations in the field and promising lines for future research.

The Home Environment and Young Children's EF

A sensible place to start is to consider the impact of *parental absence* on children's EF development. Adopting this approach, Sarsour et al. (2011) examined 60 American low-SES families with 8- to 12-year-olds and showed that children living with a single parent showed poorer inhibitory control ($\beta=0.44$) and cognitive flexibility ($\beta=0.58$) than children living with two parents. Likewise, in a study of 60 Sri Lankan 11-year-olds, Hewage, Bohlin, Wijewardena, and Lindmark (2011) found that children whose mothers had been working abroad for more than a year performed less well on tests of inhibitory control (Cohen's $d=0.54-0.58$) and verbal working memory (Cohen's $d=0.40$) than did those whose mothers were employed in Sri Lanka. Both studies also included in-depth analyses of parental behaviors that promote children's EF development. Sarsour et al. (2011) reported specific mediation effects: parental responsivity and family companionship partly mediated the relationship between SES and child inhibitory control, while enrichment activities and family companionship mediated the relationship between SES and working memory. Hewage et al. (2011) found that despite receiving more opportunities and materials for learning, children whose mothers worked abroad also received lower levels of parental responsivity (Cohen's $d=0.71$) and cognitive stimulation (Cohen's $d=0.48$), assessed using the Home Observation Measurement of Environment (HOME). Total scores on the HOME were directly and positively associated with both inhibitory control ($r=0.24$) and working memory performance ($r=0.21$). In contrast, the relationship between parental absence and poor inhibitory control appeared indirect, with emotional climate playing a mediating role.

Parental Well-Being and Young Children's EF

Also underscoring the importance of emotional climate, other researchers have investigated the effects of maternal *depression* on children's EF development. Depressive symptoms undermine parenting in several ways, including: activation of low-positive and high-negative emotion, reduced child-oriented goals and attention to child input, and increased negative appraisals of children and parenting competence, coupled with increased positive evaluations of coercive parenting (Dix & Meunier, 2009). Somewhat surprisingly then, the evidence that maternal depression adversely affects child EF is rather inconsistent, with negative findings from two studies of school-aged children and adolescents (Klimes-Dougan, Ronsaville, Wiggs, & Martinez, 2006; Micco et al., 2009) and a further study of low-income preschoolers (Rhoades, Greenberg, Lanza, & Blair, 2011). However, these null findings may reflect their common adoption of a categorical approach to depression; with one exception (Li-Grining, 2007), more positive findings have emerged from studies that adopted a continuous approach. For example, in the NICHD Early Child Care Research Network's (1999) birth-cohort study, mothers' reports of chronic depressive symptoms (obtained at 1, 6, 15, 24, and 36 months) were inversely related to cognitive-linguistic functioning ($0.25 < d < 0.43$). Likewise, in a longitudinal study that tracked children from ages 2 to 6, exposure to mothers' depressive symptoms at age 2 predicted EF at age 6 ($\beta = -0.27$), even when accounting for the effects of age 6 verbal ability (Hughes, Roman, Hart, & Ensor, 2013); in addition, child EF mediated the association between maternal depression and child conduct problems (Roman, Ensor, & Hughes, 2016).

Family Chaos and Young Children's EF

Of course, maternal depression also affects health and safety practices. For example, maternal depression was shown to relate to food insecurity ($r = 0.25$) in a low-income sample (Melchior et al., 2009). The impact of maternal depression on child EF may also include indirect effects of family chaos. Support for the importance of family chaos on child EF comes from a recent large-scale longitudinal study involving 1,292 children that showed that, over and above effects of poverty-related variables, family chaos in the first 3 years of life was associated with lower EF assessed at ages 3, 4 and 5 ($0.12 < r < 0.32$) (Vernon-Feagans, Willoughby, & Garrett-Peters, 2016). Interestingly, this association was indirect and mediated by the impact of chaos on parental responsiveness and acceptance.

Other studies have shown that children living in chaotic homes are not only deprived of a haven of support and security, but also have fewer opportunities to practice self-regulation (for a review, see Repetti, Taylor, & Seeman, 2002). As such, in chaotic environments, children's immediate goals may be to filter out the high levels of stimulation, but this may also result in the filtering out of developmentally facilitative stimulation (Matheny, Wachs, Ludwig, & Phillips, 1995). Empirical support for the negative impact of over-stimulation comes from the finding that just 9 minutes' exposure to a fast-pace cartoon (*Sponge Bob Square Pants*) impaired 4-year-old children's inhibitory control, delayed gratification, and planning abilities, compared with matched peers who watched a more gently paced educational cartoon or who spent the 9 minutes in a drawing activity ($0.65 < d < 1.30$) (Lillard & Peterson, 2011; but see also Lillard, Drell, Richey, Boguszewski, & Smith, 2015). There is also evidence that family chaos adversely affects children's EF development on a longer timescale. For example, in the same study that showed that maternal scaffolding at age 2 predicted gains in children's EF performance from ages 2 to 4 ($r = 0.31$), Hughes and Ensor (2009) also showed a specific independent predictive effect of family chaos, as indexed by mothers' responses to the CHAOS questionnaire ($r = -0.21$) (Matheny et al., 1995). Similar findings have also been reported for older children. For example, in a study of 233 low-income children who completed a delayed gratification task at age 9 and were asked to report on their own self-control 3–4 years later, Evans, Gonnella, Marcynyszyn, Gentile, and Salpekar (2005) found that the adverse effects of poverty on children's self-control were at least partly mediated by exposure to chaotic living conditions.

Observational Measures of Parental Behavior and Young Children's EF

Although empirical investigations of environmental influences on EF have only emerged in the past decade, developmental theorists have long speculated about the important role that social experiences play in the emergence of higher-order cognitive functions (Müller, Baker, & Yeung, 2013). Specifically, in the early 20th century Vygotsky (1978) speculated that children learned to master inhibition and planning skills through interactions with more skilled peers and adults. While it is tempting to treat parenting as a monolithic or unitary construct, contemporary models support a differentiated model in which different behavioral dimensions make potentially independent contributions to children's EF (Bernier, Carlson, & Whipple, 2010; Grusec & Davidov, 2010; Hughes & Ensor, 2009). Over the past decade EF in childhood has been studied in relation to a range of observable parental behaviors. These studies have focused on parental behaviors in the 'normative' range and suggest that even typical variation in parental behaviors might have developmental significance for EF in early childhood (Bernier et al., 2010). For the purposes of brevity, parental behaviors can be categorized into two types: (i) task-focused or goal-oriented parental behaviors (that is, parental behaviors that are concerned with instructing, problem-solving, or completing a task); and (ii) relationship-based parental behaviors (that is, those parental behaviors that are concerned with establishing, monitoring, or maintaining the parent–child relationship) (Zheng, Pasalich, Oberth, McMahon, & Pinderhughes, 2017). These are discussed in turn below.

Goal-Oriented Parental Behaviors and Children's EF

Both cross-sectional and longitudinal studies have examined individual differences in parental 'scaffolding' behaviors and children's EF. The concept of scaffolding was first introduced by Wood and colleagues in a series of studies of adult tutoring practices (Wood, Bruner, & Ross, 1976; Wood & Wood, 1996). Wood et al. (1976) argued that effective scaffolding (the process by which tutors cede control of a task to learners) hinged on the tutor's use of the contingency rule. That is, when children struggle the tutor should increase the level of support provided (e.g., by providing direct instructions or demonstrations) and when children succeed the tutor should decrease the level of support provided (Wood & Wood, 1996). Supporting this view, children whose parents use the contingency rule during shared problem-solving tasks are more likely to succeed on the shared task ($0.43 < r < 0.73$) (Pratt, Kerig, Cowan, & Cowan, 1988) and show better independent performance on related tasks ($0.25 < r < 0.51$) (Conner, Knight, & Cross, 1997).

The term 'scaffolding' has also been used to refer to a range of different parental behaviors in research studying the relations between parental behavior and children's EF. Some researchers have used macro ratings of parents' use of open-ended questions, praise, and elaborations (Hughes & Ensor, 2009) or parents' responsiveness, flexibility, encouragement, and perspective taking (Bernier et al., 2010). Other researchers have adopted moment-to-moment micro-level coding schemes that examine the contingency between parents and children in their verbal and non-verbal behavior (Hammond, Muller, Carpendale, Bibok, & Lieberman-Finestone, 2012; Hughes & Devine, in review). Despite these different measurement approaches, however, there is converging evidence that variation in parental scaffolding is longitudinally related to EF among preschool children (even when early EF and other factors such as verbal ability are taken into account) ($0.31 < r < 0.39$) (Hammond et al., 2012; Hughes & Ensor, 2009). Given that scaffolding has been measured in different ways in different studies, the precise mechanisms underlying the relations between parental scaffolding and children's EF remain unclear. One possibility is that both macro and micro approaches to measurement capture individual differences in parents' ability to continually challenge children's cognitive skills, as this appears to be a key ingredient for success in school-based interventions (Diamond & Ling, 2016).

Relationship-Based Parental Behaviors and Children's EF

There is now a substantial body of research showing that EF in early childhood is related to variation across a broad range of parental behaviors focused on establishing and maintaining the affective quality of parent–child interactions. These include: parental sensitivity and responsiveness (e.g., Bernier et al., 2010; Blair, Raver, & Berry, 2014), attachment security (e.g., Bernier, Beauchamp, Carlson, & Lalonde, 2015), mind-mindedness (e.g., Bernier et al., 2010), and low levels of harsh, controlling, inconsistent, or intrusive behavior (Cuevas, Deater-Deckard, Kim-Spoon, Watson, et al., 2014; Hughes & Devine, in review; Hughes & Ensor, 2009) ($0.26 < r < 0.40$). Longitudinal studies confirm that global indices of parental responsiveness (as measured by observational ratings of sensitivity, positivity, and stimulation) as well as measures of attachment security predict later EF even when the effects of prior EF and other background variables are taken into account (Blair et al., 2014; Bernier et al., 2015). Given that the majority of these studies use global composite ratings and that few studies have investigated these different constructs together (e.g., Bernier, Carlson, Deschenes, & Matte-Gagne, 2012), it is difficult to identify what mechanisms might explain the associations between these diverse relationship-based behaviors and children's EF. Some researchers have speculated that emotionally secure relationships between parents and children reduce stress and provide children with the optimal environment to regulate their own thoughts and actions (Bernier et al., 2015). Supporting this view, longitudinal data show that children's physiological responses to stress (measured via salivary cortisol) are related to EF ($r = -0.56$) and mediate the relations between parental responsiveness and later individual differences in EF (but not general intelligence) (Blair et al., 2011).

Intergenerational Resemblance in EF

While it is possible that task-oriented and relationship-based parental behaviors each provide children with the social and learning opportunities to develop their EF skills, recent work on the intergenerational links between parental EF and child EF (Cuevas, Deater-Deckard, Kim-Spoon, Wang, et al., 2014; Hughes & Ensor, 2009) suggests that the relations between different aspects of parental behavior and children's EF also reflects the genetic transmission of individual differences in EF from one generation to the next. Many aspects of parental behavior (such as, the ability to respond and adapt to a child in a contingent manner or the ability to regulate one's own emotions during a trying interaction) might reflect individual differences in parental EF (Hughes & Ensor, 2009). Supporting this perspective, a review of the available evidence indicates that measures of parental EF are moderately related to individual differences in parental sensitivity and harsh parenting (Crandall, Deater-Deckard, & Riley,

2015). That said, longitudinal data show that when individual differences in parental behavior are taken into account the association between parental EF and children's EF becomes non-significant (Cuevas, Deater-Deckard, Kim-Spoon, Watson, et al., 2014; Hughes & Devine, in review). Moreover, in a study of adopted children and their biological parents, Leve et al. (2013) found no significant correlation between individual differences in parents' EF performance and 2-year-old children's EF. Investigating the interplay between parental EF, parental behavior, and children's EF will require either genetically sensitive or within-family research designs (Jenkins, McGowan, & Knafo-Noam, 2016).

Intervention Studies

All the studies reviewed above relied upon observational longitudinal designs, which provide information about what naturally occurs during development, but cannot firmly establish the presence of causal relations (Bryant, 1990). To address this question of causal influence, future research should include genetically sensitive designs (e.g., fostered or adopted children, children of surrogates, within-family designs) (Rice et al., 2009) and intervention designs. To date, there have been a number of successful *school-based* interventions to foster young children's EF skills. In reviewing this field, Diamond and Ling (2016) have argued that narrowly focused training programs produced smaller gains than more holistic interventions designed to foster children's social and emotional development. Building on this approach, researchers have recently applied the Play and Learning Strategies (PALS) home-based intervention (Landry, Smith, Swank, & Guttentag, 2008) to examine whether increasing the responsiveness of family-based child-care providers could accelerate children's EF development (Merz, Landry, Johnson, Williams, & Jung, 2016). Although their results did not demonstrate an overall main effect, there was an age-by-intervention status interaction, such that younger (but not older) preschool children in the intervention group showed fewer attentional problems coupled with improved performance in delay inhibition (but not on other EF measures included in their assessment battery). Likewise, in a comparison of parental involvement and Early Head Start interventions, Chazan-Cohen and colleagues (2009) have reported that parental supportiveness (indexed by sensitivity, stimulation, and positive affect in video-recorded play sessions at ages 1, 2, 3, and 5 years) and the quality of the home learning environment each had a stronger influence than the center-based interventions on children's emotional regulation, early literacy, and learning motivation. Unfortunately, this study did not include EF as a child outcome measure. Given the evidence considered in this chapter for a differential model of parental influence on children's EF skills it is clear that more work is needed to identify which aspects of parenting provide the most promising focus for intervention work. At this point, however, several significant challenges deserve note. These include: (a) child-driven effects, which mean that intervention fidelity cannot be assumed but must be monitored; (b) variation in children's susceptibility to environmental influences, which requires a shift from traditional questions (e.g., 'does it work?') to more nuanced questions (e.g., 'for whom does it work?'); (c) moderating influences of other relationships, which highlight the need for multi-pronged interventions; and (d) sleeper effects, which can only be detected by including a long-term follow-up.

Future Directions

It is clear that the past decade has witnessed a growth of interest in family influences on young children's EF. There is now converging evidence from longitudinal research that, alongside general factors such as parental well-being and family chaos, individual differences in task-oriented and relationship-based parental behaviors each matter for the development of children's EF. That said, the mechanisms underpinning these associations remain unclear and at least four questions require further investigation. First, with the exception of a few studies (e.g., Bernier et al., 2010; Bernier et al., 2012; Hughes & Devine, in review; Hughes & Ensor, 2009), the majority of studies of parental influences on children's EF have focused on a single aspect of parental behavior. Hughes and Devine (in review) found that both parental scaffolding and negative control made independent contributions to individual differences in EF, suggesting that both task-oriented and relationship-based parental behaviors contribute to children's developing EF. *Future studies will benefit from investigating multiple aspects of parental behavior to examine the overlap and interplay between different dimensions of parenting and children's EF.* Second, to date, the lion's share of work on the relations between parental behavior and EF have focused solely on EF as an outcome. Consequently it is unclear whether certain parental behaviors such as scaffolding have domain-specific effects on EF in particular or more domain-general influences on children's cognition more broadly. Findings from at least three separate studies provide some insight about the specificity of the relations between parental behavior and children's EF. Longitudinal data from Blair et al. (2011) suggest that parental responsiveness and negativity are related to individual differences in both EF and general cognitive ability. In contrast, parental scaffolding appears to be specifically related to EF but not to general verbal ability (Hughes & Devine, in review; Hughes & Ensor, 2009). *Further studies will help to identify whether parental influences on children's EF are specific in their impact or contribute more widely to children's cognitive development and adjustment.*

Third, it is often assumed that parental behaviors are driving the development of children's EF, but to date, just one study has assessed the pattern of developmental relations between EF and parental behavior. Blair et al. (2014) found that the relations between parental responsiveness and children's EF show bidirectional relations. Specifically, individual differences in parental behavior at age 3 predicted growth in children's EF from age 3 to 5. In addition, individual differences in EF at age 3 predicted changes in parenting quality between ages 3 and 5. *These findings underscore the importance of examining child-driven effects in future longitudinal research.* Fourth, much of the work in this field has been restricted to relatively homogeneous Caucasian samples, such that a *further avenue for future research is to test the cultural universality of parental influences on child EF.* From a differentiated perspective of parenting, we predict that cross-cultural research will be useful not only to test the generalizability of mechanisms elucidated in this review but also to identify further ways in which variation in parenting contributes to individual differences in child EF.

Author Note

This work was funded by the UK Economic and Social Research Council (ES/L0116648/1).

References

Bernier, A., Beauchamp, M. H., Carlson, S. M., & Lalonde, G. (2015). A secure base from which to regulate: Attachment security in toddlerhood as a predictor of executive functioning at school entry. *Developmental Psychology, 51*, 1177–1189.

Bernier, A., Carlson, S. M., Deschenes, M., & Matte-Gagne, C. (2012). Social factors in the development of early executive functioning: A closer look at the caregiving environment. *Developmental Science, 15*, 12–24.

Bernier, A., Carlson, S. M., & Whipple, N. (2010). From external regulation to self-regulation: Early parenting precursors of young children's executive functioning. *Child Development, 81*, 326–339.

Blair, C., Granger, D., Willoughby, M., Mills-Koonce, R., Cox, M., Greenberg, M. T., et al. (2011). Salivary cortisol mediates effects of poverty and parenting on executive functions in early childhood. *Child Development, 82*, 1970–1984.

Blair, C., Raver, C. C., & Berry, D. J. (2014). Two approaches to estimating the effect of parenting on the development of executive function in early childhood. *Developmental Psychology, 50*, 554–565.

Bryant, P. (1990). Empirical evidence for causes in development. In G. Butterworth & P. Bryant (Eds.), *Causes of development: Interdisciplinary perspectives* (pp. 33–45). London: Harvester Wheatsheaf.

Chazan-Cohen, R., Raikes, H., Brooks-Gunn, J., Ayoub, C., Pan, B. A., Kisker, E. E., et al. (2009). Low-income children's school readiness: Parent contributions over the first five years. *Early Education and Development, 20*, 958–977.

Conner, D. B., Knight, D. K., & Cross, D. R. (1997). Mothers' and fathers' scaffolding of their 2-year-olds during problem-solving and literacy interactions. *British Journal of Developmental Psychology, 15*, 323–338.

Crandall, A. A., Deater-Deckard, K., & Riley, A. W. (2015). Maternal emotion and cognitive control capacities: A conceptual framework. *Developmental Review, 36*, 105–126.

Cuevas, K., Deater-Deckard, K., Kim-Spoon, J., Wang, Z., Morasch, K. C., & Bell, M. A. (2014). A longitudinal intergenerational analysis of executive functions during early childhood. *British Journal of Developmental Psychology, 32*, 50–64.

Cuevas, K., Deater-Deckard, K., Kim-Spoon, J., Watson, A. J., Morasch, K. C., & Bell, M. A. (2014). What's mom got to do with it? Contributions of maternal executive function and caregiving to the development of executive function across early childhood. *Developmental Science, 17*, 224–238.

Diamond, A., & Ling, D. S. (2016). Conclusions about interventions, programs, and approaches for improving executive functions that appear justified and those that, despite much hype, do not. *Developmental Cognitive Neuroscience, 18*, 34–48.

Dix, T., & Meunier, L. N. (2009). Depressive symptoms and parenting competence: An analysis of 13 regulatory processes. *Developmental Review, 29*, 45–68.

Ellis, P. D. (2010). *The essential guide to effect sizes: Statistical power, meta-analysis, and the interpretation of research results.* Cambridge: Cambridge University Press.

Evans, G. W., Gonnella, C., Marcynyszyn, L. A., Gentile, L., & Salpekar, N. (2005). The role of chaos in poverty and children's socio-emotional adjustment. *Psychological Science, 16*, 560–565.

Grusec, J. E., & Davidov, M. (2010). Integrating different perspectives on socialization theory and research: A domain-specific approach. *Child Development, 81*, 687–709.

Hammond, S. I., Muller, U., Carpendale, J. I. M., Bibok, M. B., & Lieberman-Finestone, D. P. (2012). The effects of parental scaffolding on preschoolers' executive function. *Developmental Psychology, 48*, 271–281.

Hanson, J. L., Hair, N., Shen, D., Shi, F., Gilmore, J. H., Wolfe, B., et al. (2015). Family poverty affects the rate of human infant brain growth. *PloS One, 8*, e80954.

Hewage, C., Bohlin, G., Wijewardena, K., & Lindmark, G. (2011). Executive functions and child problem behaviors are sensitive to family disruption: A study of children of mothers working overseas. *Developmental Science, 14*, 18–25.

Hill, J. (2002). Biological, psychological and social processes in conduct disorder. *Journal of Child Psychology and Psychiatry, 43*, 133–164.

Hughes, C. H., & Devine, R. T. (in review). *How do parents help or harm the development of children's EF? Let me count the ways.* Unpublished Manuscript.

Hughes, C. H., & Ensor, R. A. (2009). How do families help or hinder the emergence of early executive function? *New Directions in Child and Adolescent Development, 123*, 35–50.

Hughes, C., Roman, G., Hart, M. J., & Ensor, R. (2013). Does maternal depression predict young children's executive function? A four-year longitudinal study. *Journal of Child Psychology and Psychiatry, 54*, 169–177.

Jenkins, J. M., McGowan, P., & Knafo-Noam, A. (2016). Parent–offspring transaction: Mechanisms and the value of within family designs. *Hormones and Behavior, 77*, 53–61.

Klimes-Dougan, B., Ronsaville, D., Wiggs, E. A., & Martinez, P. E. (2006). Neuropsychological functioning in adolescent children of mothers with a history of bipolar or major depressive disorders. *Biological Psychiatry, 60*, 957–965.

Kolb, B., Myschasiuk, R., Muhammad, A., Li, Y., Frost, D., & Gibb, R. (2012). Experience and the developing prefrontal cortex. *Proceedings of the National Academy of Sciences, 109*, 17186–17193.

Landry, S. H., Smith, K. E., Swank, P. R., & Guttentag, C. (2008). A responsive parenting intervention: The optimal timing across early childhood for impacting maternal behaviors and child outcomes. *Developmental Psychology, 44*, 1335–1353.

Leve, L. D., DeGarmo, D. S., Bridgett, D. J., Neiderhiser, J. M., Shaw, D. S., Harold, G. T., et al. (2013). Using an adoption design to separate genetic, prenatal and temperament influences on toddler executive function. *Developmental Psychology, 49*, 1045–1057.

Li-Grining, C. P. (2007). Effortful control among low-income preschoolers in three cities: Stability, changes and individual differences. *Developmental Psychology*, 208–221.

Lillard, A. S., Drell, M. B., Richey, E. M., Boguszewski, K., & Smith, E. D. (2015). Further examination of the immediate impact of television on children's executive function. *Developmental Psychology, 51*, 792–805.

Lillard, A. S., & Peterson, J. (2011). The immediate impact of different types of television on young children's executive function. *Pediatrics, 128*, 1–6.

Matheny, A. P., Wachs, T. D., Ludwig, J., & Phillips, K. (1995). Bringing order out of chaos: Psychometric characteristics of the Confusion, Hubbub and Order Scale. *Journal of Applied Developmental Psychology, 16*, 429–444.

Melchior, M., Caspi, A., Howard, L. M., Ambler, A. P., Bolton, H., Mountain, N., et al. (2009). Mental health context of food insecurity: A representative cohort of families with young children. *Pediatrics, 124*, 564–572.

Merz, E. C., Landry, S. H., Johnson, U. Y., Williams, J. M., & Jung, K. (2016). Effects of a responsiveness-focused intervention in family child care homes on children's executive function. *Early Childhood Research Quarterly, 34*, 128–139.

Micco, J. A., Henin, A., Mick, E., Kim, S., Hopkins, C. A., Biederman, J., et al. (2009). Anxiety and depressive disorders in offspring at high risk for anxiety: A meta-analysis. *Journal of Anxiety Disorders, 23*, 1158–1164.

Müller, U., Baker, L., & Yeung, E. (2013). A developmental systems approach to executive function. *Advances in Child Development and Behavior, 45*, 39–66.

NICHD Early Child Care Research Network. (1999). Chronicity of maternal depressive symptoms, maternal sensitivity and child functioning at 36 months. *Developmental Psychology, 35*, 1297–1310.

Noble, K. G., Norman, M. F., & Farah, M. J. (2005). Neuro-cognitive correlates of socioeconomic status in kindergarten children. *Developmental Science, 8*, 74–87.

Pratt, M. W., Kerig, P., Cowan, P. A., & Cowan, C. P. (1988). Mothers and fathers teaching 3-year-olds: Authoritative parenting and adult scaffolding of young children's learning. *Developmental Psychology, 24*, 832–839.

Repetti, R. L., Taylor, S. E., & Seeman, T. E. (2002). Risky families: Family social environments and the mental and physical health of offspring. *Psychological Bulletin, 128*, 330–366.

Rhoades, B. L., Greenberg, M. T., Lanza, S. T., & Blair, C. (2011). Demographic and familial predictors of early executive function development: Contribution of a person-centred perspective. *Journal of Experimental Child Psychology, 108*, 638–662.

Rice, F., Harold, G. T., Boivin, J., Hay, D. F., Van Den Bree, M., & Thapar, A. (2009). Disentangling prenatal and inherited influences in humans with an experimental design. *Proceedings of the National Academy of Sciences, 106*, 2464–2467.

Roman, G. D., Ensor, R., & Hughes, C. (2016). Does executive function mediate the path from mothers' depressive symptoms to young children's problem behaviors? *Journal of Experimental Child Psychology, 142*, 158–170.

Sarsour, K., Sheridan, M., Jutte, M., Nuru-Jeter, A., Hinshaw, S., & Boyce, W. T. (2011). Family socio-economic status and child executive functions: The roles of language, home environment and single parenthood. *Journal of International Neuropsychological Society, 17*, 120–132.

Vernon-Feagans, L., Willoughby, M., & Garrett-Peters, P. (2016). Predictors of behavioral regulation in kindergarten: Household chaos, parenting and early executive functions. *Developmental Psychology, 52*, 430–441.

Vygotsky, L. S. (1978). *Mind in society: The development of higher psychological processes.* Cambridge, MA: Harvard University Press.

Wood, D., Bruner, J. S., & Ross, G. (1976). The role of tutoring in problem solving. *Journal of Child Psychology and Psychiatry, 17,* 89–100.

Wood, D., & Wood, H. (1996). Vygotsky, tutoring and learning. *Oxford Review of Education, 22,* 5–16.

Zheng, Y., Pasalich, D. S., Oberth, C., McMahon, R. J., & Pinderhughes, E. E. (2017). Capturing parenting as a multidimensional and dynamic construct with a person-oriented approach. *Prevention Science, 18,* 281–291.

11
BILINGUALISM AND THE DEVELOPMENT OF EXECUTIVE FUNCTION IN CHILDREN

The Interplay of Languages and Cognition

Gregory J. Poarch & Janet G. van Hell

Past research in the domain of bilingualism has focused on various linguistic issues such as how speakers acquire multiple languages (e.g., Hoff et al., 2012; Paradis, Genesee, & Crago, 2011), conditions that promote or constrain the co-activation of languages (e.g., Marian & Spivey, 2003; Poarch & van Hell, 2014; Thierry & Wu, 2007; van Hell & Dijkstra, 2002), and how cross-language interaction impacts lexical and syntactic processing in the first and second language (for reviews, see Caffarra, Molinaro, Davidson, & Carreiras, 2015; van Hell & Tanner, 2012; van Hell & Tokowicz, 2010). Bilingualism has proven to be an ideal testing ground of how language and cognition interact in the human brain. There is now agreement that in most language-related situations a bilingual's languages are co-activated and compete for access (e.g., Blumenfeld & Marian, 2014; Poarch & van Hell, 2012a; Thierry & Wu, 2007). Speakers who use multiple languages are often faced with situations in which one language needs to be accessed for a given context (and interlocutor) while the other language needs to be disregarded (e.g., Van Heuven, Schriefers, Dijkstra, & Hagoort, 2008; for reviews, see Kroll, Gullifer, & Rossi, 2013; van Hell & Tanner, 2012). The relevant questions here are how bilingual speakers manage to negotiate their two languages and how they are able to effortlessly switch between their languages (e.g., Kootstra, van Hell, & Dijkstra, 2010) when necessary. Thus, the cognitive and neural processes required to deal with constant language competition may reorganize brain networks in bilinguals, which has implications for how bilinguals negotiate cognitive competition (Bialystok, Craik, & Luk, 2012; Bialystok & Poarch, 2014). This review aims at presenting research in the domain of bilingualism and the effect of being bilingual on the development of executive function. The main focus will thus be on studies conducted with children, while where necessary also setting a backdrop using the breadth of research on adult populations.

Bilingual Language Control and Executive Functions

Converging evidence points to an involvement of the executive function (EF) system in multilingual language control and language switching, a system that is more generally perceived to be responsible for (non-linguistic) inhibition, shifting, and updating (see, e.g., the unity/diversity framework proposed by Miyake & Friedman, 2012; see also Chevalier & Clark, Chapter 2, this volume). The hypothesis is that bilinguals' repeated use of EF for linguistic processes to ensure context-appropriate communication may affect non-verbal cognitive tasks that engage EF (e.g., Abutalebi & Green, 2007; Green & Abutalebi, 2013; but see Branzi, Calabria, Boscarino, & Costa, 2016, for possibly differential involvement of inhibition in language control). The consequences of being bilingual, in that there is constant interaction and competition between the languages, is that dealing with this competition fine-tunes cognitive resources. In other words, bilingualism trains the ability to ignore irrelevant information, to resolve non-verbal cognitive conflict, and to switch between tasks.

EF differences between bilingual and monolingual populations have been found repeatedly using various tasks such as the Stroop task (Stroop, 1935) and the Simon task (Simon & Rudell, 1967). One prominent Stroop-like task is the flanker task (Eriksen & Eriksen, 1974) in which a central target arrow is flanked by non-target arrows either pointing in the same direction (congruent condition) or in the opposite direction (incongruent condition). Participants are required to make a directional response to the central target arrow's direction by pressing a left or a right button. Depending on whether the flankers are pointing in the same direction as the target (congruent) or not (incongruent), there is either no conflict or a conflict to be resolved. Bilinguals have been found to be faster overall in performing this task (i.e., in both conditions), which amounts to enhanced general-domain executive functions (selective attention and task monitoring) over monolinguals. Furthermore, bilinguals have also shown a smaller difference between the congruent and incongruent condition, in other words displaying a smaller effect magnitude compared to monolinguals, which amounts to enhanced domain-specific inhibitory control (e.g., Poarch & Bialystok, 2015; Poarch & van Hell, 2012b; Tao, Marzecová, Taft, Asanowicz, & Wodniecka, 2011; Yang & Yang, 2016). The Simon task has participants press the right button of a response box in response to seeing a red-color square and the left button in response to a blue-color square. Again, there is a congruent condition (red-color square on the right of the screen for which a right button press is required) and an incongruent condition (red-color square on the left for which a right button press is required), the latter of which induces conflict and requires inhibitory processes in order to perform the task correctly.

Differences in EF between Bilinguals and Monolinguals

Differences between bilinguals and monolinguals in tasks engaging EF have been found in various age groups, albeit less so in young adults (e.g., Bialystok, Craik, & Luk, 2008; Blumenfeld & Marian, 2014; Costa, Hernández, Costa-Faidella, & Sebastián-Gallés, 2009; Costa, Hernández, & Sebastián-Gallés, 2008). There are more converging findings of significant between-group differences in performing EF tasks in child populations (Bialystok & Poarch, 2014; Carlson & Meltzoff, 2008; Engel de Abreu, Cruz-Santos, Tourinho, Martin, & Bialystok, 2012; Poarch & van Hell, 2012b) and elderly adults (e.g., Bialystok, Poarch, Luo, & Craik, 2012; Salvatierra & Rosselli, 2011; Schroeder & Marian, 2012). It has been assumed that finding EF processing differences among young adult bilinguals and monolinguals may be less likely due to a ceiling effect in the cognitive development in young adults, and may also be dependent on which type of EF task is used (see Blumenfeld & Marian, 2014, for a comparison of Stroop-type and Simon-type cognitive control mechanisms). In children, on the other hand, EF development is still ongoing and their cognitive processing may draw on more brain resources than that in young adults (see Crone, Peters, & Steinbeis, Chapter 3, this volume), while in older adults, an age-related cognitive decline has been documented (e.g., Salthouse, 2009; and see Li, Vadaga, Bruce, & Lai, Chapter 4, this volume).

Turning back to research on children, in a study by Carlson and Meltzoff (2008) run in the United States, 5- to 7-year-old children enrolled in an immersion program who were either monolingual English, bilingual Spanish–English, or second-language learners performed a number of EF tasks such as the Attentional Networks task (ANT) (Rueda et al., 2004), which is a flanker task paired with a cueing task adding attentional and orienting components, and the Gift Delay task, which measures the capacity to suppress a desired action. After statistically controlling for variables such as verbal scores and socioeconomic status (SES), the authors found a significant advantage in conflict tasks but not in the delay task for the bilingual children, particularly for those tasks that required inhibitory control.

Similarly, and with a focus on exploring the trajectory of EF development, Poarch and van Hell (2012b) tested 5- to 8-year-old children in a German-speaking environment. The children were monolinguals enrolled in a German primary school as well as second-language learners, bilinguals, and trilinguals enrolled in a dual-immersion German–English school. The groups were carefully matched on age, SES, parental education, and first-language proficiency (German), but the groups differed significantly on second-language proficiency (English). The children were administered the Simon task and the ANT. Poarch and van Hell found no differences in overall processing speed between groups, which means there were no domain-general differences in EF. However, the bilinguals and trilinguals showed a smaller Simon effect and ANT conflict effect than the monolinguals, with the group of second-language learners performing partway between the other groups. A smaller effect magnitude in these tasks is an index of better conflict resolution and hence was interpreted as enhanced inhibitory control (as one of the EF components) in the multilinguals compared to the monolinguals. For the second-language learners, it was assumed that a greater extent of bilingual immersion was necessary to significantly enhance their inhibitory control over the monolinguals. This pattern of results is consistent with the notion of a continuum of language experience, experience that needs to be above a specific threshold in quantity for the emergence of any developmental differences in performance on EF tasks (see also Bialystok & Barac, 2012; van Hell & Poarch, 2014).

Corroborating evidence comes from Poarch and Bialystok (2015), who tested 8- to 11-year-old children in Canada who were monolinguals, partial bilinguals (i.e., proficient second-language learners), balanced bilinguals, or trilinguals. Again matching for background variables such as SES and English proficiency and using a modified flanker task, similar results were observed as in the study by Poarch and van Hell (2012b), with balanced bilinguals and trilinguals outperforming partial bilinguals and monolinguals. While the multilinguals did not perform faster than the other groups, they did show enhanced conflict resolution in the critical incongruent condition vs. the congruent condition, resulting in a smaller flanker effect. Recently, Sorge, Toplak, and Bialystok (2016) explored the role of (and possible interaction between) two factors that have been found to impact the development of EF: attention ability and bilingualism. Eight- to 11-year-old children performed a battery of EF tasks, among them the stop-signal task (Logan & Cowan, 1984) and the flanker task. Using hierarchical regression, the authors found that the children's attention ability best predicted their performance on the stop-signal task, while bilingualism best predicted the performance on the flanker task. Furthermore, Filippi et al. (2015) also found enhanced performance of bilingual over monolingual children during an auditory language comprehension task requiring inhibitory control due to simultaneous background noise adding irrelevant information to the task (see also Teubner-Rhodes et al., 2016).

Finally, Engel de Abreu and colleagues (2012) extended previous research by focusing on and comparing bilingual immigrant Portuguese–Luxembourgish children and monolingual Portuguese children, both of lower SES. While the monolinguals displayed higher scores on the Portuguese proficiency measure than the bilinguals (Portuguese being the dominant language for both groups), the bilinguals outperformed the monolinguals in a flanker task, again supplying evidence that irrespective of environmental factors such as SES there is a difference between bilinguals and monolinguals in the ability to resolve conflict (see also, e.g., Blom, Küntay, Messer, Verhagen, & Leseman, 2014; Crivello et al., 2016; Ladas, Carroll, & Vivas, 2015; but see studies in the section 'Studies Reporting No EF Difference between Monolingual and Bilinguals' below).

The evidence presented here so far indicates a difference between bilinguals and monolinguals in performing EF tasks that require inhibitory control. The underlying factor that is typically considered to be responsible for such a difference to emerge is that bilinguals need to constantly navigate two languages and thus inadvertently train switching between task sets (i.e., their languages), which has repercussions on their EF development. The next section focuses on evidence for differences between bilinguals and monolinguals from neurophysiological research before the research reporting no differences is reviewed.

Neurocognitive Evidence for the Effects of Bilingualism on EF

The hypothesis that bilingualism affects executive functions is also supported by neurocognitive research (albeit so far almost exclusively on adult speakers) pinpointing overlapping areas in the brain responsible for EF, language control, and language switching. This line of research asks what the neural consequences are of parallel language activation and competition across a bilingual's two languages. The repeated resolving of bilingual language competition may tune brain networks through the need to regulate the L1 to enable proficient L2 performance. Such regulatory and cognitive processes in bilinguals may have long-lasting effects on the coordination of EF components, but may also emerge differentially depending on a multitude of factors such as relative language usage and language switching behavior (see Green & Abutalebi, 2013, for their adaptive control hypothesis describing language control processes in bilinguals). Stocco, Yamasaki, Natalenko, and Prat (2014) propose a theory on the neural underpinnings supporting enhanced EF development in bilinguals. In this theory, the basal ganglia circuit (in particular the striatum) controls and routes the information to the prefrontal cortex and as such has a gating function that is utilized, for example, when bilinguals select one of their languages (see also Crone et al., Chapter 3, this volume, for the role of the striatum in working memory development). This region is thus trained through extensive language usage and language switching and, incidentally, is also called upon when performing EF tasks. Hence, repeated language switching, as a very specific form of task switching, also enhances a bilingual's performance on EF tasks.

The only study so far that has reported neurophysiological evidence for differences between monolingual and bilingual children is that by Barac, Moreno, and Bialystok (2016). Five-year-old children performed behavioral tasks with differing EF demands (gift delay, ANT flanker, go/no-go) and electrophysiological data was recorded for the task that was assumed to pose the highest EF demand. The bilingual children outperformed the monolingual children behaviorally on the ANT and the go/no-go in interference suppression and response inhibition. The authors also observed significant electrophysiological differences between the groups on the go/no-go task, with ERPs showing larger P3 amplitudes and shorter N2 and P3 peak latencies for the bilingual children than for the monolingual children. For bilingual children, but not for the monolinguals, larger P3 amplitudes and shorter N2 and P3 peak latencies were also associated with enhanced perceptual sensitivity to the go and no-go conditions. Barac et al. interpret these findings to signify earlier maturation of executive processes in bilingual children.

Pliatsikas and Luk (2016) recently reviewed functional magnetic-resonance imaging (fMRI) studies tapping cognitive processing during bilingual language processing (see also Luk & Pliatsikas, 2016). The evidence reported in the reviewed studies generally points to an overlap of brain areas in the prefrontal cortex supporting EF and bilingual language control. Furthermore, testing young adults, Abutalebi et al. (2013) found that bilingualism tunes a region in the prefrontal cortex responsible for conflict monitoring called the anterior cingulate cortex (ACC). Furthermore, bilinguals recruit less brain function and are more efficient at processing during conflict monitoring, which the authors assert is spill-over from the control of multiple languages.

In a study comparing gray matter volume (GMV) in unimodal Spanish–English bilinguals, bimodal American Sign Language–English bilinguals, and English monolinguals, Olulade et al. (2016) found greater frontal GMV in the unimodal bilinguals (two spoken languages) compared to the other two groups, who had similar GMV. This finding is in line with the prediction that benefits in EF would co-occur with anatomical differences (such as in GMV) in relevant brain areas responsible for EF and language control, which means the prefrontal cortex, but critically only in those individuals who need to choose between two spoken languages. Moving to research on older adults, Bak, Nissan, Allerhand, and Deary (2014) found enhanced cognitive abilities in a large group of older bilinguals compared to monolinguals. The authors report a positive effect induced by bilingualism in the form of a protection against cognitive decline even for those bilinguals who had learned their second language in adulthood (Alladi et al., 2013; Woumans et al., 2015; see also Cox et al., 2016, for recent evidence of better conflict processing in bilingual older adults than in monolinguals, irrespective of SES).

In summary, both behavioral and brain imaging studies supply us with evidence that the specific control processes needed to enable bilingual language processing lead to adaptive changes in the neural circuits associated with these processes – and since there is an overlap of the neural areas of language control and those of EF, it is likely that the one may influence the other (Coderre, Smith, & Van Heuven, 2016; Green & Abutalebi, 2013).

Studies Reporting No EF Difference between Monolingual and Bilinguals

In contrast to studies reporting differences in EF between bilinguals and monolinguals, recent years have seen a number of studies yielding no differences between bilinguals and monolinguals, challenging the view that bilingualism affects EF. A number of recent research studies with children and young adults has yielded no performance differences between the bilinguals and monolinguals tested, calling into question whether bilingualism per se enhances EF or whether between-group differences found in previous studies may have been driven by other variables (such as SES; see, e.g., Paap, Sawi, Dalibar, Darrow, & Johnson, 2013).

Focusing on child populations, Antón et al. (2014), for example, tested 180 bilingual and 180 monolingual children and compared their performances on the ANT task (Fan et al., 2002), finding no significant differences between groups. Similar results were reported by Duñabeitia and colleagues (2014), who investigated EF in large groups of 252 bilinguals and 252 monolinguals using a verbal and a non-verbal Stroop task. The groups performed equally in both tasks, showing no difference in favor of the bilinguals in inhibitory control. In a study by Gathercole et al. (2014) conducted in Wales, simultaneous bilinguals of various ages (3-year-olds to adults) were compared to sequential bilinguals and monolinguals. On three EF tasks, no performance differences were found between groups in all age groups, neither globally (overall speed) nor in inhibitory control (effect magnitude).

Research conducted with young adults has yielded similar results. Von Bastian, Souza, and Gade (2016), for example, administered tasks measuring nine cognitive abilities to young adults, while also assessing three continuous bilingualism variables (usage, age of acquisition, and proficiency) and participants' SES. The authors found no evidence for a cognitive bilingual benefit in inhibitory control, conflict monitoring, and shifting, and concluded that if there are cognitive benefits of being bilingual, they are not as extensive as previously assumed. They furthermore offer four explanations for their findings contrasting earlier research reporting cognitive benefits for bilinguals: the use of multiple vs. single tasks to measure cognitive abilities, the specific bilingual context, large vs. small sample sizes, and a publication bias in the field (De Bruin, Treccani, & Della Sala, 2015; but see Bak, 2016). Finally, Paap and Greenberg (2013) report a study in which, among other tasks, the flanker task and the Simon task were administered to monolingual and bilingual university students with a variety of language backgrounds and who rated themselves on first- and second-language proficiency. The groups did not differ significantly on any of the EF task performances. As such, the examples of null-result findings presented here challenge the view that there is differential development of EF in bilinguals and monolinguals. The pertinent question now is to what these mixed findings may be attributed, and to answer this question, we move to the variables at play in bilingualism research.

Variables of Relevance in Research on Bilingualism and EF

Given the mixed findings particularly in the more recently published work in the field, there is a clear need to identify the relevant factors of influence in research on EF in bilinguals. There are numerous variables that may be relevant in assessing the differential development of EF in bilinguals vs. monolinguals, and we will focus on the following: the participants' SES, the number of participants included in the studies, the participants' language backgrounds, the EF tasks employed, and the data trimming procedure in preparation for statistical analysis.

One recurrent issue brought up in the discussion on whether bilingualism may or may not affect the development of executive functions is SES. It has been found that SES is a strong predictor for the development of EF (Hackman & Farah, 2009). More generally, high-SES children commonly outperform low-SES children in EF tasks such as the flanker task (Mezzacappa, 2004). Hence, one of the claims is that differences in SES may have been driving the emergence of the effect in the studies showing EF differences between groups of monolingual and bilingual children, undermining any interpretation positing bilingualism as the influential culprit (Paap et al., 2013). However, it is very unlikely that all the studies reporting significant differences did not control for SES, while those studies that reported null-results did control for SES. Particularly the studies by Calvo and Bialystok (2014), Carlson and Meltzoff (2008), Engel de Abreu et al. (2012), and Poarch and van Hell (2012b) explicitly used SES as an independent variable or carefully matched SES across-groups. These studies all reported significant differences in EF between monolinguals and bilinguals (as did, e.g., Buac & Kaushanskaya, 2014; Ladas et al., 2015; and see Blom et al., 2014, for working memory differences between SES-matched bilingual and monolingual children). Thus, while SES can be a confounding factor if not properly controlled, there is no evidence that such a confound influenced the results only in those studies yielding an EF difference.

Another recurrent issue is the number of participants who are typically recruited for studies in this field. Paap et al. (2013) have called for studies to use a greater number of participants, without which, they claim, the studies would be statistically underpowered. However, as Hope (2015) points out, using larger numbers of participants in study design is not necessarily always better. Power may 'even be improved by reducing samples to ensure that groups are more comparable [...] and smaller studies can be more informative than larger studies, if they are better controlled' (p. 59). In other words, when comparing bilinguals and monolinguals on any task, only language background is supposed to vary across groups (e.g., 'speaks one language only' vs. 'speaks two languages fluently') with all other variables controlled. If larger numbers of participants are tested, there is potentially also more overlap across groups in terms of language background, meaning unwanted added variance in second-language proficiency in both groups – exactly what one would aim to avoid in carefully matched two-group or multiple-group designs.

For one, it is relevant to ascertain how proficient a person needs to be in one or two languages to be considered either monolingual or bilingual and how much recent exposure a person has had to one or both languages, given that it is possibly the extent of bilingual exposure that will have a critical effect on differential development of EF (see van Hell & Poarch, 2014). Future research may need to reconceptualize the possibly systematic cognitive differences between bilinguals and monolinguals (be it disadvantages or advantages) to experience related differences driving cognitive development, such as those that we see in musicians (Elbert, Pantev, Wienbruch, Rockstroh, & Taub, 1995), taxi drivers (Maguire et al., 2000), and football players (Vestberg, Gustafson, Maurex, Ingvar, & Petrovic, 2012).

Speaking of expertise, there is also recent evidence for qualitative differences between bilinguals and monolinguals in the absence of any quantitative differences. Incera and McLennan (2015) used mouse tracking to compare bilinguals and monolinguals in a classical Stroop task. Participants were instructed to click on fields on the screen that corresponded to the color of a word (and not the word itself, e.g., green in blue font). In mouse tracking, besides the overall time to click on a target, the initiation of the movement toward that target and how fast the movement proceeds to the correct response is recorded. Overall, the speed of processing did not differ across groups, but initiation times were longer and subsequent movements to the target were faster for bilinguals than for monolinguals, which the authors likened to experts in other fields (e.g., professional baseball players), who initiate a response later but then perform the relevant task faster than novices. Hence, the authors interpret their finding as bilinguals being experts in dealing with conflicting information (see also Narra, Heathcote, & Finkbeiner, 2016, for time course differences between bilinguals and monolinguals in Simon task performance).

Moreover, there is clear evidence that not all monolinguals are the same (Pakulak & Neville, 2010) and hence treating monolinguals as a homogeneous group of individuals may be just as misguided as doing so with bilinguals. This is even more relevant for children since they vary more in their performance than do adult samples, which in many studies are a highly selective and well-educated group of university students. Given the evident heterogeneity in both populations, there may be a greater overlap in linguistic and non-linguistic abilities across groups than anticipated by researchers. Hence, individual differences should be heeded to a greater extent in future research. Concurrently, using continuous predictors for EF task performance such as degree of bilingualism (which describes the relative language proficiencies in L1 and L2 and how often each language is used on a daily basis) in within-group designs can help bypass the need for matching background variables across groups (e.g., Sorge et al., 2016; Thomas-Sunesson, Hakuta, & Bialystok, 2016).

Finally, future research may also want to aim at using longitudinal designs to examine the development of EF over time and the relationship between language experience and EF development. Poarch and van Hell (2012b), for example, reported a cross-sectional study showing bilingual experience effects emerging after 2.5 years of dual language immersion (see also Crivello et al., 2016; Guerrero, Smith, & Luk, 2016).

We now move away from variables pertaining to participants to considerations in bilingual research pertaining to the EF tasks employed and how the collected data is handled in preparation for statistical analysis. Concerning the tasks used in this line of research, the studies on bilingualism have relied heavily on tasks such as the flanker and the Simon task, which come in variants that are either cognitively very simple or more complex (see, e.g., Costa et al., 2009, for two variants of the flanker task differing in cognitive load). From this follows that the cognitive demands a specific EF task poses on participants will inevitably tap cognitive processes to differing degrees – there may be a need for tasks to be sufficiently cognitively demanding, otherwise the cognitive load may be insufficient to tax the system for EF differences to emerge (Macnamara & Conway, 2014; Qu, Low, Zhang, Li, & Zelazo, 2016). Concerning data trimming procedures, Zhou and Krott (2016) conducted a meta-analysis and found that differences in EF tasks between groups may also be driven by responses that are in the tail of the response time distribution. Hence, identifying slow responses as outliers and trimming them may eliminate the potential effect. Given that children's performances in EF tasks are more variable and responses are typically slower, this point is all the more relevant as inclusion or exclusion of slow reaction times depending on the setting of a usually arbitrary cut-off threshold may obscure or eliminate possible differences between groups of monolinguals and bilinguals (see also Grundy & Timmer, 2016, for a similar line of argumentation).

In conclusion, at present the picture of across-group differences in comparisons between bilinguals and monolinguals is not fully converging. Hence, there is a need to specify more accurately and strictly what type of participants take part in studies in this line of research and under which conditions differential EF emerge. It is particularly relevant to capture and assess as thoroughly as possible the diversity in participants in terms of background variables such as socio-economic status, age of acquisition, length of immersion, relative language proficiencies, language experience (Yang, Hartanto, & Yang, 2016), frequency of usage (De Leeuw & Bogulski, 2016), context and availability of the languages (Ye, Mo, & Wu, 2016), heritage vs. minority vs. majority language, recency of language usage, and language switching behavior (Verreyt, Woumans, Vandelanotte, Szmalek, & Duyck, 2015), to name just a few. While this may be too time-consuming, as can be evidenced by the relative heterogeneity in the background measures reported across studies, determining patterns of language usage and language switching may be a viable manner of ascertaining in which way populations do differ and whether such differences affect the development (and possible enhancement) of EF. Furthermore, the tasks that have been used to tap EF will need to be re-assessed as to whether they are sufficiently cognitively demanding and, critically, which EF processes they actually measure (see also Crone et al., Chapter 3, this volume). The same goes for seemingly inconsequential procedures such as data trimming, where slow reaction times may carry very relevant information and, hence, should not be disregarded per se.

Having said that, future research may also need to explore individual differences in groups of bilinguals to a greater extent (Baum & Titone, 2014; Dong & Li, 2015). In other words, it may answer the question as to which particular language experiences are at the root of the effects repeatedly but not universally found.

Author Note

The writing of this paper was partially supported by NSF grants OISE-0968369, BCS-1349110, and OISE-1545900 to Janet van Hell.

References

Abutalebi, J., Della Rosa, P. A., Ding, G., Weekes, B., Costa, A., & Green, D. W. (2013). Language proficiency modulates the engagement of cognitive control areas in multilinguals. *Cortex, 49*, 905–911.

Abutalebi, J., & Green, D. W. (2007). Bilingual language production: The neurocognition of language representation and control. *Journal of Neurolinguistics, 20*, 242–275.

Alladi, S., Bak, T. H., Duggirala, V., Surampudi, B., Shailaja, M., Shukla, A. K., & Kaul, S. (2013). Bilingualism delays age at onset of dementia, independent of education and immigration status. *Neurology, 81*, 1938–1944.

Antón, E., Duñabeitia, J. A., Estévez, A., Hernández, J. A., Castillo, A., Fuentes, L. J., et al. (2014). Is there a bilingual advantage in the ANT task? Evidence from children. *Frontiers in Psychology, 5*, 398.

Bak, T. H. (2016). Cooking pasta in La Paz: Bilingualism, bias and the replication crisis. *Linguistic Approaches to Bilingualism, 6*, 699–717.

Bak, T. H., Nissan, J. J., Allerhand, M. M., & Deary, I. J. (2014). Does bilingualism influence cognitive aging? *Annals of Neurology, 75*, 959–963.

Barac, R., Moreno, S., & Bialystok, E. (2016). Behavioral and electrophysiological differences in executive control between monolingual and bilingual children. *Child Development, 87*, 1277–1290.

Baum, S., & Titone, D. (2014). Moving toward a neuroplasticity view of bilingualism, executive control, and aging. *Applied Psycholinguistics, 35*, 857–894.

Bialystok, E., & Barac, R. (2012). Emerging bilingualism: Dissociating advantages for metalinguistic awareness and executive control. *Cognition, 122*, 67–73.

Bialystok, E., Craik, F. I. M., & Luk, G. (2008). Cognitive control and lexical access in younger and older bilinguals. *Journal of Experimental Psychology: Learning, Memory, and Cognition, 34*, 859–873.

Bialystok, E., Craik, F. I. M., & Luk, G. (2012). Bilingualism: Consequences for mind and brain. *Trends in Cognitive Sciences, 16*, 240–250.

Bialystok, E., & Poarch, G. J. (2014). Language experience changes language and cognitive ability. *Zeitschrift für Erziehungswissenschaft, 17*, 433–446.

Bialystok, E., Poarch, G. J., Luo, L., & Craik, F. I. M. (2014). Effects of bilingualism and aging on executive function and working memory. *Psychology and Aging, 29*, 696–705.

Blom, W. B. T., Küntay, A. C., Messer, M. H., Verhagen, J., & Leseman, P. P. M. (2014). The benefits of being bilingual: Working memory in bilingual Turkish-Dutch children. *Journal of Experimental Child Psychology, 128*, 105–119.

Blumenfeld, H. K., & Marian, V. (2014). Cognitive control in bilinguals: Advantages in Stimulus-Stimulus inhibition. *Bilingualism: Language and Cognition, 17*, 610–629.

Branzi, F. M., Calabria, M., Boscarino, M. L., & Costa, A. (2016). On the overlap between bilingual language control and domain-general executive control. *Acta Psychologica, 166*, 21–30.

Buac, M., & Kaushanskaya, M. (2014). The relationship between linguistics and non-linguistic cognitive control skills in bilingual children from low socio-economic backgrounds. *Frontiers in Psychology, 5*, 1098.

Caffarra, S., Molinaro, N., Davidson, D., & Carreiras, M. (2015). Second language syntactic processing revealed through event-related potentials: An empirical review. *Neuroscience & Biobehavioral Reviews, 51*, 31–47.

Calvo, A., & Bialystok, E. (2014). Independent effects of bilingualism and socioeconomic status on language ability and executive functioning. *Cognition, 130*, 278–288.

Carlson, S., & Meltzoff, A. N. (2008). Bilingual experience and executive functioning in young children. *Developmental Science, 11*, 282–298.

Coderre, E. L., Smith, J. F., & Van Heuven, W. J. B. (2016). The functional overlap of executive control and language processing in bilinguals. *Bilingualism: Language and Cognition, 19*, 471–488.

Costa, A., Hernández, M., Costa-Faidella, J., & Sebastián-Gallés, N. (2009). On the bilingual advantage in conflict processing: Now you see it, now you don't. *Cognition, 113*, 135–149.

Costa, A., Hernández, M., & Sebastián-Gallés, N. (2008). Bilingualism aids conflict resolution: Evidence from the ANT task. *Cognition, 106*, 59–86.

Cox, S. R., Bak, T. H., Allerhand, M., Redmond, P., Starr, J. M., Deary, I. J., & MacPherson, S. E. (2016). Bilingualism, social cognition and executive functions: A tale of chickens and eggs. *Neuropyschologica, 42*, 1029–1040.

Crivello, C., Kuzyk, O., Rodrigues, M., Friend, M., Zesiger, P., & Poulin-Dubois, D. (2016). The effects of bilingual growth on toddlers' executive function. *Journal of Experimental Child Psychology, 141*, 121–132.

De Bruin, A., Treccani, B., & Della Sala, S. (2015). Cognitive advantage in bilingualism: An example of publication bias? *Psychological Science, 26*, 99–107.

De Leeuw, E., & Bogulski, C. A. (2016). Frequent L2 language use enhances executive control in bilinguals. *Bilingualism: Language and Cognition, 19*, 907–913.

Dong, Y., & Li, P. (2015). The cognitive science of bilingualism. *Language and Linguistics Compass, 9*, 1–13.

Duñabeitia, J. A., Hernández, J. A., Antón, E., Macizo, P., Estévez, A., Fuentes, L. J., & Carreiras, M. (2014). The inhibitory advantage in bilingual children revisited: Myth or reality? *Experimental Psychology, 61*, 234–251.

Elbert, T., Pantev, C., Wienbruch, C., Rockstroh, B., & Taub, E. (1995). Increased cortical representation of the fingers of the left hand in string players. *Science, 270,* 305–307.

Engel de Abreu, P. M. J., Cruz-Santos, A., Tourinho, C. J., Martin, A., & Bialystok, E. (2012). Bilingualism enriches the poor: Enhanced cognitive control in low-income minority children. *Psychological Science, 23,* 1364–1371.

Eriksen, B. A., & Eriksen, C. W. (1974). Effects of noise letters upon the identification of a target letter in a nonsearch task. *Perception Psychophysics, 16,* 143–149.

Fan, J., McCandliss, B. D., Sommer, T., Raz, A., & Posner, M. I. (2002). Testing the efficiency and independence of attentional networks. *Journal of Cognitive Neuroscience, 14,* 340–347.

Filippi, R., Morris, J., Richardson, F. M., Bright, P., Thomas, M. S. C., Karmiloff-Smith, A., & Marian, V. (2015). Bilingual children show an advantage in controlling verbal interference during spoken language comprehension. *Bilingualism: Language and Cognition, 18,* 490–501.

Gathercole, V. C. M., Thomas, E. M., Kennedy, I., Prys, C., Young, N., Viñas Guasch, N., et al. (2014). Does language dominance affect cognitive performance in bilinguals? Lifespan evidence from preschoolers through older adults on card sorting, Simon, and metalinguistic tasks. *Frontiers in Psychology, 5,* 11.

Green, D. W., & Abutalebi, J. (2013). Language control in bilinguals: The adaptive control hypothesis. *Journal of Cognitive Psychology, 25,* 143–149.

Grundy, J. G., & Timmer, K. (2016). Cognitive mechanisms underlying performance differences between monolinguals and bilinguals. In J. W. Schwieter (Ed.), *Cognitive control and consequences of multilingualism* (pp. 375–396). Amsterdam: John Benjamins.

Guerrero, S. L., Smith, S., & Luk, G. (2016). Home language usage and executive function in bilingual preschoolers. In J. W. Schwieter (Ed.), *Cognitive control and consequences of multilingualism* (pp. 351–374). Amsterdam: John Benjamins.

Hackman, D. A., & Farah, M. J. (2009). Socioeconomic status and the developing brain. *Trends in Cognitive Sciences, 13,* 65–73.

Hoff, E., Core, C., Place, S., Rumiche, R., Señor, M., & Parra, M. (2012). Dual language exposure and early bilingual development. *Journal of Child Language, 39,* 1–27.

Hope, T. M. H. (2015). The bilingual cognitive advantage: No smoke without fire? *AIMS Neuroscience, 2,* 58–65.

Incera, S., & McLennan, C. T. (2015). Mouse tracking reveals that bilinguals behave like experts. *Bilingualism: Language and Cognition, 19,* 610–620.

Kootstra, G. J., van Hell, J. G., & Dijkstra, T. (2010). Syntactic alignment and shared word order in code-switched sentence production: Evidence from bilingual monologue and dialogue. *Journal of Memory and Language, 63,* 210–231.

Kroll, J. F., Gullifer, J. W., & Rossi, E. (2013). The multilingual lexicon: The cognitive and neural basis of lexical comprehension and production in two or more languages. *Annual Review of Applied Linguistics, 33,* 102–127.

Ladas, A. I., Carroll, D. J., & Vivas, A. B. (2015). Attentional processes in low-socioeconomic status bilingual children: Are they modulated by the amount of bilingual experience? *Child Development, 86,* 557–578.

Logan, G. D., & Cowan, W. B. (1984). On the ability to inhibit thought and action: A theory of an act of control. *Psychological Review, 91,* 295–327.

Luk, G., & Pliatsikas, C. (2016). Converging diversity to unity: Commentary on the neuroanatomy of bilingualism. *Language, Cognition and Neuroscience, 31,* 349–352.

Macnamara, B. N., & Conway, A. R. A. (2014). Novel evidence in support of the bilingual advantage: Influences of task demands and experience on cognitive control and working memory. *Psychonomic Bulletin & Review, 21*, 520–525.

Maguire, E. A., Gadian, D. G., Johnsrude, I. S., Good, C. D., Ashburner, J., Frackowiak, R. S., & Frith, C. D. (2000). Navigation-related structural changes in the hippocampi of taxi drivers. *Proceedings of the National Academy of the United States of America, 97*, 4398–4403.

Marian, V., & Spivey, M. (2003). Competing activation in bilingual language processing: Within- and between-language competition. *Bilingualism: Language and Cognition, 6*, 97–115.

Mezzacappa, E. (2004). Alerting, orienting, and executive attention: Developmental properties and sociodemographic correlates in an epidemiological sample of young, urban children. *Child Development, 75*, 1373–1386.

Miyake, A., & Friedman, N. P. (2012). The nature and organization of individual differences in executive functions: Four general conclusions. *Current Directions in Psychological Science, 21*, 8–14.

Narra, M., Heathcote, A., & Finkbeiner, M. (2016). Time course differences between bilinguals and monolinguals in the Simon task. In J. W. Schwieter (Ed.), *Cognitive control and consequences of multilingualism* (pp. 397–426). Amsterdam: John Benjamins.

Olulade, O. A., Jamal, N. I., Koo, D. S., Perfetti, C. A., LaSasso, C., & Eden, G. F. (2016). Neuroanatomical evidence in support of the bilingual advantage theory. *Cerebral Cortex, 26*, 3196–3204.

Paap, K. R., & Greenberg, Z. I. (2013). There is no coherent evidence for a bilingual advantage in executive processing. *Cognitive Psychology, 66*, 232–258.

Paap, K. R., Sawi, O., Dalibar, C., Darrow, J., & Johnson, H. A. (2013). The brain mechanisms underlying the cognitive benefits of bilingualism may be extraordinarily difficult to discover. *AIMS Neuroscience, 1*, 245–256.

Pakulak, E., & Neville, H. J. (2010). Proficiency differences in syntactic processing of monolingual native speakers indexed by event-related potentials. *Journal of Cognitive Neuroscience, 22*, 2728–2744.

Paradis, J., Genesee, F., & Crago, M. B. (2011). *Dual language development and disorders: A handbook on bilingualism and second language learning*. Baltimore, MD: Brookes Publishing.

Pliatsikas, C., & Luk, G. (2016). Executive control in bilinguals: A concise review on fMRI studies. *Bilingualism: Language and Cognition, 19*, 699–705.

Poarch, G. J., & Bialystok, E. (2015). Bilingualism as a model for multitasking. *Developmental Review, 35*, 113–124.

Poarch, G. J., & van Hell, J. G. (2012a). Cross-language activation in children's speech production: Evidence from second language learners, bilinguals, and trilinguals. *Journal of Experimental Child Psychology, 111*, 419–438.

Poarch, G. J., & van Hell, J. G. (2012b). Executive functions and inhibitory control in multilingual children: Evidence from second language learners, bilinguals, and trilinguals. *Journal of Experimental Child Psychology, 113*, 535–551.

Poarch, G. J., & van Hell, J. G. (2014). Cross-language activation in same-script and different-script trilinguals. *International Journal of Bilingualism, 18*, 693–716.

Qu, L., Low, J. J. W., Zhang, T., Li, H., & Zelazo, P. D. (2016). Bilingual advantage in executive control when task demands are considered. *Bilingualism: Language and Cognition, 19*, 277–293.

Rueda, M., Fan, J., McCandliss, B., Halparin, J., Gruber, D., Lercari, L., & Posner, M. (2004). Development of attentional networks in children. *Neuropyschologica, 42*, 1029–1040.

Salthouse, T. A. (2009). When does age-related cognitive decline begin? *Neurobiology of Aging, 30*, 507–514.

Salvatierra, J. L., & Rosselli, M. (2011). The effect of bilingualism and age on inhibitory control. *International Journal of Bilingualism, 15*, 26–37.

Schroeder, S., & Marian, V. (2012). A bilingual advantage for episodic memory in older adults. *Journal of Cognitive Psychology, 24*, 591–601.

Simon, J. R., & Rudell, A. P. (1967). Auditory S–R compatibility: The effect of an irrelevant cue on information processing. *Journal of Applied Psychology, 51*, 300–304.

Sorge, G. B., Toplak, M. E., & Bialystok, E. (2016). Interactions between levels of attention ability and levels of bilingualism in children's executive functioning. *Developmental Science*. Advance online publication. doi:10.1111/desc.12408.

Stocco, A., Yamasaki, B., Natalenko, R., & Prat, C. S. (2014). Bilingual brain training: A neurobiological framework of how bilingual experience improves executive function. *International Journal of Bilingualism, 18*, 67–92.

Stroop, J. R. (1935). Studies of interference in serial verbal reactions. *Journal of Experimental Psychology, 18*, 643–662.

Tao, L., Marzecová, A., Taft, M., Asanowicz, D., & Wodniecka, Z. (2011). The efficiency of attentional networks in early and late bilinguals: The role of age of acquisition. *Frontiers in Psychology, 2*, 123.

Teubner-Rhodes, S. E., Mishler, A., Corbett, R., Andreu, L., Sanz-Torrent, M., Trueswell, J. C., & Novick, J. M. (2016). The effects of bilingualism on conflict monitoring, cognitive control, and garden-path recovery. *Cognition, 150*, 213–231.

Thierry, G., & Wu, Y. J. (2007). Brain potentials reveal unconscious translation during foreign-language comprehension. *Proceedings of the National Academy of the United States of America, 104*, 12530–12535.

Thomas-Sunesson, D., Hakuta, K., & Bialystok, E. (2016). Degree of bilingualism modifies executive control in Hispanic children in the USA. *International Journal of Bilingual Education and Bilingualism*. Advance online publication. doi:10.1080/13670050.2016.1148114.

van Hell, J. G., & Dijkstra, T. (2002). Foreign language knowledge can influence native language performance in exclusively native contexts. *Psychonomic Bulletin and Review, 9*, 780–789.

van Hell, J. G., & Poarch, G. J. (2014). How much bilingual experience is needed to affect executive control? *Applied Psycholinguistics, 35*, 925–928.

van Hell, J. G., & Tanner, D. (2012). Second language proficiency and cross-language lexical activation. *Language Learning, 62*, 148–171.

van Hell, J. G., & Tokowicz, N. (2010). Event-related brain potentials and second language learning: Syntactic processing in late L2 learners at different L2 proficiency levels. *Second Language Research, 26*, 43–74.

Van Heuven, W. J. B., Schriefers, H., Dijkstra, T., & Hagoort, P. (2008). Language conflict in the bilingual brain. *Cerebral Cortex, 18*, 2706–2716.

Verreyt, N., Woumans, E., Vandelanotte, D., Szmalek, A., & Duyck, W. (2015). The influence of language-switching experience on the bilingual executive control advantage. *Bilingualism: Language and Cognition, 19*, 181–190.

Vestberg, T., Gustafson, R., Maurex, L., Ingvar, M., & Petrovic, P. (2012). Executive functions predict the success of top soccer players. *PLoS ONE, 7*, e34731.

Von Bastian, C. C., Souza, A. S., & Gade, M. (2016). No evidence for bilingual cognitive advantages: A test of four hypotheses. *Journal of Experimental Psychology: General, 145*, 246–258.

Woumans, E., Santens, P., Sieben, A., Versijpt, J., Stevens, M., & Duyck, W. (2015). Bilingualism delays clinical manifestation of Alzheimer's disease. *Bilingualism: Language and Cognition, 18*, 568–574.

Yang, H., Hartanto, A., & Yang, S. (2016). The importance of bilingual experience in assessing bilingual advantages in executive functions. *Cortex, 75*, 237–240.

Yang, S., & Yang, H. (2016). Bilingual effects on deployment of the attention system in linguistically and culturally homogeneous children and adults. *Journal of Experimental Child Psychology, 146*, 121–136.

Ye, Y., Mo, L., & Wu, Q. (2016). Mixed cultural context brings out bilingual advantage on executive function. *Bilingualism: Language and Cognition.* Advance online publication. doi:10.1017/S1366728916000481.

Zhou, B., & Krott, A. (2016). Data trimming procedure can eliminate bilingual cognitive advantage. *Psychonomic Bulletin & Review, 23*, 1221–1230.

12
PHYSICAL ACTIVITY, EXERCISE, AND EXECUTIVE FUNCTIONS

Nicolas Berryman, Kristell Pothier, & Louis Bherer

Introduction

Mens Sana In Corpore Sano

This famous quotation carries the essence of a whole emerging field of research that has become increasingly important over the last few decades, supporting the notion that physical activity, exercise, and fitness are associated with neurocognitive and brain functions. Physical activity is defined as any bodily movement produced by skeletal muscles that result in energy expenditure above basal metabolic rate. Exercise is a subtype of physical activity that is planned, structured, repetitive, and purposive, whereas physical fitness is a set of measurable health- and skill-related attributes that include cardiorespiratory (aerobic) performance, muscle strength, body composition, balance, agility, reaction time, and power (Garber et al., 2011).

In 2003, a seminal research article revealed that physical exercise interventions were associated with improvements in cognition in older adults (Colcombe & Kramer, 2003). Particularly, this meta-analysis showed that beneficial effects were more substantial on executive functions (EF) than other cognitive domains (e.g., memory, speed of processing, etc.). Moreover, the meta-analysis highlighted the importance of combining both strength and aerobic training. Mechanisms supporting the long-term (chronic) relationship between exercise interventions and cognition are still under investigation. Regarding EF, a review (Barenberg, Berse, & Dutke, 2011) indicated the involvement of three potential physiological explanations: (1) an increased cerebral oxygenation found in the prefrontal cortex, known to be associated with EF (Dupuy et al., 2015); (2) the plasticity of brain architecture, modulated by neurotrophic factors (brain derived neurotrophic factors – BDNF, vascular endothelial growth factor – VEGF, insulin-like growth factor – IGF-1), which play an important role in neurogenesis and angiogenesis (Cassilhas et al., 2012; Voss, Nagamatsu, Liu-Ambrose, & Kramer, 2011); and (3) the role of brain neurotransmitters, in particular norepinephrine and dopamine, both potentially upregulated by physical activity and known to play a role in executive control processes (Stroth et al., 2010). Recent studies also suggest that acute effects of physical activity on cognition could be associated with increases in catecholamine levels, sympathetic nervous system activity, central nervous system arousal (Brisswalter, Collardeau, & Rene, 2002), and acute modifications in cerebral blood flow and oxygenation (Mekari et al., 2015; but see Ando, 2016). These observations bring hope that physical activity and exercise could be an efficient way to enhance neurocognitive functions throughout life and help reduce age-associated cognitive decline. While this field of investigation is rapidly gaining popularity, there is already a substantial amount of research dedicated to uncovering its basic mechanisms and its potential as a non-pharmaceutics in numerous medical conditions. However, there is still much to know about the impact of physical activity and exercise on EF in healthy populations. This chapter is an attempt to provide a comprehensive overview of the literature investigating the acute and chronic effects of physical activity and exercise on EF across the life span of healthy individuals.

Effects of Acute Physical Activity and Exercise on Executive Functions throughout the Life Span

An increasing number of studies have tried to better understand the link between physical activity and cognition by studying its acute effect. According to different literature reviews and meta-analyses (Bangsbo et al., 2016; Chang, Labban, Gapin, & Etnier, 2012; Donnelly et al., 2016; Ludyga, Gerber, Brand, Holsboer-Trachsler, & Puhse, 2016; Verburgh, Konigs, Scherder, & Oosterlaan, 2014), it is well accepted that acute physical activity has a beneficial effect on cognition. However, the overall effects are sometimes rather small and it appears that multiple variables could influence this relationship.

Effects of Acute Physical Activity and Exercise on Executive Functions in Childhood and Early Adulthood

Most studies on the acute effect of physical activity and exercise on cognition have focused on its impact on EF. A recent meta-analysis on 21 studies with children, adolescents, and young adults included a total of 586 participants (Verburgh et al., 2014). It was shown that acute exercise bouts were associated with some beneficial effects on EF in different age groups, when assessed immediately after exercise (note that studies in which EFs were assessed during exercise were excluded). While a moderate global effect was found ($d = 0.52$), no significant differences were reported between age groups (children: 6–12 years old, adolescents: 13–17 years old, young adults: 18–35 years old). Moreover, it was found that inhibition and interference are particularly sensitive to acute exercise while a moderate effect size was reported (d = 0.46). Furthermore, a sub-analysis with only 48 participants showed non-significant acute effects of exercise on working memory (d = 0.05) (Verburgh et al., 2014). Importantly, regarding potential confounding moderators, the meta-analysis showed that exercise duration (typically between 10 and 40 minutes) was not significantly associated with executive performances. Exercise intensity was not reported with consistency in the included studies and was, therefore, not included in the analyses.

A recent position stand of the American College of Sports Medicine (ACSM) (Donnelly et al., 2016) supports a beneficial effect of acute physical activity on cognition. The focus of this review was on elementary school children (5–13 years old) and it is important to mention that the included studies had to present fitness and/or physical activity levels data. This report identified a total of 17 studies investigating the effects of acute physical activity bouts on cognition with samples sizes ranging from 20 to 1,274 students. The authors reported that 14 studies were supportive of a beneficial relationship between acute physical activity and cognition and more particularly on executive control. Exercise duration varied between 4 and 42 minutes at a moderate intensity of 60–70% of maximal heart rate. Importantly, in studies that failed to support the positive acute effect of physical activity on cognition, no deleterious effects were reported.

One should be careful when interpreting these results of acute exercise on EF due to numerous methodological issues. It seems that the timing of the assessment of EF and the exercise intensity are key factors. As a matter of fact, in a dual-task paradigm in which both motor and cognitive task are performed simultaneously (a modified-Stroop task while pedaling on a cycle ergometer at 40%, 60%, and 80% of peak power output), it was shown that the transition from moderate to high aerobic intensity (60 to 80–85% of maximal oxygen uptake – VO_2max) was accompanied by a decrease in switching performance during exercise in young adults (Labelle, Bosquet, Mekary, & Bherer, 2013; Mekari et al., 2015). These cognitive performance drops at higher exercise intensities were associated with a reduction in cerebral oxygenation in the prefrontal cortex (Mekari et al., 2015). These results fit well with the hypofrontality hypothesis, according to which exercise at higher intensities creates higher neural demands in some regions of the brain. In response to that phenomenon, a shift in the allocation of oxygen and glucose is observed in order to support those regions associated with high-intensity exercise. The prefrontal cortex, which does not play a crucial role for exercise maintenance at higher intensities, will be negatively affected by this resource shift (Dietrich & Sparling, 2004). The observation that higher-fit individuals seem to particularly benefit from an acute bout of exercise to improve cognition when assessments are made during exercise (dual-task) could also be related to this hypofrontality hypothesis. Indeed, it could be argued that these participants have more resources available to execute a cognitive task while doing physical activity, since a lower portion of these resources is necessary for the exercise task in comparison to lower-fit individuals (Chang et al., 2012).

Taken together, these results suggest that there is a beneficial acute effect of physical activity and exercise on EF in children, adolescents, and young adults when EF are assessed immediately after the exercise bout. However, it appears that fitness level, exercise intensity, and the timing of EF assessment, at least with aerobic exercise bouts in young adults, are crucial variables that could interact with this effect. Furthermore, it seems that more investigations are necessary to identify which type of exercise provides the most beneficial acute effect on EF. Whereas a majority of studies focused on aerobic exercise (Chang et al., 2012; Donnelly et al., 2016; Ludyga et al., 2016), it was recently argued that strength training or integrative neuromuscular training should also be considered for an optimal brain development (Myer et al., 2015). Accordingly, it is suggested that more original research is needed to increase our understanding of different interactions between variables, which could influence this acute physical activity/EF relationship in children and young adults.

Effects of Acute Physical Activity and Exercise on Executive Functions in Older Adults

Studies on the effects of an acute bout of physical activity on cognition in middle-aged and older adults are still scarce. However, recent studies support the positive impact of single bouts of different types of physical activities on EF.

Using an acute resistance exercise (submaximal strength), a modified Stroop task was administered to 30 community-dwelling older adults (55–70 years old) before (pre) and following exercise (post). Compared to a control (reading) group, the exercise treatment resulted in significantly enhanced performance from pre- to post-test across all Stroop conditions, with larger effects in the executive component (Chang, Tsai, Huang, Wang, & Chu, 2014). More recently, these results were extended to another exercise type. In this study (Chu, Chen, Hung, Wang, & Chang, 2015), 70 healthy older adults (60–70 years old) were instructed to complete an acute aerobic exercise (a 30-minute cycling ergometer protocol) and to perform the Stroop test within 5 minutes of exercise cessation. Again, compared to a control-reading group, results showed that acute exercise led to benefits in EF, with a smaller difference between the congruent and incongruent conditions of the Stroop task after exercise. In 2012, Alves and colleagues (Alves et al., 2012) compared the effects of acute moderate-intensity aerobic (walking) and strength exercises (submaximal strength) on selected EF (inhibition and flexibility, assessed by the Stroop and the Trail Making tests, respectively) in 42 middle-aged women (52 ± 7.3 years). Results showed that both acute aerobic and strength exercises equally and selectively improved inhibition (Stroop task). Taken together, these results highlight the positive impact of single bouts of moderate to vigorous physical activities on inhibitory control when assessed after the exercise. However, as for younger adults, exercise intensity can also represent a key factor for older adults. Indeed, using the Stroop task while biking at different intensities, it was recently shown that the transition between 60 and 80% of VO_2max negatively impacted EF performance in older adults (60–70 years old) (Labelle et al., 2014).

Future studies are needed to help understand in more detail the impact of acute bouts of different exercise modalities on specific cognitive functions in adults and seniors.

Effects of Chronic Physical Activity and Exercise on Executive Functions throughout the Life Span

Observational Studies in Childhood and Young Adults

In a recent observational study with 74 elementary school children (mean age: 8.64 +/− 0.58 years old), a positive relationship was found between aerobic fitness and inhibitory control (modified Eriksen flanker task). Interestingly, no significant associations were found between cognition and moderate-to-vigorous physical activity assessed during 7 days with an accelerometer (Pindus et al., 2016). These results suggest that aerobic fitness, rather than the physical activity level per se, represent a key factor in this relationship with cognitive performances.

These results are in line with a recent review of cross-sectional studies in children (5–13 years old) (Donnelly et al., 2016), which revealed a beneficial relationship between aerobic fitness and cognitive performances. Interestingly, brain function was assessed with tasks mainly related to inhibition. This positive relationship between fitness and executive functioning was maintained even after inclusion of potential confounding variables such as gender, pubertal stage, socioeconomic status, body composition, age, grade, and IQ. The review (Donnelly et al., 2016) also included two longitudinal studies (Chaddock et al., 2012; Niederer et al., 2011), allowing assessment of the long-term relationship between aerobic fitness and performances in inhibition or working memory. It was observed that children with higher aerobic fitness level performed better and improved more during the follow-up period than their less fit colleagues. Interestingly, it appears that brain structures and neural networks that support EF are particularly sensitive to the benefits of exercise (Donnelly et al., 2016).

In young and middle-aged adults, similar observations were made. Indeed, a total of 1,803 healthy participants (24.7 years old in average, range from 15 to 55 years old) from 14 studies were included in a systematic review of the literature investigating the relationship between long-term physical activity and cognition (Cox et al., 2016). The authors reported that 12 studies found a positive association between physical activity, assessed through questionnaires or accelerometers, and at least one domain of cognition. In line with the results described in children, this positive association was more pronounced with EF. However, results for fitness levels were not reported in this analysis.

Taken together, these results in children and young adults support a positive association between chronic physical activity and cognition, with a particular emphasis on EF. However, more research is needed to understand and disentangle the specific roles of fitness and lifestyles.

Observational Studies in Older Adults

In older adults, it is well accepted that long-term physical activity and exercise are associated with better EF. Indeed, it was reported in a study including 349 adults over 55 years old that baseline VO_2max remained significantly associated with all three measures of EF (switching, inhibition, attentional control) during a 6-year follow-up (Barnes, Yaffe, Satariano, & Tager, 2003). Whereas aerobic fitness seems to be associated with better cognitive performances in this study, it is important to mention that muscular strength is also a key fitness marker related to cognition. Indeed, in a cohort of 970 older participants included in a 6-year follow-up, participants who did not develop Alzheimer's disease (AD; n = 832) showed greater strength performances (both lower and upper body) at baseline than the participants who developed AD (Boyle, Buchman, Wilson, Leurgans, & Bennett, 2009). Again, these results tend to support the benefits of neuromuscular fitness, along with cardiovascular health, on cognition. Although these results are rather convincing, recent intervention studies including various exercises and the use of neuroimaging techniques have opened new research avenues.

Intervention Studies in Childhood and Young Adults

In its latest position stand the ACSM reported the results of 14 intervention studies, with 11 of them being randomized controlled trial (RCT; SMART Trial, Georgia Trial, FIT Kids Trial) (Donnelly et al., 2016). Results were in line with previous studies suggesting selective benefits of chronic physical activity and exercise on EF. Interestingly, body mass appeared as a significant confounding variable in one cohort study that reported greater improvements in inhibition for overweight children involved in an enhanced physical education program in school (Crova et al., 2014). Moreover, a dose-response relationship is noticeable as greater attendance in exercise sessions is associated with greater benefits on EF. However, a meta-analysis by Verburgh and colleagues (Verburgh et al., 2014) recently showed that chronic exercise yielded a non-significant effect (d = 0.16) on EF in a sample of 337 pre-adolescent children. Considering that RCT studies were mainly focused on aerobic fitness and motor skills development and taking into account previous positive reports on the effects of strength training, we recommend that future research should emphasize comparisons between exercises modalities. Clearly, more research is needed to identify the key variables that could play a role in moderating the relationship between exercise and EF in children.

In young adults, it seems that the research is much less abundant. To our knowledge, only a few intervention studies were published in this age group (Masley, Weaver, Peri, & Phillips, 2008; Stroth, Hille, Spitzer, & Reinhardt, 2009) and results regarding cognition and more specifically EF are not as convincing (mitigated effects and rather low overall sample size) as was found for children or older adults. More research is needed in this population.

Intervention Studies in Older Adults

In a key article, Kramer and colleagues showed that older adults (60–75 years old) who completed a 6-month aerobic training program had greater improvements than a control group (stretching exercises) in three tasks (task switching, response compatibility, and stopping) that involved executive control processes (Kramer et al., 1999). In line with these results, the cardiovascular hypothesis was proposed and suggested that improvement in aerobic fitness, as measured by VO_2max, was necessary to induce cognitive enhancements after an exercise intervention (Etnier, Nowell, Landers, & Sibley, 2006). This theoretical model was challenged when improvements in cognition were reported without significant changes in aerobic fitness (Smiley-Oyen, Lowry, Francois, Kohut, & Ekkekakis, 2008). Importantly, a meta-analysis of the literature revealed that combined resistance and aerobic training was a more efficient training regimen than aerobic training alone to improve cognition and, more specifically, EF in older adults (Colcombe & Kramer, 2003).

Recent studies suggest that strength training alone can also have beneficial effects on older adults' executive performance (Liu-Ambrose & Donaldson, 2009). It was demonstrated that a 12-month high-intensity resistance exercise intervention (60 minutes, three times a week) could decrease reaction times during a discrimination task (the oddball task) in 48 healthy elderly males (65–79 years old). These results, associated with changes in P3 amplitudes (a sensitive biomarker for normal aging) in the oddball condition, suggest that this type of training could also delay the decline in EF (Tsai, Wang, Pan, & Chen, 2015). Other studies offer convincing support for the benefit of strength training exercise on brain plasticity. Among them, in a neuroimaging study, improved flanker task scores (evaluating selective attention and conflict resolution) were reported in 52 healthy older women (65–75 years old) who underwent a 12-month resistance program (60 minutes, twice a week) (Liu-Ambrose,

Nagamatsu, Voss, Khan, & Handy, 2012). Again, these behavioral data were in conjunction with increased hemodynamic activity in brain regions associated with EF, suggesting functional plasticity after resistance training.

It was recently suggested that gross motor skills training can also lead to improvements in cognition in older adults independently of aerobic fitness or strength training. Voelcker-Rehage and colleagues (Voelcker-Rehage, Godde, & Staudinger, 2011) performed a 12-month neuroimaging study to investigate the effects of aerobic and coordination (designed to improve fine- and gross-motor body coordination) training on executive control (flanker task) and perceptual speed (Visual Search Task) in 44 healthy older adults (aged 63–79 years). Compared to a control group (relaxation and stretching), both experimental groups improved in executive functioning and perceptual speed. They also observed that the neurocognitive mechanisms that underlie cognitive changes could differ depending on training intervention, with cardiovascular training being associated with an increased activation of the sensorimotor network and coordination training leading to an increased activation in the visual–spatial network. Two other research groups using the Random Number Generation task (inhibition and working memory) recently supported these results and showed the beneficial effects of gross motor skills training on improving EF in older adults (Berryman et al., 2014; Forte et al., 2013).

Recently, tai chi interventions have shown beneficial effects on cognition and EF for the elderly (Kelly et al., 2014; Wayne et al., 2014). However, it was found in a meta-analysis that the cognitive task could be a key factor (Miller & Taylor-Piliae, 2014). Besides the variety in tai chi types or doses, a lot of different tasks (from the Trail Making Test to the Clock Drawing Test) were used in the literature, making the relationship between tai chi and EF hard to interpret. Nevertheless, authors did find statistically significant improvements in EF in 10 studies among the 12 included. Thus, these new popular physical activities could be considered as an example of a multi-component exercise that has the potential to enhance cognitive performance in older adults.

Conclusion

This review of the literature on the relationship between physical activity and exercise on EF across the life span is rather convincing. Both acute and chronic exercises, from childhood to the elderly, seem to have beneficial effects on EF. Overall, the studies reported here also suggest that, besides cardiovascular training, other types of physical activity can have a beneficial effect on cognitive functions in healthy populations. Taken together, these results suggest that an individual has multiple avenues to get involved in some sort of physical activities potentially leading to improvements in EF, which reinforces this crucial public health message: get active, stay active!

As described in a previous chapter (Li, Vadaga, Bruce, & Lai, Chapter 4, this volume), mobility performances could be associated to EF in older adults. Moreover, according to the Central Benefit Model (Liu-Ambrose, Nagamatsu, Hsu, & Bolandzadeh, 2013), it is suggested that exercise interventions could reduce falls in older adults through retention and improvements in EF. Recently, a cross-sectional analysis of 48 healthy older adults (70.5 +/− 5.3 years old) showed that faster participants in mobility tests demonstrated higher levels of aerobic fitness (VO_2max), lower energy cost of locomotion and greater neuromuscular performances in comparison to slower individuals. Interestingly, this study also revealed significant positive relationships between mobility performances and EF as measured by the Stroop task (Berryman et al., 2013). Considering the latter observations and the results presented in the previous section, it is suggested that maintaining an active lifestyle represents a crucial healthy life habit. Indeed, physical activity could be seen as a privileged entry into a virtuous circle where fitness, mobility, and cognition interact.

However, some areas need more investigation. Indeed, whereas previous chapters highlighted the importance of EF for academic performance (see Chevalier & Clark, Chapter 2, and Crone, Peters, & Steinbeis, Chapter 3, this volume), it appears that more intervention studies are necessary before firmly concluding as to the potential beneficial effects of physical activity on academic achievement (Donnelly et al., 2016).

References

Alves, C. R., Gualano, B., Takao, P. P., Avakian, P., Fernandes, R. M., Morine, D., & Takito, M. Y. (2012). Effects of acute physical exercise on executive functions: A comparison between aerobic and strength exercise. *Journal of Sport and Exercise Psychology, 34*(4), 539–549.

Ando, S. (2016). Acute exercise and cognition. In T. McMorris (Ed.), *Exercise-cognition interaction: Neuroscience perspectives* (pp. 131–145). London: Elsevier.

Bangsbo, J., Krustrup, P., Duda, J., Hillman, C., Andersen, L. B., Weiss, M., et al. (2016). The Copenhagen Consensus Conference 2016: Children, youth, and physical activity in schools and during leisure time. *British Journal of Sports Medicine*.

Barenberg, J., Berse, T., & Dutke, S. (2011). Executive functions in learning processes: Do they benefit from physical activity? *Educational Research Review, 6*(3), 208–222.

Barnes, D. E., Yaffe, K., Satariano, W. A., & Tager, I. B. (2003). A longitudinal study of cardiorespiratory fitness and cognitive function in healthy older adults. *Journal of the American Geriatrics Society, 51*(4), 459–465.

Berryman, N., Bherer, L., Nadeau, S., Lauziere, S., Lehr, L., Bobeuf, F., et al. (2013). Executive functions, physical fitness and mobility in well-functioning older adults. *Experimental Gerontology, 48*(12), 1402–1409.

Berryman, N., Bherer, L., Nadeau, S., Lauziere, S., Lehr, L., Bobeuf, F., et al. (2014). Multiple roads lead to Rome: Combined high-intensity aerobic and strength training vs. gross motor activities leads to equivalent improvement in executive functions in a cohort of healthy older adults. *Age (Dordr), 36*(5), 9710.

Boyle, P. A., Buchman, A. S., Wilson, R. S., Leurgans, S. E., & Bennett, D. A. (2009). Association of muscle strength with the risk of Alzheimer disease and the rate of cognitive decline in community-dwelling older persons. *Archives of Neurology, 66*(11), 1339–1344.

Brisswalter, J., Collardeau, M., & Rene, A. (2002). Effects of acute physical exercise characteristics on cognitive performance. *Sports Medicine, 32*(9), 555–566.

Cassilhas, R. C., Lee, K. S., Fernandes, J., Oliveira, M. G., Tufik, S., Meeusen, R., & de Mello, M. T. (2012). Spatial memory is improved by aerobic and resistance exercise through divergent molecular mechanisms. *Neuroscience, 202,* 309–317.

Chaddock, L., Hillman, C. H., Pontifex, M. B., Johnson, C. R., Raine, L. B., & Kramer, A. F. (2012). Childhood aerobic fitness predicts cognitive performance one year later. *Journal of Sports Science, 30*(5), 421–430.

Chang, Y. K., Labban, J. D., Gapin, J. I., & Etnier, J. L. (2012). The effects of acute exercise on cognitive performance: A meta-analysis. *Brain Research, 1453,* 87–101.

Chang, Y. K., Tsai, C. L., Huang, C. C., Wang, C. C., & Chu, I. H. (2014). Effects of acute resistance exercise on cognition in late middle-aged adults: General or specific cognitive improvement? *Journal of Science and Medicine in Sport, 17*(1), 51–55.

Chu, C. H., Chen, A. G., Hung, T. M., Wang, C. C., & Chang, Y. K. (2015). Exercise and fitness modulate cognitive function in older adults. *Psychology and Aging, 30*(4), 842–848.

Colcombe, S., & Kramer, A. F. (2003). Fitness effects on the cognitive function of older adults: A meta-analytic study. *Psychological Science, 14*(2), 125–130.

Cox, E. P., O'Dwyer, N., Cook, R., Vetter, M., Cheng, H. L., Rooney, K., & O'Connor, H. (2016). Relationship between physical activity and cognitive function in apparently healthy young to middle-aged adults: A systematic review. *Journal of Science and Medicine in Sport, 19*(8), 616–628.

Crova, C., Struzzolino, I., Marchetti, R., Masci, I., Vannozzi, G., Forte, R., & Pesce, C. (2014). Cognitively challenging physical activity benefits executive function in overweight children. *Journal of Sports Science, 32*(3), 201–211.

Dietrich, A., & Sparling, P. B. (2004). Endurance exercise selectively impairs prefrontal-dependent cognition. *Brain and Cognition, 55*(3), 516–524.

Donnelly, J. E., Hillman, C. H., Castelli, D., Etnier, J. L., Lee, S., Tomporowski, P., et al. (2016). Physical activity, fitness, cognitive function, and academic achievement in children: A systematic review. *Medicine & Science in Sports & Exercise, 48*(6), 1197–1222.

Dupuy, O., Gauthier, C. J., Fraser, S. A., Desjardins-Crepeau, L., Desjardins, M., Mekary, S., et al. (2015). Higher levels of cardiovascular fitness are associated with better executive function and prefrontal oxygenation in younger and older women. *Frontiers in Human Neuroscience, 9,* 66.

Etnier, J. L., Nowell, P. M., Landers, D. M., & Sibley, B. A. (2006). A meta-regression to examine the relationship between aerobic fitness and cognitive performance. *Brain Research Reviews, 52*(1), 119–130.

Forte, R., Boreham, C. A., Leite, J. C., De Vito, G., Brennan, L., Gibney, E. R., & Pesce, C. (2013). Enhancing cognitive functioning in the elderly: Multicomponent vs resistance training. *Clinical Interventions in Aging, 8,* 19–27.

Garber, C. E., Blissmer, B., Deschenes, M. R., Franklin, B. A., Lamonte, M. J., Lee, I. M., et al. (2011). American College of Sports Medicine position stand. Quantity and quality of exercise for developing and maintaining cardiorespiratory, musculoskeletal, and neuromotor fitness in apparently healthy adults: Guidance for prescribing exercise. *Medicine & Science in Sports & Exercise, 43*(7), 1334–1359.

Kelly, M. E., Loughrey, D., Lawlor, B. A., Robertson, I. H., Walsh, C., & Brennan, S. (2014). The impact of exercise on the cognitive functioning of healthy older adults: A systematic review and meta-analysis. *Ageing Research Reviews, 16*, 12–31.

Kramer, A. F., Hahn, S., Cohen, N. J., Banich, M. T., McAuley, E., Harrison, C. R., et al. (1999). Ageing, fitness and neurocognitive function. *Nature, 400*(6743), 418–419.

Labelle, V., Bosquet, L., Mekary, S., & Bherer, L. (2013). Decline in executive control during acute bouts of exercise as a function of exercise intensity and fitness level. *Brain and Cognition, 81*(1), 10–17.

Labelle, V., Bosquet, L., Mekary, S., Vu, T. T., Smilovitch, M., & Bherer, L. (2014). Fitness level moderates executive control disruption during exercise regardless of age. *Journal of Sport and Exercise Psychology, 36*(3), 258–270.

Liu-Ambrose, T., & Donaldson, M. G. (2009). Exercise and cognition in older adults: Is there a role for resistance training programmes? *British Journal of Sports Medicine, 43*(1), 25–27.

Liu-Ambrose, T., Nagamatsu, L. S., Hsu, C. L., & Bolandzadeh, N. (2013). Emerging concept: "Central benefit model" of exercise in falls prevention. *British Journal of Sports Medicine, 47*(2), 115–117.

Liu-Ambrose, T., Nagamatsu, L. S., Voss, M. W., Khan, K. M., & Handy, T. C. (2012). Resistance training and functional plasticity of the aging brain: A 12-month randomized controlled trial. *Neurobiology of Aging, 33*(8), 1690–1698.

Ludyga, S., Gerber, M., Brand, S., Holsboer-Trachsler, E., & Puhse, U. (2016). Acute effects of moderate aerobic exercise on specific aspects of executive function in different age and fitness groups: A meta-analysis. *Psychophysiology*.

Masley, S. C., Weaver, W., Peri, G., & Phillips, S. E. (2008). Efficacy of lifestyle changes in modifying practical markers of wellness and aging. *Alternative Therapies in Health and Medicine, 14*(2), 24–29.

Mekari, S., Fraser, S., Bosquet, L., Bonnery, C., Labelle, V., Pouliot, P., et al. (2015). The relationship between exercise intensity, cerebral oxygenation and cognitive performance in young adults. *European Journal of Applied Physiology, 115*(10), 2189–2197.

Miller, S. M., & Taylor-Piliae, R. E. (2014). Effects of tai chi on cognitive function in community-dwelling older adults: A review. *Geriatric Nursing, 35*(1), 9–19.

Myer, G. D., Faigenbaum, A. D., Edwards, N. M., Clark, J. F., Best, T. M., & Sallis, R. E. (2015). Sixty minutes of what? A developing brain perspective for activating children with an integrative exercise approach. *British Journal of Sports Medicine, 49*(23), 1510–1516.

Niederer, I., Kriemler, S., Gut, J., Hartmann, T., Schindler, C., Barral, J., & Puder, J. J. (2011). Relationship of aerobic fitness and motor skills with memory and attention in preschoolers (Ballabeina): A cross-sectional and longitudinal study. *BMC Pediatrics, 11*, 34.

Pindus, D. M., Drollette, E. S., Scudder, M. R., Khan, N. A., Raine, L. B., Sherar, L. B., et al. (2016). Moderate-to-vigorous physical activity, indices of cognitive control, and academic achievement in preadolescents. *Journal of Pediatrics, 173*, 136–142.

Smiley-Oyen, A. L., Lowry, K. A., Francois, S. J., Kohut, M. L., & Ekkekakis, P. (2008). Exercise, fitness, and neurocognitive function in older adults: The "selective improvement" and "cardiovascular fitness" hypotheses. *Annals of Behavioral Medicine, 36*(3), 280–291.

Stroth, S., Hille, K., Spitzer, M., & Reinhardt, R. (2009). Aerobic endurance exercise benefits memory and affect in young adults. *Neuropsychological Rehabilitation, 19*(2), 223–243.

Stroth, S., Reinhardt, R. K., Thone, J., Hille, K., Schneider, M., Hartel, S., et al. (2010). Impact of aerobic exercise training on cognitive functions and affect associated to the COMT polymorphism in young adults. *Neurobiology of Learning and Memory, 94*(3), 364–372.

Tsai, C. L., Wang, C. H., Pan, C. Y., & Chen, F. C. (2015). The effects of long-term resistance exercise on the relationship between neurocognitive performance and GH, IGF-1, and homocysteine levels in the elderly. *Frontiers in Behavioral Neuroscience, 9*, 23.

Verburgh, L., Konigs, M., Scherder, E. J., & Oosterlaan, J. (2014). Physical exercise and executive functions in preadolescent children, adolescents and young adults: A meta-analysis. *British Journal of Sports Medicine, 48*(12), 973–979.

Voelcker-Rehage, C., Godde, B., & Staudinger, U. M. (2011). Cardiovascular and coordination training differentially improve cognitive performance and neural processing in older adults. *Frontiers in Human Neuroscience, 5*, 26.

Voss, M. W., Nagamatsu, L. S., Liu-Ambrose, T., & Kramer, A. F. (2011). Exercise, brain, and cognition across the life span. *Journal of Applied Physiology, 111*(5), 1505–1513.

Wayne, P. M., Walsh, J. N., Taylor-Piliae, R. E., Wells, R. E., Papp, K. V., Donovan, N. J., & Yeh, G. Y. (2014). Effect of tai chi on cognitive performance in older adults: Systematic review and meta-analysis. *Journal of the American Geriatrics Society, 62*(1), 25–39.

13
COGNITIVE TRAINING TO PROMOTE EXECUTIVE FUNCTIONS

Matthias Kliegel, Alexandra Hering, Andreas Ihle, & Sascha Zuber

Introduction

At least three perspectives have been employed to motivate and frame the study of cognitive training that may promote executive functions across the life span: a developmental, an applied and a conceptual perspective. The first motivation for this line of research concerns a core question of developmental psychology itself and directly refers to prominent life-span theories, especially their notion of plasticity. According to life-span theories of development, plasticity is a central aspect of lifelong development because it encompasses the adaptive potential of an individual to develop (i.e., change and retain stability) through experience (Baltes, Lindenberger, & Staudinger, 2006; Willis & Schaie, 2009). Plasticity of cognitive functions has been examined in psychology mostly by employing systematic intervention or training approaches exploring the amount of plasticity in different domains of cognitive functioning and possible age-related differences in plasticity, executive functioning being one of the more recent domains targeted in this regard. The second perspective focuses on the intervention side of cognitive training and adopts a preventive or even clinical approach. Here, the aim is to prevent a specific population potentially at risk, such as older adults (e.g., Hertzog, Kramer, Wilson, & Lindenberger, 2009), from falling below a critical threshold of functioning in executive control (considering its general importance for a broad variety of important life outcomes, such as goal-directed behavior, socioemotional development, or physical health). In addition, in case (specific) executive dysfunctions have been identified as a core marker of a clinically relevant impairment, such as inhibitory problems in children with attention deficit hyperactivity disorder (ADHD), in this framework one aims at improving or even restoring the efficiency of this compromised neurocognitive process through process-based cognitive training (e.g., Klingberg et al., 2005). In both cases, the study goals are twofold: restoring the trained executive function (training effects) and aiming for generalized improvements of those training effects in other cognitive or life domains (transfer effects). Transfer effects can be differentiated in near and far transfer effects. Near transfer refers to improvements in untrained tasks of the same cognitive ability, whereas far transfer refers to improvements in untrained tasks of different cognitive abilities than the trained one (e.g., Morrison & Chein, 2011). Somewhat more recently, transfer effects have become a key conceptual target in a third perspective taken in the literature on executive function training across the life span. Here, cognitive training to promote executive functions in developmental populations are employed as experimental manipulations and used to test and further shape theories and models of cognitive and socioemotional development that propose (specific) executive functions as developmental mechanisms underlying the trajectories in higher-order cognitive processes such as reasoning or prospective memory (e.g., Kliegel, Altgassen, Hering, & Rose, 2011). Complementing traditional experimental approaches that have mostly manipulated cognitive control or specific executive functions via adding executive task load (and thereby reducing the efficiency of cognitive control) on top of the cognitive process of interest (e.g., Voigt et al., 2014), the conceptual training approach opts for the other direction and aims at increasing capacity in specific executive functions through training (mimicking developmental progress) and then examining (age-related) transfer effects in the target constructs testing for a causal impact.

All three perspectives are of course complementary approaches, but are partly discussed in distinct literature families – thus, the present chapter aims at bridging these traditions and systematically reviewing available evidence across the life span, discussing literature on children and young and older adults.

Childhood

Although the first cognitive training studies in children emerged in the 1980s, applying a training approach with the goal to enhance executive functions in children seems to be a relatively novel approach. In 2000, Dowsett and Livesey showed that healthy preschool children aged three to five were able to improve their inhibitory capacities by repeatedly working on a set of executive tasks. These findings were among the first to illustrate that even relatively short interventions (Dowsett & Livesey's study applied a total of approximately 60 minutes of training, spread over three sessions) may be able to enhance children's executive functions (see also Chevalier & Clark, Chapter 2, and Moriguchi, Chapter 5, this volume, on developmental aspects of executive functions during childhood). Following these promising results, a multitude of subsequent studies have demonstrated that cognitive training interventions can promote executive functioning in healthy children. Taken together, the literature has found to date that by repeatedly training on executively demanding tasks, children from a very young to an adolescent age can show (albeit sometimes small) improvements in working memory (e.g., Jaeggi, Buschkuehl, Jonides, & Perrig, 2008; Jaeggi et al., 2010; Karbach, Strobach, & Schubert, 2015; Klingberg, 2010; Thorell, Lindqvist, Nutley, Bohlin, & Klingberg, 2009), in task switching (e.g., Karbach & Kray, 2009; Kray, Eber, & Karbach, 2008; Zinke, Einert, Pfennig, & Kliegel, 2012), as well as in inhibitory control (e.g., Rueda, Rothbart, McCandliss, Saccomanno, & Posner, 2005; Thorell et al., 2009).

After establishing the possible trainability and plasticity of executive functions, many studies aimed at examining the broader benefit of cognitive training interventions, by investigating whether a transfer of the improvements could be seen in other, non-trained domains of more general importance, such as children's intelligence or their academic achievement. Jaeggi, Buschkuehl, Jonides, and Shah (2011), for example, found that after completing their cognitive training, children had not only increased their performance in a working memory task, but also showed improvements on a fluid intelligence test (for similar findings showing a positive effect of training on tests measuring intelligence, see, e.g., Ang, Lee, Cheam, Poon, & Koh, 2015; Karbach & Kray, 2009; Rueda et al., 2005). Furthermore, Jaeggi et al. observed a mid-term benefit of their intervention, as children maintained their improvements on a 3-month follow-up assessment.

Besides the possible benefits of cognitive training for healthy populations, as indicated above, many researchers were interested in using executively demanding interventions to help children with specific executive dysfunctions. As, for example, symptoms of ADHD are frequently associated with executive, especially inhibitory, deficits (see, e.g., Brown, 2013), different studies have looked into the efficiency of executive functioning training for those children. Klingberg, Forssberg, and Westerberg (2002), for example, applied working memory training with children diagnosed with ADHD and found that the intervention increased 7- to 15-year-olds' performance on a working memory and an intelligence test. Moreover, Klingberg and colleagues (2005) found that ADHD-diagnosed children improved working memory, inhibitory control as well as reasoning, and furthermore that participants' parents reported their children to demonstrate fewer symptoms of inattention following the intervention (for similar findings, see Kray, Karbach, Haenig, & Freitag, 2012).

Although this illustrates that training interventions can be efficient (in healthy as well as clinical samples) and that more general benefits than only task-related improvements are possible, a recent wave of reviews and meta-analyses that aggregated several cognitive training studies and analyzed the effects and possible influence factors, challenges the early optimism especially concerning possible (clinically relevant) transfer effects. This critical view is particularly motivated by the current emergence of commercially available, computer-based training applications, which promise "significant and lasting improvement in attention, impulse control, social functioning, academic performance, and complex reasoning skills for children with ADHD" (Rapport, Orban, Kofler, & Friedman, 2013, p. 1238). For example, Melby-Lervag and Hulme (2013) argued that instead of targeting the executive resources, some of the commercial interventions are more likely to work on the short-term memory component of a task, which is not considered to be one of the main deficits in ADHD. Furthermore, the authors stated that although participants increase their performance on different cognitive tasks, the training intervention does not seem to reflect on either children's behavior or their academic performance. This suggests that the conventional training approaches – although they may improve performance in some cognitive domains – may not be specific enough to decrease symptoms related to ADHD. Rapport and colleagues (2013), who came to the same conclusion, argued that "collectively, meta-analytic results indicate that claims regarding the academic, behavioral, and cognitive benefits associated with extant cognitive training programs are unsupported in ADHD" (p. 1249). Furthermore, Rapport and colleagues underlined that when assessing the efficiency of a training program, researchers should be cautious of the Hawthorne effect: usually, raters who know which participants did and did not undergo the training (*unblinded raters*) report larger changes in symptom-related behaviors for the training group than raters who are not aware of group affiliation (*blinded raters*), which indicates that many studies might overestimate training benefits.

Taken together, the current state of research seems to indicate that although in some cases executive function training can show long-term, non-trained, and real-life transfer (e.g., in healthy children), future studies will have to develop what the exact mechanisms behind these effects are, and how training approaches have to be modified to benefit specific populations (such as children with ADHD).

As an example of the third experimental approach to executive function training in children, we briefly present an ongoing research project from our own group. The conceptual target in this project is the development of prospective memory: the ability to remember to perform previously planned activities in the future such as remembering to take home a signed letter from school (Kliegel et al., 2013). Descriptively, prospective memory has been reported to increase during childhood (Kvavilashvili, Kyle, & Messer, 2008; Mahy & Moses, 2011) but little is known about the mechanisms that drive its development. One mechanism under debate is executive functioning, as individual differences in executive functions have been found to correlate with prospective memory performance in children (e.g., Ford, Driscoll, Shum, & Macaulay, 2012). Further, studies have documented a similar developmental trajectory of prospective memory and executive functions that show marked increases in performance during the early childhood years (Carlson & Moses, 2001; Mahy & Moses,

2011) and continue to relate to prospective memory performance in adulthood (Kliegel, Mackinlay, & Jäger, 2008; West & Craik, 1999). Yet, so far, all available evidence is purely correlational. Thus, in an ongoing study, we directly train the executive functions assumed to underlie prospective memory over 4 weeks using established training procedures for working memory, inhibition, and task switching and compare their relevance for prospective memory development. In a classical pre-post randomized controlled trial (RCT) design, we test for transfer effects on different markers of prospective memory functioning across childhood. Initial evidence suggests an important effect of especially inhibitory control and task-switching training for age-related prospective memory performance. It supports evidence from studies using structural equation modeling that suggest these executive processes to be particularly related to prospective memory development (Schnitzspahn, Stahl, Zeintl, Kaller, & Kliegel, 2013).

Young Adulthood

The majority of studies on executive functioning training has been conducted in young adults and trained mainly working memory. In general, these studies reported improvements for the trained tasks (typically n-back training tasks; see, e.g., Enriquez-Geppert, Huster, & Herrmann, 2013; Klingberg, 2010; Morrison & Chein, 2011; Strobach, Salminen, Karbach, & Schubert, 2014, for overviews); yet, as in children, results regarding possible transfer effects to untrained tasks are mixed. For example, studies using extensive training procedures (usually including 8 to 20 sessions) revealed transfer effects to untrained working memory tasks (near transfer) as well as to other tasks of cognitive domains (far transfer) such as fluid intelligence and task switching (e.g., Dahlin, Nyberg, Bäckman, & Neely, 2008; Jaeggi et al., 2008; Salminen, Strobach, & Schubert, 2012), while other studies did not find far transfer effects (see, e.g., Lilienthal, Tamez, Shelton, Myerson, & Hale, 2013; Morrison & Chein, 2011; Shipstead, Redick, & Engle, 2012). However, Au et al. (2015) recently conducted a meta-analysis on n-back trainings in young adults and found small but significant far transfer effects to fluid intelligence. Melby-Lervag and Hulme (2016) raised methodological issues on this meta-analysis (i.e., not accounting for baseline differences and failing to emphasize the difference between studies with treated vs. untreated control groups) and instead reported an absence of transfer effects in their own meta-analysis (see also Melby-Lervag & Hulme, 2013). In response to that, Au, Buschkuehl, Duncan, and Jaeggi (2016) meta-analytically demonstrated that the type of control group per se does not moderate observed transfer effects of working memory training on measures of fluid intelligence and thereby reaffirm the initial conclusions of Au et al. (2015; see also Bogg & Lasecki, 2015). This debate underlines the requirement for more systematic studies on potential moderators such as intensity, duration, or adaptivity of the training procedure (Enriquez-Geppert et al., 2013; Schwaighofer, Fischer,

& Buhner, 2015). Likewise, further research is needed to disentangle different possible mechanisms underlying transfer effects: for example, in the case of working memory training whether the training enhances working memory capacity (training-related increase of amount that can be held in working memory) and/or efficiency to better use the available working memory capacity (e.g., von Bastian & Oberauer, 2014).

While training of working memory has been extensively evaluated, experimental studies investigating the trainability of other executive functioning are relatively sparse (see, e.g., Enriquez-Geppert et al., 2013, for an overview). For inhibition, most studies focused on the Stroop effect (i.e., a reduced difference between congruent and incongruent trial response times; e.g., Davidson, Zacks, & Williams, 2003; Dotson, Sozda, Marsiske, & Perlstein, 2013) and showed a reduced Stroop effect with more pronounced practice-related gains on incongruent compared to congruent trials, which is consistent with the view that training enhances interference processing by improving the suppression of reading processes in the Stroop task (e.g., Dulaney & Rogers, 1994; MacLeod, 1998; see, e.g., Strobach et al., 2014, for an overview). Besides the Stroop task, there are a few studies that investigated trainability of inhibition with other tasks in young adults, such as the Stop-Signal training by Noel et al. (2016). The authors found that stop performance improved if stimuli were consistently associated with the need to stop (in contrast to stop stimuli without consistent association). Furthermore, training effects in inhibition were observed for Go/No-Go training (e.g., Manuel, Grivel, Bernasconi, Murray, & Spierer, 2010). Notably, while these studies showed that inhibition of irrelevant stimuli may be enhanced in the trained task, Strobach et al. (2014) concluded that there is no convincing evidence yet for near or far transfer. For example, in one of our own studies, Enge et al. (2014) trained inhibitory control using the Go/No-Go and Stop-Signal task, but did not find evidence for near transfer to an untrained Stroop task nor far transfer to measures of fluid intelligence.

For task switching, evidence showed that already relatively short task-switching training (e.g., six practice sessions) can produce a reduction of switch costs (e.g., Buchler, Hoyer, & Cerella, 2008; Cepeda, Kramer, & Gonzalez de Sather, 2001; Karbach & Kray, 2009; Kray & Eppinger, 2006; Kray & Lindenberger, 2000; Minear & Shah, 2008; Tayeb & Lavidor, 2016). In addition, there is evidence suggesting that task-switching training can lead to performance improvements in new (untrained) task-switching tasks and thus evoke near transfer (e.g., Karbach & Kray, 2009; Pereg, Shahar, & Meiran, 2013). There is also evidence of far transfer with enhanced performance in other tasks of cognitive functioning such as working memory, inhibition, and fluid intelligence (e.g., Karbach & Kray, 2009; see also, e.g., Enriquez-Geppert et al., 2013; Strobach et al., 2014, for overviews).

Thus, while the trainability of different executive functions seems indeed promising, far transfer may become evident only on specific conditions.

In general, it has been discussed that the prerequisites for transfer effects of executive functioning training may be an overlap of neural and cognitive processes involved in the trained task and in the transfer tasks (see, e.g., Enriquez-Geppert et al., 2013, for an overview). Yet, despite growing efforts to experimentally evaluate the trainability of executive functioning in younger adults during the last decade – particularly for inhibition and task switching – research is still in its infancy and detailed investigations are needed to help understand the mechanisms underlying the observed outcomes of different training procedures for executive functions (see, e.g., Enriquez-Geppert et al., 2013; Hsu, Novick, & Jaeggi, 2014; Strobach et al., 2014, for overviews). In this avenue, investigating training-related neural activation changes (e.g., Li et al., 2015) and intraindividual fluctuations during training (Könen & Karbach, 2015) may be fruitful methodological advances.

Late Adulthood

Based on the promising patterns of early studies suggesting the possibility of enhancing efficiency in executive functions, older adults have very quickly become an important target in this training literature because decline of executive functions is one of the first cognitive losses in late adulthood (see Li, Vadaga, Bruce, & Lai, Chapter 4, this volume; McFall, Sapkota, Thibeau, & Dixon, Chapter 17, this volume; West, Chapter 6, this volume). As with children or younger adults, training studies in older adults have focused on the full spectrum of executive functions such as working memory, inhibition, and task switching and examined both training and transfer effects.

Again, early studies in this literature have reported very promising results. For example, Borella, Carretti, Riboldi, and De Beni (2010) trained healthy older adults on working memory training for three sessions and showed that the training group clearly benefited from the verbal working memory training. Trained participants improved in working memory, fluid intelligence, short-term memory, and inhibition compared to the control group. Furthermore, the training gains and the improvement in the fluid intelligence measure were maintained even 8 months after the training. The results of this study are impressive and show the usefulness of cognitive training in promoting older adults' executive functioning. Considering the evidence from meta-analyses, the efficiency of cognitive training in older adults has received further support (but see Melby-Lervag & Hulme, 2016, for a contradictory view). Karbach and Verhaeghen (2014) compared 49 different studies on working memory and executive function training in older adults. They showed large improvements in performance of the trained tasks as well as large near transfer effects for tasks of the same cognitive domain as the trained task. Furthermore, Karbach and Verhaeghen (2014) found small but significant far transfer effects for task performance in cognitive domains other than the trained one (e.g., reasoning for working memory training). In older adults, similarly to research on clinical populations such as children with ADHD, transfer of working memory or executive function training to markers of everyday life functioning are considered to be especially crucial. Yet, so far, only a few studies have targeted this topic and found transfer to real-world tasks such as instrumental activities of daily living, everyday-like multi-tasking scenarios, reading comprehension, or self-reports on everyday functioning (e.g., Brehmer, Westerberg, & Backman, 2012; Carretti, Borella, Zavagnin, & de Beni, 2013; McDaniel et al., 2014; Rose et al., 2015).

To better understand which cognitive training promotes executive functions and is in fact beneficial for older adults also, gerontological training research has started to systematically examine moderators that influence the training efficacy. In late adulthood, age seems to play an especially important role. There is a differentiation between young-olds aged 60 to 75 and old-olds aged 75 and above. Whereas cognitive training benefits working memory and executive functions as reported in the meta-analyses of Karbach and Verhaeghen (2014), the picture is less clear in old-old adults. There exist only a few studies investigating this age group (e.g., Buschkuehl et al., 2008; Zinke, Zeintl, Eschen, Herzog, & Kliegel, 2012). For example, the two mentioned studies both used working memory training in old-old adults aged 80 years and above. Buschkuehl et al. (2008) found not only training gains but also near transfer effects. Only the training group improved in a not-trained working memory task. In a study from our own group, Zinke et al. (2012) investigated the impact of working memory training on executive functions. Here, we showed clear training improvements but did not find any transfer effects to executive functions. Furthermore, we differentiated the training group according to their working memory capacity at the beginning of the training into high performer vs. low performer. Separate analyses for each group revealed that the low-capacity training group benefited the most from the working memory training. In a follow-up working memory study from our group on young-olds and old-olds, Zinke et al. (2014) confirmed their previous finding and showed again that baseline performance, that is, performance before the training in the training tasks, impacts training gains. Specifically, we trained old adults in four different working memory tasks for nine sessions over three weeks and tested for transfer to executive functions such as inhibition and reasoning. Old adults with lower performance at the beginning improved more over the course of the training. Additionally, we showed that age and the actual training gains influenced the transfer to executive functions. Transfer effects decreased with increasing age but participants with higher training gains also showed higher transfer effects.

Although these studies considered already important influencing factors for cognitive training, there are still lots of open questions on other mediating or moderating factors such as motivation, lifestyle differences, and leisure activities. In the same context of discovering mechanisms of training and transfer effects in old age, more and more studies on neural changes due to cognitive interventions have emerged. Studies using neuroscientific methods such as neuroimaging techniques (fMRI) or electroencephalography (EEG) to evaluate the influence of training gains on the brain reported in general more efficient neural processing. For example, Heinzel et al. (2014) found a more "youth-like" brain response in older adults after their working memory training. Gajewski and Falkenstein (2012) used cognitive training to promote task-switching performance. As neural indicators they assessed specific event-related potentials for maintaining and coordinating multiple tasks (e.g., N2, P3b) and could show that after the cognitive training these indicators were enhanced, indicating better response selection.

Overall, there is a significant potential in cognitive training to promote older adults' executive functioning. Behavioral training studies show training gains and some transfer effects in young-olds and old-olds. However, there are also important factors that critically influence the efficacy of cognitive training, such as age of participants, with increasing age reducing the training gains and transfer effects.

Conclusions

Taken together, across all age groups and across healthy and clinical populations, first results on cognitive training to promote executive functions across the life span, indeed, appeared promising as they (i) revealed plasticity in trained tasks from childhood into old age. Moreover, (ii) some of those interventions also revealed transfer to non-trained tasks in the laboratory. More recently, however, the initial enthusiasm has been challenged by several studies not or only marginally revealing training effects for some domains, failing to generally find far transfer at all, or suggesting partly limited benefits for non-trained tasks. Moreover, evidence on possible effects in everyday life is virtually non-existent. Thus, to date, three core conclusions seem to be justified: (a) executive functions can be trained; (b) training and even more so transfer effects are critically moderated; and (c) those moderators are largely unknown. In consequence, further systematic research is urgently needed to test possible moderators of training and transfer effects in cognitive functioning across the life span – given the heterogeneity of study protocols and populations included, a coordinated effort across research groups and traditions seems more than warranted.

References

Ang, S. Y., Lee, K., Cheam, F., Poon, K., & Koh, J. (2015). Updating and working memory training: Immediate improvement, long-term maintenance, and generalisability to non-trained tasks. *Journal of Applied Research in Memory and Cognition, 4*(2), 121–128.

Au, J., Buschkuehl, M., Duncan, G. J., & Jaeggi, S. M. (2016). There is no convincing evidence that working memory training is NOT effective: A reply to Melby-Lervag and Hulme (2015). *Psychonomic Bulletin & Review, 23*(1), 331–337.

Au, J., Sheehan, E., Tsai, N., Duncan, G. J., Buschkuehl, M., & Jaeggi, S. M. (2015). Improving fluid intelligence with training on working memory: A meta-analysis. *Psychonomic Bulletin & Review, 22*(2), 366–377.

Baltes, P. B., Lindenberger, U., & Staudinger, U. M. (2006). Lifespan theory in developmental psychology. In W. Damon & R. M. Lerner (Eds.), *Handbook of child psychology: Vol. 1. Theoretical models of human development* (6th ed., pp. 569–664). New York: Wiley.

Bogg, T., & Lasecki, L. (2015). Reliable gains? Evidence for substantially underpowered designs in studies of working memory training transfer to fluid intelligence. *Frontiers in Psychology, 5*(1589).

Borella, E., Carretti, B., Riboldi, F., & De Beni, R. (2010). Working memory training in older adults: Evidence of transfer and maintenance effects. *Psychology and Aging, 25*(4), 767–778.

Brehmer, Y., Westerberg, H., & Backman, L. (2012). Working-memory training in younger and older adults: Training gains, transfer, and maintenance. *Frontiers in Human Neuroscience, 6*(63).

Brown, E. T. (2013). *A new understanding of ADHD in children and adults: Executive function impairments*. New York: Routledge.

Buchler, N. G., Hoyer, W. J., & Cerella, J. (2008). Rules and more rules: The effects of multiple tasks, extensive training, and aging on task-switching performance. *Memory & Cognition, 36*(4), 735–748.

Buschkuehl, M., Jaeggi, S. M., Hutchison, S., Perrig-Chiello, P., Däpp, C., Müller, M., et al. (2008). Impact of working memory training on memory performance in old-old adults. *Psychology and Aging, 23*(4), 743–753.

Carlson, S. M., & Moses, L. J. (2001). Individual differences in inhibitory control and children's theory of mind. *Child Development, 72*, 1032–1053.

Carretti, B., Borella, E., Zavagnin, M., & de Beni, R. (2013). Gains in language comprehension relating to working memory training in healthy older adults. *International Journal of Geriatric Psychiatry, 28*(5), 539–546.

Cepeda, N. J., Kramer, A. F., & Gonzalez de Sather, J. C. M. (2001). Changes in executive control across the life span: Examination of task-switching performance. *Developmental Psychology, 37*(5), 715–730.

Dahlin, E., Nyberg, L., Bäckman, L., & Neely, A. S. (2008). Plasticity of executive functioning in young and older adults: Immediate training gains, transfer, and long-term maintenance. *Psychology and Aging, 23*(4), 720–730.

Davidson, D. J., Zacks, R. T., & Williams, C. C. (2003). Stroop interference, practice, and aging. *Aging, Neuropsychology, and Cognition, 10*(2), 85–98.

Dotson, V. M., Sozda, C. N., Marsiske, M., & Perlstein, W. M. (2013). Within-session practice eliminates age differences in cognitive control. *Aging, Neuropsychology, and Cognition, 20*(5), 522–531.

Dowsett, S. M., & Livesey, D. J. (2000). The development of inhibitory control in preschool children: Effects of "executive skills" training. *Developmental Psychobiology, 36*(2), 161–174.

Dulaney, C. L., & Rogers, W. A. (1994). Mechanisms underlying reduction in Stroop interference with practice for young and old adults. *Journal of Experimental Psychology: Learning, Memory, and Cognition, 20*(2), 470–484.

Enge, S., Behnke, A., Fleischhauer, M., Kuttler, L., Kliegel, M., & Strobel, A. (2014). No evidence for true training and transfer effects after inhibitory control training in young healthy adults. *Journal of Experimental Psychology: Learning, Memory, and Cognition, 40*(4), 987–1001.

Enriquez-Geppert, S., Huster, R. J., & Herrmann, C. S. (2013). Boosting brain functions: Improving executive functions with behavioral training, neurostimulation, and neurofeedback. *International Journal of Psychophysiology, 88*(1), 1–16.

Ford, R. M., Driscoll, T., Shum, D., & Macaulay, C. E. (2012). Executive and theory-of-mind contributions to event-based prospective memory in children: Exploring the self-projection hypothesis. *Journal of Experimental Child Psychology, 111*, 468–489.

Gajewski, P. D., & Falkenstein, M. (2012). Training-induced improvement of response selection and error detection in aging assessed by task switching: Effects of cognitive, physical, and relaxation training. *Frontiers in Human Neuroscience, 6*(130).

Heinzel, S., Lorenz, R. C., Brockhaus, W. R., Wustenberg, T., Kathmann, N., Heinz, A., & Rapp, M. A. (2014). Working memory load-dependent brain response predicts behavioral training gains in older adults. *Journal of Neuroscience, 34*(4), 1224–1233.

Hertzog, C., Kramer, A. F., Wilson, R. S., & Lindenberger, U. (2009). Enrichment effects on adult cognitive development. *Psychological Science in the Public Interest, 9*, 1–65.

Hsu, N. S., Novick, J. M., & Jaeggi, S. M. (2014). The development and reliability of executive control abilities. *Frontiers in Behavioral Neuroscience, 8*(221).

Jaeggi, S. M., Buschkuehl, M., Jonides, J., & Perrig, W. J. (2008). Improving fluid intelligence with training on working memory. *Proceedings of the National Academy of Sciences of the United States of America, 105*(19), 6829–6833.

Jaeggi, S. M., Buschkuehl, M., Jonides, J., & Shah, P. (2011). Short- and long-term benefits of cognitive training. *Proceedings of the National Academy of Sciences of the United States of America, 108*(25), 10081–10086.

Jaeggi, S. M., Studer-Luethi, B., Buschkuehl, M., Su, Y. F., Jonides, J., & Perrig, W. J. (2010). The relationship between n-back performance and matrix reasoning: Implications for training and transfer. *Intelligence, 38*(6), 625–635.

Karbach, J., & Kray, J. (2009). How useful is executive control training? Age differences in near and far transfer of task-switching training. *Developmental Science, 12*(6), 978–990.

Karbach, J., Strobach, T., & Schubert, T. (2015). Adaptive working-memory training benefits reading, but not mathematics in middle childhood. *Child Neuropsychology, 21*(3), 285–301.

Karbach, J., & Verhaeghen, P. (2014). Making working memory work: A meta-analysis of executive-control and working memory training in older adults. *Psychological Science, 25*(11), 2027–2037.

Kliegel, M., Altgassen, M., Hering, A., & Rose, N. (2011). A process-model based approach to prospective memory impairment in Parkinson's disease. *Neuropsychologia, 49*, 2166–2177.

Kliegel, M., Mackinlay, R., & Jäger, T. (2008). Complex prospective memory: Development across the lifespan and the role of task interruption. *Developmental Psychology, 44*(2), 612–617.

Kliegel, M., Mahy, C. E. V., Voigt, B., Henry, J. D., Rendell, P. G., & Aberle, I. (2013). The development of prospective memory in young schoolchildren: The impact of ongoing task absorption, cue salience, and cue centrality. *The Journal of Experimental Child Psychology, 116*, 792–810.

Klingberg, T. (2010). Training and plasticity of working memory. *Trends in Cognitive Sciences, 14*(7), 317–324.

Klingberg, T., Fernell, E., Olesen, P. J., Johnson, M., Gustafsson, P., Dahlstrom, K., et al. (2005). Computerized training of working memory in children with ADHD:

A randomized, controlled trial. *Journal of the American Academy of Child and Adolescent Psychiatry, 44*(2), 177–186.

Klingberg, T., Forssberg, H., & Westerberg, H. (2002). Training of working memory in children with ADHD. *Journal of Clinical and Experimental Neuropsychology, 24*(6), 781–791.

Könen, T., & Karbach, J. (2015). The benefits of looking at intraindividual dynamics in cognitive training data. *Frontiers in Psychology, 6*(615).

Kray, J., Eber, J., & Karbach, J. (2008). Verbal self-instructions in task switching: A compensatory tool for action-control deficits in childhood and old age? *Developmental Science, 11*(2), 223–236.

Kray, J., & Eppinger, B. (2006). Effects of associative learning on age differences in task-set switching. *Acta Psychologica, 123*(3), 187–203.

Kray, J., Karbach, J., Haenig, S., & Freitag, C. (2012). Can task-switching training enhance executive control functioning in children with attention deficit/-hyperactivity disorder? *Frontiers in Human Neuroscience, 5*(180).

Kray, J., & Lindenberger, U. (2000). Adult age differences in task switching. *Psychology and Aging, 15*(1), 126–147.

Kvavilashvili, L., Kyle, F., & Messer, D. J. (2008). The development of prospective memory in children: Methodological issues, empirical findings and future directions. In M. Kliegel, M. A. McDaniel, & G. O. Einstein (Eds.), *Prospective memory: Cognitive, neuroscience, developmental, and applied perspectives* (pp. 115–140). Mahwah, NJ: Erlbaum.

Li, X., Xiao, Y.-H., Zhao, Q., Leung, A. W. W., Cheung, E. F. C., & Chan, R. C. K. (2015). The neuroplastic effect of working memory training in healthy volunteers and patients with schizophrenia: Implications for cognitive rehabilitation. *Neuropsychologia, 75*, 149–162.

Lilienthal, L., Tamez, E., Shelton, J. T., Myerson, J., & Hale, S. (2013). Dual n-back training increases the capacity of the focus of attention. *Psychonomic Bulletin & Review, 20*(1), 135–141.

MacLeod, C. M. (1998). Training on integrated versus separated Stroop tasks: The progression of interference and facilitation. *Memory & Cognition, 26*(2), 201–211.

Mahy, C. E. V., & Moses, L. J. (2011). Executive functioning and prospective memory in young children. *Cognitive Development, 26*, 269–281.

Manuel, A. L., Grivel, J., Bernasconi, F., Murray, M. M., & Spierer, L. (2010). Brain dynamics underlying training-induced improvement in suppressing inappropriate action. *Journal of Neuroscience, 30*(41), 13670–13678.

McDaniel, M. A., Binder, E. F., Bugg, J. M., Waldum, E. R., Dufault, C., Meyer, A., et al. (2014). Effects of cognitive training with and without aerobic exercise on cognitively demanding everyday activities. *Psychology and Aging, 29*(3), 717–730.

Melby-Lervag, M., & Hulme, C. (2013). Is working memory training effective? A meta-analytic review. *Developmental Psychology, 49*(2), 270–291.

Melby-Lervag, M., & Hulme, C. (2016). There is no convincing evidence that working memory training is effective: A reply to Au et al. (2014) and Karbach and Verhaeghen (2014). *Psychonomic Bulletin & Review, 23*(1), 324–330.

Minear, M., & Shah, P. (2008). Training and transfer effects in task switching. *Memory & Cognition, 36*(8), 1470–1483.

Morrison, A. B., & Chein, J. M. (2011). Does working memory training work? The promise and challenges of enhancing cognition by training working memory. *Psychonomic Bulletin & Review, 18*(1), 46–60.

Noel, X., Brevers, D., Hanak, C., Kornreich, C., Verbanck, P., & Verbruggen, F. (2016). On the automaticity of response inhibition in individuals with alcoholism. *Journal of Behavior Therapy and Experimental Psychiatry, 51*, 84–91.

Pereg, M., Shahar, N., & Meiran, N. (2013). Task switching training effects are mediated by working-memory management. *Intelligence, 41*(5), 467–478.

Rapport, M. D., Orban, S. A., Kofler, M. J., & Friedman, L. M. (2013). Do programs designed to train working memory, other executive functions, and attention benefit children with ADHD? A meta-analytic review of cognitive, academic, and behavioral outcomes. *Clinical Psychology Review, 33*(8), 1237–1252.

Rose, N. S., Rendell, P. G., Hering, A., Kliegel, M., Bidelman, G., & Craik, F. I. M. (2015). Cognitive and neural plasticity in older adults' prospective memory following training with the Virtual Week Computer Game. *Frontiers in Human Neuroscience, 9*(592).

Rueda, M. R., Rothbart, M. K., McCandliss, B. D., Saccomanno, L., & Posner, M. I. (2005). Training, maturation, and genetic influences on the development of executive attention. *Proceedings of the National Academy of Sciences of the United States of America, 102*(41), 14931–14936.

Salminen, T., Strobach, T., & Schubert, T. (2012). On the impacts of working memory training on executive functioning. *Frontiers in Human Neuroscience, 6*(166).

Schnitzspahn, K. M., Stahl, C., Zeintl, M., Kaller, C. P., & Kliegel, M. (2013). The role of shifting, updating, and inhibition in prospective memory performance in young and older adults. *Developmental Psychology, 49*, 1544–1553.

Schwaighofer, M., Fischer, F., & Buhner, M. (2015). Does working memory training transfer? A meta-analysis including training conditions as moderators. *Educational Psychologist, 50*(2), 138–166.

Shipstead, Z., Redick, T. S., & Engle, R. W. (2012). Is working memory training effective? *Psychological Bulletin, 138*(4), 628–654.

Strobach, T., Salminen, T., Karbach, J., & Schubert, T. (2014). Practice-related optimization and transfer of executive functions: A general review and a specific realization of their mechanisms in dual tasks. *Psychological Research, 78*(6), 836–851.

Tayeb, Y., & Lavidor, M. (2016). Enhancing switching abilities: Improving practice effect by stimulating the dorsolateral pre frontal cortex. *Neuroscience, 313*, 92–98.

Thorell, L. B., Lindqvist, S., Nutley, S. B., Bohlin, G., & Klingberg, T. (2009). Training and transfer effects of executive functions in preschool children. *Developmental Science, 12*(1), 106–113.

Voigt, B., Mahy, C. E. V., Ellis, J., Schnitzspahn, K., Krause, I., Altgassen, M., & Kliegel, M. (2014). The development of time-based prospective memory in childhood: The role of working memory updating. *Developmental Psychology, 50*, 2393–2404.

von Bastian, C. C., & Oberauer, K. (2014). Effects and mechanisms of working memory training: A review. *Psychological Research, 78*(6), 803–820.

West, R., & Craik, F. I. M. (1999). Age-related decline in prospective memory: The roles of cue accessibility and cue sensitivity. *Psychology & Aging, 14*, 264–272.

Willis, S. L., & Schaie, K. W. (2009). Cognitive training and plasticity: Theoretical perspective and methodological consequences. *Restorative Neurology and Neuroscience, 27*(5), 375–389.

Zinke, K., Einert, M., Pfennig, L., & Kliegel, M. (2012). Plasticity of executive control through task switching training in adolescents. *Frontiers in Human Neuroscience, 6*(41).

Zinke, K., Zeintl, M., Eschen, A., Herzog, C., & Kliegel, M. (2012). Potentials and limits of plasticity induced by working memory training in old-old age. *Gerontology, 58*(1), 79–87.

Zinke, K., Zeintl, M., Rose, N. S., Putzmann, J., Pydde, A., & Kliegel, M. (2014). Working memory training and transfer in older adults: effects of age, baseline performance, and training gains. *Developmental Psychology, 50*(1), 304–315.

Part IV
Atypical Patterns of Executive Function Development Across the Life Span

14

EXECUTIVE DYSFUNCTION IN VERY PRETERM CHILDREN AND ASSOCIATED BRAIN PATHOLOGY

Elisha K. Josev & Peter J. Anderson

Very Preterm Birth and Neurodevelopmental Impairment

Very preterm birth (< 32 weeks' completed gestation) represents a biological immaturity for extrauterine life (Behrmann & Butler, 2007). Compared with term birth, it is associated with increased rates of mortality and major long-term morbidity such as brain injury and cognitive disability. Advances in medical technology, obstetric management, and intensive neonatal care have led to an improvement in survival rates of preterm infants over the past 3 decades (Saigal & Doyle, 2008). Nevertheless, the rate and severity of longer-term cognitive and neurological problems remains high. Indeed, it has been estimated that more than 70% of extremely preterm children (< 28 weeks' completed gestation) have at least one neurodevelopmental impairment, and 50% exhibit multiple areas of concern (Saigal & Doyle, 2008).

Cognitive impairment is the most common neurodevelopmental disability identified among preterm survivors (Wood, Marlow, Costeloe, Gibson, & Wilkinson, 2000; Vohr et al., 2004; Vohr et al., 2000). Children born very preterm show lower scores on global measures of IQ, and higher rates of developmental delay, learning difficulty, grade repetition, remedial schooling, and intellectual disability, compared with term-born peers (Bhutta, Cleves, Casey, Cradock, & Anand, 2002; Roberts, Anderson, & Doyle, 2009; Saigal et al., 2003; Saigal, Hoult, Streiner, Stoskopf, & Rosenbaum, 2000; Vohr et al., 2000). In particular, cognitive deficits in the area of executive functioning are common in very preterm cohorts, which may at least partly explain concerns with academic achievement and social-behavioural functioning (Burnett, Scratch, & Anderson, 2013). Longitudinal studies of extremely preterm cohorts suggest that these children show slower skill development and little 'catch-up' growth over time, and persistence of executive dysfunction and other cognitive impairment into adolescence and adulthood has been reported (Nosarti et al., 2007; Rose, Feldman, Jankowski, & Van Rossem, 2005; Saigal et al., 2003; Taylor, Minich, Klein, & Hack, 2004).

Very preterm children with neurological insults and comorbidities tend to be most at risk for serious cognitive impairment, like executive dysfunction. This is because preterm birth is associated with both direct and indirect brain injury. Indeed, brain development is particularly rapid and complex during the 20–40-week gestational period, and the neuropathology seen in very preterm infants is due to multiple neurobiological processes that are occurring during this period (Volpe, 2009). The main forms of direct brain injury in the preterm infant include diffuse noncystic white matter abnormalities (i.e., delayed myelination characterised by microgliosis and astrogliosis, which can appear as reduced white matter volume, enlarged lateral ventricles, thinning of the corpus callosum, and immature gyral formation on magnetic resonance imaging), periventricular leukomalacia (i.e., focal areas of necrosis and a loss of all cellular elements, leading to softening of the periventricular white matter), intraventricular haemorrhage (i.e., bleeding at the immature and thin-walled germinal matrix ventrolateral to the lateral ventricles, causing toxicity, inflammation, and oxidative stress), and axonal and neuronal disease (Dobbing, 1974; Volpe, 2009).

These brain pathologies are thought to be related to the preterm infant's immature cerebral circulation system that is often unable to respond to fluctuations in blood flow. Term infants show greater maturity in their vascular systems at birth, and as such, are capable of autoregulation of their cerebral blood flow despite fluctuations in blood pressure. In contrast, the preterm vascular system is 'pressure-passive' in that cerebral blood flow can fall in conjunction with small drops in blood pressure (Volpe, 1997), putting the preterm infant at further risk of haemorrhage or hypoxic-ischemic injury. Indeed, haemorrhagic infarction within the parenchyma of the periventricular white matter represents a significant complication of intraventricular haemorrhage, which has disastrous effects on cerebral perfusion and cerebral venous pressure during the perinatal period (Wyatt, 2010). Hypoxic-ischaemic and inflammatory pathways are proposed as two main aetiological mechanisms for direct brain injury in preterm infants. In the former, multiple and possibly co-occurring processes are thought to be involved: namely, *hypoxia* in cerebral tissues, which leads to reduced protein synthesis and neuronal death; *anoxia*, which triggers excessive release of excitatory neurotransmitters and free radicals leading to cell damage; and *apoptosis*, which may be exacerbated by and further influence ischaemic effects leading to tissue necrosis (Arpino et al., 2005). In the latter, brain injury and preterm birth itself may be the consequence of direct microbial infection of the foetal membrane and indirect inflammatory responses of the mother or foetus (Arpino et al., 2005).

Indirect brain injury may also contribute to this picture, with primary injury (i.e., acute vascular events such as intraventricular haemorrhage, ischaemic episodes, and inflammatory reactions) causing secondary damage or disruption to the development and integrity of other brain regions. For instance, intraventricular haemorrhage can spread into the ventricular system, disrupting neuroblast formation and future glial cell proliferation, which may result in white matter damage, or loss of white and grey matter tissue. Altered neurodevelopmental trajectories may also be a product of delayed maturational processes, most commonly in those born at earlier gestational age, beyond that of the direct brain injury associated with preterm birth (Bhutta & Anand, 2002). For example, researchers undertaking allometric analysis of human brain development have demonstrated that greater prematurity is associated with a greater amount of disruption to the normal growth pattern of cortical surface area relative to cerebral brain volume from birth to term-equivalent age, independent of whether preterm infants possess major destructive brain lesions (Kapellou et al., 2006). The amount of growth disruption was also found to directly predict measures of neurodevelopmental delay at age 2 years. Cerebral white matter injury may also lead to downstream effects such as cortical grey and white matter reduction in preterm infants at term-equivalent age, secondary to direct brain injury (Andiman et al., 2010; Huppi et al., 2001; Inder, Warfield, Wang, Huppi, & Volpe, 2005).

The following sections of this chapter will describe a conceptual framework for understanding executive functioning in the developing preterm infant, the types of executive dysfunction experienced by very preterm individuals across the life span, and neurological predictors of this impairment.

A Model for Executive Functioning in the Developing Preterm Infant

Executive functioning is a broad term that refers to a range of higher-order cognitive processes that guide goal-directed behaviour. These processes are still in development in childhood and adolescence. As explained in Cuevas, Rajan, and Bryant (Chapter 1, this volume), many models exist that characterise executive functioning as either a single construct (Duncan, Johnson, Swales, & Freer, 1997) or separable components (Miyake et al., 2000; Stuss & Alexander, 2000). Here, we divide executive functioning into discrete but interrelated domains according to the conceptual framework of the Executive Control System (Anderson, 2002; Anderson, Anderson, Northam, Jacobs, & Catroppa, 2001; Lezak, 1995). These domains are attentional control (i.e., selective attention, self-regulation, self-monitoring, inhibition), information processing (i.e., speed of processing, efficiency, fluency), cognitive flexibility (i.e., divided attention, working memory, conceptual transfer), and goal setting (i.e., planning, conceptual reasoning, initiative), as shown in Figure 14.1. *Attentional control* involves the ability to regulate and monitor attention to stimuli, either selectively or across divided attentional resources. *Information processing* refers to the efficiency,

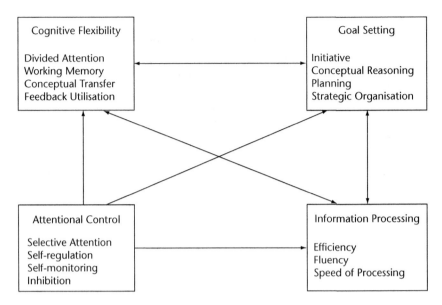

FIGURE 14.1 The four main interrelated domains of higher-order cognitive processes that form executive function in childhood and adolescence, as proposed by Anderson (2002) (adapted from Anderson, 2002, p. 71).

fluency, and speed with which information is processed, which can directly and indirectly affect goal-directed behaviour. *Cognitive flexibility* represents the ability to divide attention, as well as maintain and process multiple sources of information concurrently within one's working memory. This domain also includes the ability to learn from errors, formulate alternative strategies, and shift between response sets (Anderson, 2002). This requires adequate adaptive behaviour to accommodate changing demands in one's environment. *Goal setting* involves anticipating, planning, organising, and formulating actions until an end goal is achieved.

The Executive Control System framework shares some features of other models of executive functioning. For example, the Miyake et al. (2000) model proposes three executive factors, namely updating, shifting, and inhibition, which overlap with the attentional control and cognitive flexibility domains of the Executive Control System. It should be noted that many recent theories of executive function, including that of Miyake and colleagues, assume that while specific executive abilities are separable and distinguishable from one another, they are not completely independent and share some overlapping processes.

While deficits within a specific executive domain are more common than global executive dysfunction (Grattan & Eslinger, 1991), both types of difficulties can influence a child's academic achievement, as well as impact on social and emotional behaviour. In novel situations, the executive ability to formulate, apply, monitor, and evaluate new strategies is also dependent on the functioning of lower-order memory, language, perceptual, and visuospatial processes. Therefore, when executive functions are measured via neuropsychological tests, the end point score is not only a reflection of a child's capacity within that specific executive domain; it is also an indication of how well they can integrate isolated, lower-level non-executive skills to efficiently achieve a goal (Anderson, Anderson, Jacobs, & Spencer-Smith, 2008).

Executive processes emerge during infancy (see Cuevas et al., Chapter 1, this volume) and continue to develop during childhood (see Chevalier & Clark, Chapter 2, this volume) and well into late adolescence and early adulthood (Anderson, 1998; Crone, Peters, & Steinbeis, Chapter 3, this volume). Certain executive skills come 'online' and mature earlier than others. This is thought to reflect the underlying stage-like maturation of neural and white matter networks in the developing brain (Anderson, 2002; Welsh, Pennington, & Groisser, 1991). It is well established that the prefrontal lobes of the brain are important for executive functioning due to numerous lesion studies in brain-damaged populations (Benton, 1991; Luria, 1973; Stuss & Alexander, 2000) and the observation that the prefrontal region is generally the last to become myelinated and mature, as explained in Chevalier and Clark (Chapter 2, this volume) (i.e., the posterior to anterior gradient of brain growth; Gogtay et al., 2004). The protracted development of frontal regions is thought to put children at particular risk of executive deficits (Klingberg, Vaidya, Gabrieli, Moseley, & Hedehus, 1999). However, with the aid of advanced MRI techniques, it is now recognised that executive dysfunction in children is more likely to be the result of altered development, damage, and de-activation of multiple brain areas, and in particular, *disconnection* between these areas. As such, the prevalent injuries and abnormalities in white matter that are noted in very preterm children are thought to be the most likely cause of executive deficits in this population (Boardman & Dyet, 2007). The following sections provide a review of research that has examined executive functioning in very preterm children.

Attentional Control

Very preterm children who lack *attentional control* have difficulty focusing and sustaining their attention. Compared with term-born controls, studies have reported deficits in very preterm children's ability to selectively attend to stimuli while ignoring distracting stimuli, and to maintain their concentration and performance during long, monotonous tasks (Bohm, Smedler, & Forssberg, 2004; Elgen, Lundervold, & Sommerfelt, 2004; Mulder, Pitchford, Hagger, & Marlow, 2009; Murray et al., 2014; Taylor, Minich, Klein et al., 2004; Vicari, Caravale, Carlesimo, Casadei, & Allemand, 2004). For example, 8-year-old extremely preterm children have been reported to be 2.4 times more likely to have these attentional deficits than term controls (Anderson et al., 2011). While attention deficits have been reported to persist in childhood and into adolescence (Katz et al., 1996; Taylor, Minich, Klein et al., 2004; Vicari et al., 2004; Wilson-Ching et al., 2013), other studies have observed no impairments (Taylor, Hack, & Klein, 1998), or have found that group differences diminish after controlling for IQ or parental factors (Bohm et al., 2004; Elgen et al., 2004). Yet other literature demonstrates that lower birth weight and neonatal medical complications (i.e., bronchopulmonary dysplasia) confer greater risk of attentional impairment, and that deficits may persist even after adjustment for neurosensory and overall intellectual impairment, and sociodemographic factors (Breslau, Chilcoat, DelDotto, Andreski, & Brown, 1996; Taylor, Minich, Bangert, Filipek, & Hack, 2004; Taylor, Klein, Minich, & Hack, 2000). Variation in the selection of assessment tools used, characteristics of the sample, and definitions of severity of impairment may account for these inconsistent results.

Self-monitoring and inhibition are other elements of attentional control, which include the capacity to monitor one's own behaviour in response to environmental cues (i.e., to detect errors and correct them accordingly), and inhibit previously learned responses when required (i.e., exercising impulse control). Self-control deficits are commonly reported by parents and teachers of very preterm children, who describe impulsivity, distractibility, carelessness, difficulty following instructions, or difficulty remaining seated for long periods of time (Anderson, Doyle, & Victorian Infant Collaborative Study Research Group, 2004; Howard, Anderson, & Taylor, 2008). Research has documented more impulsive responses and poorer inhibitory control in very preterm children compared with term-born peers, with deficits persisting after adjusting for IQ and language skills (Bohm et al., 2004; Harvey, O'Callaghan, & Mohay, 1999). While more longitudinal studies are needed, there is some suggestion that inhibitory control deficits may recede with age, with some studies showing no group differences in impulsivity and vigilance between preterm and term cohorts in early and later adolescence (Elgen et al., 2004; Kulseng et al., 2006). However, it is possible that the tools used to assess these areas in older children are insensitive to mild impairments and show ceiling effects, which may explain a lack of differentiation between these groups.

Brain abnormalities on neonatal MRI have been found to predict attention in very preterm 7-year-olds (Murray et al., 2014). More specifically, increased severity of deep grey matter abnormalities (i.e., basal ganglia and thalamus) was predictive of selective attention impairments in this cohort of children, while white matter abnormalities and cerebellar abnormalities were predictive of sustained attentional deficits. Later in development, it appears that very preterm individuals may employ alternative strategies and different brain regions for response inhibition compared with their term-born peers. Nosarti and colleagues (2006) found that very preterm adolescents demonstrated reduced BOLD signal responses on a functional MRI response inhibition task in the bilateral cerebellum, right caudate nucleus, thalamus and prefrontal regions, and increased responses in temporal regions, which differed from healthy controls who typically show an 'inhibitory' cognitive network in predominantly left orbital, medial and inferior frontal and parietal cortices, and right basal ganglia, cerebellum, and temporo-occipital cortices (Garavan, Ross, Murphy, Roche, & Stein, 2002; Watanabe et al., 2002). This difference in activation and inactivation of networks occurred despite very preterm adolescents showing similar task performance to controls. The authors hypothesised that these differences in frontal–striatal–cerebellar circuitry may account for the engagement of alternative neural pathways for response inhibition processes (Nosarti et al., 2006), which may have occurred as a result of prematurity-related early brain damage that led to plasticity and reorganisation of this circuitry.

Information Processing

Very preterm children are often characterised by slower reaction times, longer time required to interpret and respond to stimuli, and reduced output (Anderson, 2014). These deficits may be present in the 1st year of life, and continue through school-age into adulthood (de Kieviet, van Elburg, Lafeber, & Oosterlaan, 2012; Nosarti et al., 2007; Rose, Feldman, & Jankowski, 2002; Rose & Feldman, 1996; Strang-Karlsson et al., 2010). Compared with full-term controls, very preterm infants may require 20% more trials and up to 30% more inspection time to identify a novel stimulus (Rose et al., 2002), while very preterm school-age children may require significantly greater decision time on choice reaction tasks with increasing task complexity (Rose & Feldman, 1996). Similar results have been noted in older very preterm cohorts. Finnish very preterm adults aged 18–27 without neurosensory impairments displayed significantly slower psychomotor speed and lower accuracy on multiple computerised tasks of reaction time, divided attention, learning capacity, and working memory (Strang-Karlsson et al., 2010). These results suggest that very preterm individuals have difficulty with sustaining efficient processing of information, especially for more complicated tasks, which can impact on goal-directed behaviour.

Given that diffuse white matter loss is a common consequence of preterm birth, reduced neural transmission and whole-brain connectivity is a plausible mechanism by which reduced output and slowed information processing may occur in the preterm population. Severity of brain injury of neonatal MRI, particularly abnormalities within white matter and deep grey matter, has been shown to predict processing speed deficits in 7-year-old very preterm children (Murray et al., 2014). In very preterm adolescents, processing speed deficits have been shown to correlate with reductions in white matter volumes but not grey matter volume (Soria-Pastor et al., 2008).

Cognitive Flexibility

The ability to shift between response sets, and the fluency with which this is achieved, typically emerges in early childhood and develops throughout middle childhood (Anderson, 2002; Anderson et al., 2001). However, the rate at which this ability develops across middle childhood has been shown to be slower in children born <750 g birth weight compared to children born between 750 g and 1499 g or full-term controls (Taylor, Minich, Klein et al., 2004; Taylor et al., 2000). Further, impairment in the ability to shift from or divide one's attention between different activities has been found to occur 3 times as often in extremely preterm children than term controls, with extremely preterm children performing between 0.3 and 0.6 SD below their term peers (Anderson et al., 2011). There is a suggestion that these deficits may be mediated by prematurity-related brain injury; for example very preterm children with low-grade intraventricular haemorrhage have been found to perform less efficiently on tasks involving shifting response set compared with very preterm children without intraventricular haemorrhage or term-born controls (Ross, Boatright, Auld, & Nass, 1996). IQ and maternal education may also partly explain and contribute to the prediction of the deficits in mental flexibility and problem-solving ability observed in preterm children compared with term controls (Taylor, Klein, Schatschneider, & Hack, 1998; Tideman, 2000).

Few studies have investigated the neurological basis of higher-order divided and shifting attentional abilities in the preterm population. Nevertheless, recent research suggests that increased severity of brain injury on neonatal MRI, particularly white matter, deep grey matter and cerebellar abnormalities, may be predictive of greater deficits in cognitive flexibility at 4 (Woodward, Clark, Bora, & Inder, 2012) and 7 (Murray et al., 2014) years of age.

Deficits in working memory are one of the most well-studied and common problems identified in preterm children because of their devastating impact for classroom education and learning, academic attainment, and social behaviour (Howard et al., 2008). In comparison with term-born controls, survivors of preterm birth demonstrate poorer working memory performance on various working memory measures, at toddler age, school age, during adolescence (Luu, Ment, Allan, Schneider, & Vohr, 2011; Taylor, Minich, Klein et al., 2004), and even as adults (Strang-Karlsson et al., 2010). Group differences tend to persist even after exclusion of individuals with neurosensory impairment or IQ below 70 (Luu et al., 2011).

A meta-analysis from Mulder and colleagues (2009) concluded that preterm children at toddler or pre-school age performed particularly poorly on visuospatial working memory tasks involving memory for locations, compared with term-born and normal birth weight peers (Mulder et al., 2009). In particular, the analysis showed large effect sizes to describe the difference in performance between term controls and 2-year-old preterm children born with gestational ages of 28 to 36 weeks (effect size of 0.73) (Espy et al., 2002), and 3-year-old preterm children born with gestational ages of 30 to 34 weeks (effect size of 1.09) (Caravale, Tozzi, Albino, & Vicari, 2005). Further, analysis of school-age cohorts demonstrates that preterm children still perform below their term-born peers. This is evident in the findings from a long-term follow-up study of an Australian cohort of very preterm children, who, at 7 years of age, performed significantly worse on tasks of verbal short-term and working memory, and visuospatial short-term memory, compared with term controls (Omizzolo et al., 2014). Perhaps more revealing in this study were the rates of impairment (1 SD below the term control group mean), with the very preterm children 2.9 and 3.2 times more likely to exhibit impairment on verbal and visuospatial short-term memory tasks, respectively. On verbal working memory tasks, very preterm children were 3 times more likely to show impairment, with rates of 36.2% impairment compared with 16.2% in controls (Omizzolo et al., 2014). Of the limited number of studies performed in adolescents and young adults born preterm, evidence suggests that visuospatial working memory deficits remain prominent in this population. A significantly larger number of recall errors were noted on the CANTAB visuospatial working memory task in 17-year-old VLBW adolescents (effect size 0.34) and ELBW adolescents (effect size 1.08), compared with term-born controls (Taylor, Minich, Klein et al., 2004). Very preterm 16-year-olds have also been shown to display significantly lower performance on visuospatial working memory tasks compared with term-born controls (Luu et al., 2011), with moderate effect sizes (0.7) (Saavalainen et al., 2007).

Considerable research has investigated the potential neuropathological substrates of working memory deficits in the preterm population. Significant linear associations have been observed between the severity of white matter abnormalities in the neonatal period and later performance on object working memory measures at 2 and 4 years of age (Edgin et al., 2008; Woodward, Anderson, Austin, Howard, & Inder, 2006; Woodward, Edgin, Thompson, & Inder, 2005). Similarly, Clark and Woodward (2010) found a trend in very preterm children, where (a) an increasing severity of white matter abnormalities at term-equivalent age was associated with poorer verbal and visuospatial working memory at age 6, and (b) very preterm children without neonatal white matter abnormalities demonstrated similar working memory profiles to term controls. More recent research has focused on a regional approach to white matter abnormalities and connectivity in relation to working memory dysfunction, with working memory thought to rely on distributed cortical and subcortical networks (Collete & Van der Linden, 2002). Interestingly, preterm birth is also postulated to disturb cortical-subcortical development and connectivity, particularly in circuits that connect frontal, striatal, and thalamic regions, and this disturbance is inversely proportional to gestational age and birth weight (Ball et al.,

2013; Cornelieke, Hanan, Weisglas-Kuperus, van Goudoever, & Oosterlaan, 2009; Counsell et al., 2003; Pandit et al., 2014; Partridge et al., 2004; Pavlova & Krageloh-Mann, 2013).

Goal Setting

Very preterm children with impairments in this domain have poor conceptual reasoning and difficulties in developing effective and efficient strategies. These difficulties may also be associated with a poor capacity to remember and retrieve plans (Anderson, 2002). Planning deficits have been noted in preterm individuals in early to middle childhood and adolescence compared to term-born controls (Harvey et al., 1999; Luciana, Lindeke, Georgieff, Mills, & Nelson, 1999; Rickards, Kelly, Doyle, & Callanan, 2001; Taylor, Minich, Klein et al., 2004), with some but not all studies (Luciana et al., 1999; Rickards et al., 2001) suggesting that lower gestational age and birth weight infers a greater risk of impairment (Anderson et al., 2004; Harvey et al., 1999; Taylor, Minich, Bangert et al., 2004; Taylor et al., 2000). In their meta-analysis on executive functioning in preterm children, Mulder and colleagues (2009) found that preterm children performed 0.4 SD below controls on tests of planning ability, with extremely preterm children performing the worst (Mulder et al., 2009). Importantly, deficits in planning and goal setting appear to remain after adjusting for intellectual impairment, neurosensory abnormalities, social risk, verbal functioning, and illness-related factors (Anderson et al., 2004; Harvey et al., 1999; Luciana et al., 1999; Taylor, Minich, Bangert et al., 2004; Taylor et al., 2000). Research examining the neurobiological bases of goal-setting deficits in very preterm survivors are lacking.

Clinical Implications

While more longitudinal studies are needed, there is evidence to suggest that poorer executive performance either increases with age, or persists over time, which has the potential effect of widening the gap between preterm children and their peers (Gray, Indurkhya, & McCormick, 2004; Guzzetta, Mercuri, & Spano, 2000; Sansavini et al., 2014; Taylor, Minich, Klein et al., 2004). These children may fail to make appropriate gains in the acquisition of executive skill sets, such as attentional control, information processing, cognitive flexibility, and goal setting, at critical stages during cognitive and behavioural maturation. This may give the appearance of preterm individuals 'growing into' their deficit over time. In some instances, executive deficits observed at earlier stages of development may dissipate, or resolve with time (Ment et al., 2003), with potential for plasticity and reorganisation of the preterm brain's circuitry which may manifest as the employment of alternative cognitive strategies when challenged (Nosarti et al., 2006). In order to identify preterm children most vulnerable to executive impairment and associated brain injury, it is clear that studies must involve detailed assessment of executive abilities within a state-of-the-art executive functioning framework, as well as a comprehensive evaluation and monitoring of neurological risk factors. Indeed, the high rates of executive impairment and neurological injury in this population emphasises the importance of neuroprotective measures that promote normal development and optimise the health and maturation of preterm infants.

References

Anderson, P. J. (2002). Assessment and development of executive function (EF) during childhood. *Child Neuropsychology, 8*(2), 71–82.

Anderson, P. J. (2014). Neuropsychological outcomes of children born very preterm. *Seminars in Fetal & Neonatal Medicine, 19*, 90–96.

Anderson, P. J., De Luca, C. R., Hutchinson, E., Spencer-Smith, M. M., Roberts, G., Doyle, L. W., and the Victorian Infant Collaborative Study Research Group. (2011). Attention problems in a representative sample of extremely preterm/extremely low birth weight children. *Developmental Neuropsychology, 36*, 57–73.

Anderson, P. J., Doyle, L., & the Victorian Infant Collaborative Study Research Group. (2004). Executive functioning in school-aged children who were born very preterm or with extremely low birth weight in the 1990s. *Pediatrics, 114*, 50–57.

Anderson, V. A. (1998). Assessment of executive function in children: An introduction. *Child Neuropsychology, 8*, 69–70.

Anderson, V. A., Anderson, P. J., Jacobs, R., & Spencer-Smith, M. (2008). In V. Anderson, R. Jacobs, & P. J. Anderson (Eds.), *Executive functions and the frontal lobes* (pp. 123–155). New York: Taylor & Francis.

Anderson, V. A., Anderson, P. J., Northam, E., Jacobs, R., & Catroppa, C. (2001). Developmental of executive functions through late childhood and adolescence in an Australian sample. *Developmental Neuropsychology, 20*, 385–406.

Andiman, S. E., Haynes, R. L., Trachtenberg, F. L., Billiards, S. S., Folkerth, R. S., & Volpe, J. J. (2010). The cerebral cortex overlying periventricular leukomalacia: Analysis of pyramidal neurons. *Brain Pathology, 20*, 803–814.

Arpino, C., D'Argenzio, L., Ticconi, C., Di Paolo, A., Stellin, V., Lopez, L., & Curatolo, P. (2005). Brain damage in preterm infants: Etiological pathways. *Annali dell'Istituto Superiore di Sanità, 41*, 229–237.

Ball, G., Boardman, J., Aljabar, P., Pandit, A., Arichi, T., Merchant, N., et al. (2013). The influence of preterm birth on the developing thalamocortical connectome. *Cortex, 49*, 1711–1721.

Behrmann, R. E., & Butler, A. S. (Eds.). (2007). *Preterm birth: Causes, consequences, and prevention*. Washington, DC: National Academies Press (US).

Benton, A. L. (1991). Prefrontal injury and behaviour in children. *Developmental Neuropsychology, 7*, 275–281.

Bhutta, A. T., & Anand, K. J. S. (2002). Vulnerability of the developing brain. *Clinics in Perinatology, 29*(3), 357–372.

Bhutta, A. T., Cleves, M. A., Casey, P. H., Cradock, M. M., & Anand, K. J. (2002). Cognitive and behavioral outcomes of school-aged children who were born preterm: A meta-analysis. *Journal of the American Medical Association, 288*, 728–737.

Boardman, J. P., & Dyet, L. E. (2007). Recent advances in imaging preterm brain injury. *Minerva Pediatrica, 59*, 349–368.

Bohm, B., Smedler, A. C., & Forssberg, H. (2004). Impulse control, working memory, and other executive functions in preterm children when starting school. *Acta Paediatrica, 93*(10), 1363–1371.

Breslau, N., Chilcoat, H., DelDotto, J., Andreski, P., & Brown, G. (1996). Low birth weight and neurocognitive status at six years of age. *Biological Psychiatry, 40*(5), 389–397.

Burnett, A. C., Scratch, S. E., & Anderson, P. J. (2013). Executive function outcome in preterm adolescents. *Early Human Development, 89*, 215–220.

Caravale, B., Tozzi, C., Albino, G., & Vicari, S. (2005). Cognitive development in low risk preterm infants at 3–4 years of life. *Archives of Disease in Childhood: Fetal and Neonatal Edition, 90*(6), F474–F479.

Clark, C. A. C., & Woodward, L. J. (2010). Neonatal cerebral abnormalities and later verbal and visuospatial working memory abilities of children born very preterm. *Developmental Neuropsychology, 35*, 622–642.

Collete, F., & Van der Linden, M. (2002). Brain imaging of the central executive component of working memory. *Neuroscience Biobehavioural Reviews, 26*, 105–125.

Cornelieke, S., Hanan, A. M., Weisglas-Kuperus, N., van Goudoever, J. B., & Oosterlaan, J. (2009). Meta-analysis of neurobehavioral outcomes in very preterm and/or very low birth weight children. *Pediatrics, 124*, 717–728.

Counsell, S., Allsop, J., Harrison, M., Larkman, D., Kennea, N., Kapellou, O., et al. (2003). Diffusion-weighted imaging of the brain in preterm infants with focal and diffuse white matter abnormality. *Pediatrics, 112*, 176–180.

de Kieviet, J. F., van Elburg, R. M., Lafeber, H. N., & Oosterlaan, J. (2012). Attention problems of very preterm children compared with age-matched term controls at school age. *Journal of Pediatrics, 161*, 824–829.

Dobbing, J. (1974). The later growth of the brain and its vulnerability. *Pediatrics, 53*(1), 2–6.

Duncan, J., Johnson, R., Swales, M., & Freer, C. (1997). Frontal lobe deficits after head injury: Unity and diversity of function. *Cognitive Neuropsychology, 14*, 713–741.

Edgin, J. O., Inder, T. E., Anderson, P., Hood, K. M., Clark, C. A. C., & Woodward, L. J. (2008). Executive functioning in preschool children born very preterm: Relationship with early white matter pathology. *Journal of the International Neuropsychological Society, 14*, 90–101.

Elgen, I., Lundervold, A. J., & Sommerfelt, K. (2004). Aspects of inattention in low birth weight children. *Pediatric Neurology, 30*(2), 92–98.

Espy, K. A., Stalets, M. M., McDiarmid, M. M., Senn, T. E., Cwik, M. F., & Hamby, A. (2002). Executive functions in preschool children born preterm: Application of cognitive neuroscience paradigms. *Child Neuropsychology, 8*(2), 83–92.

Garavan, H., Ross, T. J., Murphy, K., Roche, R. A., & Stein, E. A. (2002). Dissociable executive functions in the dynamic control of behavior: Inhibition, error detection, and correction. *NeuroImage, 17*, 1820–1829.

Gogtay, N., Giedd, J. N., Lusk, L., Hayashi, K. M., Greenstein, D., Vaituzis, A. C., et al. (2004). Dynamic mapping of human cortical development during childhood through early adulthood. *Proceedings of the National Academy of Sciences, 101*, 8174–8179.

Grattan, L., & Eslinger, P. (1991). Frontal lobe damage in children and adults: A comparative review. *Developmental Neuropsychology, 7*(3), 283–326.

Gray, R. F., Indurkhya, A., & McCormick, M. C. (2004). Prevalence, stability, and predictors of clinically significant behaviour problems in low birth weight children at 3, 5, and 8 years of age. *Pediatrics, 114*, 736–743.

Guzzetta, F., Mercuri, E., & Spano, M. (2000). Congenital lesions of the cerebellum. In A. Benton, E. De Renti & D. Riva (Eds.), *Localization of brain lesions and development functions* (pp. 147–152). London: John Libbey.

Harvey, J. M., O'Callaghan, M. J., & Mohay, H. (1999). Executive function of children with extremely low birth weight: A case controlled study. *Developmental Medicine and Child Neurology, 41*(5), 292–297.

Howard, K. H., Anderson, P. J., & Taylor, G. H. (2008). Executive functioning and attention in children born preterm. In V. Anderson, R. Jacobs & P. J. Anderson, *Executive functions and the frontal lobes: A lifespan perspective* (pp. 219–243). New York: Taylor & Francis.

Huppi, P. S., Murphy, B., Maier, S. E., Zientara, G., Inder, T. E., & Barnes, P. D. (2001). Microstructural brain development after perinatal cerebral white matter injury assessed by diffusion tensor magnetic resonance imaging. *Pediatrics, 107*, 455–460.

Inder, T. W., Warfield, S. K., Wang, H. X., Huppi, P. S., & Volpe, J. J. (2005). Abnormal cerebral structure is present at term in premature infants. *Pediatrics, 115*, 286–294.

Kapellou, O., Counsell, S. J., Kennea, N., Dyet, L., Saeed, N., & Stark, J. (2006). Abnormal cortical development after premature birth shown by altered allometric scaling of brain growth. *PLoS Medicine, 3*, e265.

Katz, K. S., Dubowitz, L. M. S., Henderson, S., Jongmans, M., Kay, G. G., Nolte, C. A., & De Vries, L. (1996). Effect of cerebral lesions on continuous performance test responses of school age children born prematurely. *Journal of Pediatric Psychology, 21*(6), 841–855.

Klingberg, T., Vaidya, C. J., Gabrieli, J. D., Moseley, M. E., & Hedehus, M. (1999). Myelination and organization of the frontal white matter in children: A diffusion tensor MRI study. *NeuroReport, 10*(13), 2817–2821.

Kulseng, S., Jennekens-Schinkel, A., Naess, P., Romundstad, P., Indredavik, M., Vik, T., & Brubakk, A. M. (2006). Very-low birthweight and term small-for-gestational-age adolescents: Attention revisited. *Acta Paediatrica, 95*(2), 224–230.

Lezak, M. D. (1995). *Neuropsychological assessment*. New York: Oxford University Press.

Luciana, M., Lindeke, L., Georgieff, M., Mills, M., & Nelson, C. (1999). Neurobehavioral evidence for working memory deficits in school-aged children with histories of prematurity. *Developmental Medicine and Child Neurology, 41*, 521–533.

Luria, A. R. (1973). *The working brain*. New York: Basic Books.

Luu, T. M., Ment, L., Allan, W., Schneider, K., & Vohr, B. R. (2011). Executive and memory function in adolescents born very preterm. *Pediatrics, 127*, 639–646.

Ment, L. R., Vohr, B., Allan, W., Katz, K. H., Schneider, K. C., Westerveld, M., et al. (2003). Change in cognitive function over time in very low-birth-weight infants. *Journal of the American Medical Association, 289*(6), 705–711.

Miyake, A., Friedman, N. P., Emerson, M. J., Witzki, A. H., Howerter, A., & Wager, T. D. (2000). The unity and diversity of executive functions and their contributions to complex 'frontal lobe' tasks: A latent variable analysis. *Cognitive Psychology, 41*, 49–100.

Mulder, H., Pitchford, N. J., Hagger, M. S., & Marlow, N. (2009). Development of executive function and attention in preterm children: A systematic review. *Developmental Neuropsychology, 34*(4), 393–421.

Murray, A. L., Scratch, S., Thompson, D., Inder, T. E., Doyle, L. W., Anderson, J. F., & Anderson, P. J. (2014). Neonatal brain pathology predicts adverse attention and processing speed outcomes in very preterm and/or very low birth weight children. *Neuropsychology, 28*(4), 552–562.

Nosarti, C., Giouroukou, E., Micali, N., Rifkin, L., Morris, R. G., & Murray, R. M. (2007). Impaired executive functioning in young adults born very preterm. *Journal of International Neuropsychological Society, 13*, 571–581.

Nosarti, C., Rubia, K., Smith, A., Frearson, S., Williams, S., Rifkin, L., & Murray, R. (2006). Altered functional neuroanatomy of response inhibition in adolescent males who were born very preterm. *Developmental Medicine & Child Neurology, 48*(4), 265–271.

Omizzolo, C., Scratch, S. E., Stargatt, R., Kidokoro, H., Thompson, D. K., Lee, K. J., et al. (2014). Neonatal brain abnormalities and memory and learning outcomes at 7 years in children born preterm. *Memory, 22*(6), 605–615.

Pandit, A. S., Robinson, E., Aljabar, P., Ball, G., Gousias, I. S., Wang, Z., et al. (2014). Whole-brain mapping of structural connectivity in infants reveals altered connectivity strength associated with growth and preterm birth. *Cerebral Cortex, 24*(9), 2324–2333.

Partridge, S. C., Mukherjee, P., Henry, R. G., Miller, S. P., Berman, J. I., Jin, H., et al. (2004). Diffusion tensor imaging: Serial quantitation of white matter tract maturity in premature newborns. *NeuroImage, 22*, 1302–1314.

Pavlova, M. A., & Krageloh-Mann, I. (2013). Limitations of the developing preterm brain: Impact of periventricular white matter lesions on brain connectivity and cognition. *Brain, 136*, 998–1011.

Rickards, A. L., Kelly, E. A., Doyle, L. W., & Callanan, C. (2001). Cognition, academic progress, behaviour, and self-concept at 14 years of very low birth weight children. *Journal of Developmental and Behavioural Pediatrics, 22*(1), 11–18.

Roberts, G., Anderson, P. J., & Doyle, L. W. (2009). Neurosensory disabilities at school age in geographic cohorts of extremely low birth weight children born between the 1970s and the 1990s. *Journal of Pediatrics, 154*, 829–834.

Rose, S. A., & Feldman, J. F. (1996). Memory and processing speed in preterm children at eleven years: A comparison with full-terms. *Child Development, 67*, 2005–2021.

Rose, S. A., Feldman, J. F., & Jankowski, J. J. (2002). Processing speed in the 1st year of life: A longitudinal study of preterm and full-term infants. *Developmental Psychology, 38*(6), 895–902.

Rose, S. A., Feldman, J. F., Jankowski, J. J., & Van Rossem, R. (2005). Pathways from prematurity and infant abilities to later cognition. *Child Development, 76*, 1172–1184.

Ross, G., Boatright, S., Auld, P. A. M., & Nass, R. (1996). Specific cognitive abilities in 2-year-old children with subependymal and mild intraventricular haemorrhage. *Brain and Cognition, 32*(1), 1–13.

Saavalainen, P., Luoma, L., Bowler, D., Määttä, S., Kiviniemi, V., Laukkanen, E., et al. (2007). Spatial span in very prematurely born adolescents. *Developmental Neuropsychology, 32*(3), 769–785.

Saigal, S., den Ouden, L., Wolke, D., Hoult, L., Paneth, N., Streiner, D. L., et al. (2003). School-age outcomes in children who were extremely low birth weight from four international population-based cohorts. *Pediatrics, 112*, 943–950.

Saigal, S., & Doyle, L. (2008). An overview of mortality and sequelae of preterm birth from infancy to adulthood. *The Lancet, 371*, 261–269.

Saigal, S., Hoult, L. A., Streiner, D. L., Stoskopf, B. L., & Rosenbaum, P. L. (2000). School difficulties at adolescence in a regional cohort of children who were extremely low birth weight. *Pediatrics, 105*, 325–331.

Sansavini, A., Pentimonti, J., Justice, L., Guarini, A., Savini, S., Alessandroni, R., & Faldella, G. (2014). Language, motor and cognitive development of extremely preterm children: Modelling individual growth trajectories over the first three years of life. *Journal of Communication Disorders, 49*, 55–68.

Soria-Pastor, S., Gimenez, M., Narberhaus, A., Falcon, C., Botet, F., Bargallo, N., et al. (2008). Patterns of cerebral white matter damage and cognitive impairment in adolescents born very preterm. *International Journal of Developmental Neuroscience, 26*(7), 647–654.

Strang-Karlsson, S., Andersson, S., Paile-Hyvarinen, M., Darby, D., Hovi, P., Raikkonrn, K., et al. (2010). Slower reaction times and impaired learning in young adults with birth weight < 1500 g. *Pediatrics, 125*(1), 74–82.

Stuss, D., & Alexander, M. (2000). Executive functions and the frontal lobes: A conceptual view. *Psychological Research, 63*, 289–298.

Taylor, H. G., Hack, M., & Klein, N. K. (1998). Attention deficits in children with <750 gm birth weight. *Child Neuropsychology, 4*, 21–34.

Taylor, H. G., Klein, N., Minich, N. M., & Hack, M. (2000). Middle-school-age outcomes in children with very low birth weight. *Child Development, 71*, 1495–1511.

Taylor, H. G., Klein, N., Schatschneider, C., & Hack, M. (1998). Predictors of early school age outcomes in very low birth weight children. *Journal of Developmental and Behavioural Pediatrics, 19*(4), 235–243.

Taylor, H. G., Minich, N., Bangert, B., Filipek, P. A., & Hack, M. (2004). Long-term neuropsychological outcomes of very low birth weight: Associations with early risks for periventricular brain insults. *Journal of the International Neuropsychological Society, 10*, 987–1004.

Taylor, H. G., Minich, N. M., Klein, N., & Hack, M. (2004). Longitudinal outcomes of very low birth weight: Neuropsychological findings. *Journal of the International Neuropsychological Society, 10*, 149–163.

Tideman, E. (2000). Longitudinal follow-up of children born preterm: Cognitive development at age 19. *Early Human Development, 58*(2), 81–90.

Vicari, S., Caravale, B., Carlesimo, G. A., Casadei, A. M., & Allemand, F. (2004). Spatial working memory deficits in children at ages 3–4 who were low birth weight, preterm infants. *Neuropsychology, 18*(4), 673–678.

Vohr, B. R., Wright, L. L., Dusick, A. M., Mele, L., Verter, J., Seichen, J. J., et al. (2000). Neurodevelopmental and functional outcomes of extremely low birth weight infants in the National Institute of Child Health and Human Development Neonatal Research Network, 1993–1994. *Pediatrics, 105*(6), 1216–1226.

Vohr, B. R., Wright, L. L., Dusick, A. M., Perritt, R., Poole, W. K., Tyson, J. E., et al. (2004). Center differences and outcomes of extremely low birth weight infants. *Pediatrics, 113*, 781–789.

Volpe, J. J. (1997). Brain injury in the premature infant: From pathogenesis to prevention. *Brain & Development, 19*(8), 519–534.

Volpe, J. J. (2009). Brain injury in premature infants: A complex amalgam of destructive and developmental disturbances. *Lancet Neurology, 8*, 110–124.

Watanabe, J., Sugiura, M., Sato, K., Sato, Y., Maeda, Y., Matsue, Y., et al. (2002). The human prefrontal and parietal association cortices are involved in NO-GO performances: An event-related fMRI study. *NeuroImage, 17*, 1207–1216.

Welsh, M. C., Pennington, B. F., & Groisser, D. B. (1991). A normative-developmental study of executive function: A window on prefrontal function in children. *Developmental Neuropsychology, 7*(2), 131–149.

Wilson-Ching, M., Molloy, C. S., Anderson, V. A., Burnett, A., Roberts, G., Cheong, J. L. Y., et al. (2013). Attention difficulties in a contemporary geographic cohort of adolescents born extremely preterm/extremely low birth weight. *Journal of the International Neuropsychological Society, 19*, 1–12.

Wood, N. S., Marlow, N., Costeloe, K., Gibson, A. T., & Wilkinson, A. R. (2000). Neurologic and developmental disability after extremely preterm birth: EPICure Study group. *New England Journal of Medicine, 343*, 378–384.

Woodward, L. J., Anderson, P. J., Austin, N. C., Howard, K., & Inder, T. E. (2006). Neonatal MRI to predict neurodevelopmental outcomes in preterm infants. *New England Journal of Medicine, 355*, 685–694.

Woodward, L. J., Clark, C. A. C., Bora, S., & Inder, T. E. (2012). Neonatal white matter abnormalities an important predictor of neurocognitive outcomes for very preterm children. *PLOS One, 7*(12), e51879.

Woodward, L. J., Edgin, J. O., Thompson, D., & Inder, T. E. (2005). Object working memory deficits predicted by early brain injury and development in the preterm infant. *Brain, 128*, 2578–2587.

Wyatt, J. (2010). The changing face of intensive care for preterm newborns. In C. Nosarti, R. M. Murray, & M. Hack (Eds.), *Neurodevelopmental outcomes of preterm birth: From childhood to adult life*. Cambridge: Cambridge University Press.

15

EXECUTIVE FUNCTIONS AND DEVELOPMENTAL PSYCHOPATHOLOGY

Neurobiology of Emotion Regulation in Adolescent Depression and Anxiety

Kristina L. Gelardi, Veronika Vilgis, & Amanda E. Guyer

Introduction

Humans have an exquisite set of psychological processes available to direct their attention and actions in order to prioritize, organize, and target their emotions and behaviors in a goal-oriented manner. Executive functions, which include working memory, inhibitory control, and attentional shifting (Miyake et al., 2000), are fundamental to cognitive strategies supporting effective self-regulation of emotions and behaviors (Hofmann, Schmeichel, & Baddeley, 2012; Ochsner & Gross, 2008). The development of executive functions that support emotion regulation, and related cognitive strategies such as reappraisal and attentional shifting, follow a protracted trajectory, taking many years to become more automated, fine-tuned, and nuanced (Best & Miller, 2010; Blakemore & Choudhury, 2006; Chevalier & Clark, Chapter 2, this volume). Indeed, learning to control and manage one's emotions and behaviors is viewed as a principal task of child and adolescent development, and relies on the maturation of cognitive and social-emotional skills. The progression of emotion regulation across development is further supported by socialization experiences and fine-tuning of the connections among brain circuits that underlie affective and cognitive functions.

When executive functioning does not mature as expected or becomes maladaptive in response to environmental demands, it may produce dysregulated emotions, impair daily functioning, and lead to psychopathology (Aldao, Nolen-Hoeksema, & Schweizer, 2010; Snyder, Miyake, & Hankin, 2015). Risk for emotion regulation difficulties is particularly heightened in adolescence, attributed in part to the onset of puberty, continued development of brain regions needed to support executive functions (Crone, Peters, & Steinbeis, Chapter 3, this volume; Luciana & Ewan, Chapter 16, this volume), and changing social demands (Guyer, Silk, & Nelson, 2016). Indeed, a substantial increase in the onset of mood and anxiety disorders, which involve emotion dysregulation, occurs in adolescence (Avenevoli, Swendsen, He, Burstein, & Merikangas, 2015; Paus, Keshavan, & Giedd, 2008). Therefore, the negative impact of emotion dysregulation during adolescence vs. childhood may be more impactful for long-term psychopathology (DeSteno, Gross, & Kubzansky, 2013). A domain general perspective of neural functioning supporting these strategies emphasizes the universality of neural mechanisms associated with top-down cognitive control (Johnson, 2011), evidenced in executive functions and cognitive emotion regulation strategies, and subsequently mental health (Cole, Repovš, & Anticevic, 2014).

The goal of this chapter is to discuss emotion regulation, which relies on a composite of executive functions, that, when compromised or dysregulated, relates to risk for mood and anxiety problems, with a focus on adolescence and findings from neuroscience. First, key issues regarding definitions and developmental trajectories of executive functions, psychopathology, and emotion regulation are presented. Second, we review brain regions involved in executive functions and emotion regulation. Third, we present central neuroscience research findings on emotion regulation and adolescent mood and anxiety disorders. Finally, the chapter closes with a discussion of future directions for and conclusions about the importance of integrating research on cognitive control and psychopathology to better meet the needs of children and adolescents at risk for psychopathology.

Definitions and Developmental Trajectories

Executive Functions and Psychopathology

Executive functions include cognitive processes associated with working memory, inhibitory control, and attentional shifting (Miyake et al., 2000). Deficits in these cognitive processes are found along with the dysregulated emotionality (e.g., negative affect, mood disturbance) characteristic of anxiety disorders and depression. For example, anxiety disorders are associated with attentional biases and reduced ability to disengage from threat, deficits in working memory, and reduced inhibitory control (Berggren & Derakshan, 2013; Eysenck, Derakshan, Santos, & Calvo, 2007). Similarly, depression is associated with greater attention to negative stimuli, rumination, reduced working memory, and impulsive behavior (Joormann & Gotlib, 2008; Kyte, Goodyer, & Sahakian, 2005), each reflecting deficits in executive functions.

The risk of developing psychopathology increases in middle childhood and adolescence given that many anxiety disorders (e.g., social, panic, and generalized) onset during childhood and adolescence (Beesdo, Knappe, & Pine, 2011) and rates of depression double in adolescence compared to childhood (Kessler,

Avenevoli, & Ries Merikangas, 2001). The effects of these disorders are widespread as nearly a third of adolescents report anxiety symptoms and over 10% experience severe impairment and/or distress as a result of mood disorders, such as depression (Avenevoli et al., 2015). Substance use experimentation also often begins in adolescence and may disrupt executive function via neural changes, as discussed elsewhere in this volume (Luciana & Ewan, Chapter 16, this volume).

Emotion Regulation and Dysregulation

Effective emotion regulation is associated with prosocial behavior (Eisenberg, Fabes, Guthrie, & Reiser, 2000), social connections (Butler et al., 2003), strong working memory (Evans & Fuller-Rowell, 2013), and better physical (DeSteno et al., 2013) and mental health (Aldao, et al., 2010). Executive functions – attentional shifting, working memory, and inhibitory control – are the underlying cognitive processes supporting effective self-regulation of emotions and behaviors (Hofmann et al., 2012; Ochsner & Gross, 2008). Emotion regulation relies on these cognitive processes to alter the onset of and recovery from one's emotional state or behavior – impacting when and which emotions we have and how we express or experience emotions (Gross, 2014; Thompson, 1994).

A number of emotion regulation strategies in response to negative events or aversive stimuli have been identified and studied in relation to mood and anxiety problems. Two in particular, attentional shifting and cognitive reappraisal, have been shown to be effective at attenuating negative emotional reactivity in both behavioral and developmental neuroscience studies. For example, attentional shifting involves directing attention away from an aversive stimulus to reduce the emotional salience of the stimulus (Ferri, Schmidt, Hajcak, & Canli, 2013). Cognitive restructuring or reappraisal is another widely studied strategic form of emotion regulation. Reappraisal involves changing one's thoughts to reframe the stimulus or experience, such as an aversive one, with the goal of altering one's emotional response (McRae, Ciesielski, & Gross, 2012). Cognitive reappraisal integrates attentional shifting with the other core executive functions in emotion regulation, such as representing goals in working memory (Hofmann et al., 2012; Pe, Raes, & Kuppens, 2013).

The utilization of emotion regulation strategies changes across development, as children and adolescents begin to differentiate the efficacy of discrete regulatory strategies. Indeed, children's understanding about the effectiveness of using thoughts to reduce emotional distress increases linearly from preschool age to early adolescence (Pons, Harris, & De Rosnay, 2004) and becomes more effective during middle and late adolescence (McRae et al., 2012; Silvers et al., 2012). While younger children often use behavioral strategies such as distraction from an emotion-eliciting stimulus, adolescents tend to employ cognitive skills such as taking the perspective of another person and attentional control to regulate their emotional reaction to a stimulus or situation (Cohen Kadosh,

Heathcote, & Lau, 2014). The increased use of cognitive emotion regulation strategies in adolescence is accompanied by maturation of other executive functions such as goal-directed behavior, working memory, and response inhibition, which have been linked with brain maturation (Blakemore & Choudhury, 2006; Luna, Marek, Larsen, Tervo-Clemmens, & Chahal, 2015).

Although there is normative variability in individuals' ability to control or direct their emotional experiences, for some, emotion regulation deficits can lead to increased dysfunction in other domains (Gross & Jazaieri, 2014). More specifically, emotion dysregulation involves patterns of emotional experience or expression that interfere with appropriate goal-directed behaviors (Beauchaine & Gatzke-Kopp, 2012). For mood and anxiety disorders, experiencing fear, anxiety, panic, or sadness may become pathological if they are felt too intensely or for too long and cease to be adaptive (Beauchaine, Gatzke-Kopp, & Mead, 2007), preventing children and adolescents from participating in social events or concentrating at school. Negative emotions may also be experienced in unpredictable and extreme forms relative to situational or contextual demands if efforts to appropriately control these emotions are not executed or are poorly executed.

Neural Correlates of Executive Functions and Emotion Regulation

Executive functions and emotion regulation skills rely heavily on input from regions within the prefrontal cortex (PFC; Moriguchi, Chapter 5, this volume), an area of the brain that develops in a protracted fashion during childhood and into early adulthood (Giedd et al., 2015; Mills et al., 2016). Maturation of the PFC is associated with cortical thinning in lateral regions and increases in white matter, which serve to strengthen connections between the PFC and other areas of the brain. Neuroimaging studies show that executive functions and cognitive emotion regulation rely on the function of specific PFC subregions, including the ventrolateral, dorsolateral, and medial PFC as well as the anterior cingulate cortex (ACC) and parts of the orbitofrontal cortex (OFC) (Blakemore & Choudhury, 2006; Kohn et al., 2014). The majority of these regions are part of a domain general cognitive control network spanning primarily the frontal and parietal regions (Kohn et al., 2014).

Prefrontal regions have reciprocal connections with subcortical regions such as the amygdala, insula, and hippocampus, which are associated with the detection of emotionally salient stimuli and memory, and influence both executive functions and emotion regulation (Moriguchi, Chapter 5, this volume). These subcortical regions mature early in development relative to their prefrontal counterparts, resulting in heightened emotional salience and reduced cognitive control in the presence of emotions during adolescence (Casey, 2015; Guyer et al., 2016). The prevailing literature suggests that this typical, but asynchronous, development of brain regions associated with attending to, interpreting, and responding to emotionally salient stimuli may create a window of vulnerability for psychopathology during adolescence (Casey, 2015; Guyer et al., 2016). In the following sections we consider evidence from neuroscience that illustrate the imbalance of top-down and bottom-up emotion processing in the context of anxiety and depression.

Psychopathology, Emotion Regulation, and the Brain

Extant behavioral and neuroimaging research provides clear links between psychopathology, emotion regulation (including related executive functions), and brain function. To probe the brain's response to emotionally salient stimuli (i.e., stimuli that elicit the need for regulation), studies often use pictures of negative facial expressions and threatening or aversive scenes. This body of empirical research shows that the use of regulatory strategies, such as attentional shifting and cognitive reappraisal to down-regulate negative affect, is associated with reduced amygdala activity and increased PFC activation (Ferri et al., 2013; Kohn et al., 2014; Nelson, Fitzgerald, Klumpp, Shankman, & Phan, 2015). These patterns are also evident outside of lab tasks and associated with the use of habitual reappraisal strategies in daily life manifest as reduced amygdala and increased PFC activity when viewing negative facial expressions (Drabant, McRae, Manuck, Hariri, & Gross, 2009). Similarly, individuals scoring high in reappraisal ability are found to exhibit greater amygdala attenuation during a reappraisal task in a longitudinal study, suggesting some stability to this pattern given its persistence over time (Lee, Heller, van Reekum, Nelson, & Davidson, 2012). Finally, studies of neural connectivity, which more directly measure coordination between amygdala and PFC activity, demonstrate that reappraisal success is associated with greater inverse connectivity between the amygdala and ACC, OFC, and dlPFC (Lee et al., 2012) and greater reductions of negative affect as predicted by connectivity strength between the amygdala and the OFC (Banks, Eddy, Angstadt, Nathan, & Luan Phan, 2007).

Anxiety Disorders

Anxiety disorders are associated with deficits in each of the traditional constructs of executive functions (i.e., attentional shifting, working memory, inhibitory control) as well as cognitive reappraisal (Amstadter, 2008). Some work suggests that this is the result of a disruption in the balance between top-down and bottom-up neural systems (Eysenck et al., 2007; Mueller et al., 2012). Indeed, there is evidence of disrupted executive function and emotion regulation through attenuated top-down control, implicit in executive functions and emotion regulation.

Developmental behavioral studies have examined anxiety and emotion regulation primarily through how attention is impacted, with less focus on the use of cognitive reappraisal strategies among individuals with anxiety disorders. Evidence suggests associations between anxiety and attention, with developmental differences emerging from middle childhood to young adulthood. For example, children and adolescents with anxiety disorders show increased attention bias to threatening stimuli (Roy et al., 2008), whereas young adults with anxiety take longer to disengage and shift attention from negative stimuli than their non-anxious peers (Fox, Russo, Bowles, & Dutton, 2001). Individual differences also exist, with evidence of executive functions, such as attentional control (Derryberry & Reed, 2002) and working memory (Visu-Petra, Stanciu, Benga, Miclea, & Cheie, 2014), predicting emotion regulation deficits. Furthermore, anxiety may not actually reduce effective cognitive control, but rather might attenuate the efficiency of cognitive efforts (Berggren & Derakshan, 2013), as evidenced by longer reaction times but similar accuracy (Cohen Kadosh et al., 2014).

Neuroimaging findings also indicate an important link between emotion regulation and anxiety. Atypical patterns of brain activation have been found in association with anxiety (Sylvester et al., 2012), as well as generalized anxiety and emotion dysregulation (Etkin, Prater, Hoeft, Menon, & Schatzberg, 2010), including in youth (Pine, Guyer, & Leibenluft, 2008). Important roles for the dorso- and ventrolateral PFC and ACC have been identified in the cognitive control of emotions, with reduced activation in these regions accompanied by elevated amygdala activation in individuals with anxiety. Specifically, atypical amygdala responses are found during attention bias to threat faces and anticipation of peer evaluation in adolescents with anxiety compared to healthy controls. For example, adolescents with general anxiety disorder showed greater right amygdala activation, and weaker negative coupling between the right amygdala and right ventrolateral PFC than did healthy controls when subconsciously exposed to angry faces (Monk et al., 2008). Similarly, social anxiety disorder was related to elevated amygdala activation during the anticipation of peer evaluation (Guyer et al., 2008), but in contrast to Monk et al. (2008), positive coupling between the ventrolateral PFC and the amygdala was found among the socially anxious vs. healthy adolescents (Guyer et al., 2008). These discrepancies may be associated with symptom differences between general and social anxiety, as well as the attentional processes engaged by the different functional magnetic resonance imaging (fMRI) tasks used.

Another index of brain activation relevant to emotion regulation among anxious youth is the time course that neural activation takes in response to fearful stimuli. The time course reflects the degree to which the amygdala is engaged, at what point, and for how long, and whether cognitive control regions are disproportionately recruited to manage amygdala engagement. For example, amygdala activation has shown a differential time course for child and adolescent participants with anxiety disorders compared to healthy youth (Swartz et al., 2014). Whereby amygdala activation was elevated at the start of the task and became significantly attenuated later in the task for anxious youth, healthy youth showed no significant difference in amygdala activation across the course of the task. Furthermore, anxiety status was associated with reduced amygdala–dorsal PFC connectivity for youth with anxiety disorders.

While extant research does not allow for directional interpretations, evidence of similar dysregulated brain activation can be found for children at risk of developing clinical levels of anxiety. For example, in anticipation of threatening stimuli, temperamentally inhibited children aged 8–10 showed attenuated PFC engagement and reduced connectivity between the PFC and limbic regions (Clauss, Benningfield, Rao, & Blackford, 2016). Reduced limbic–PFC connectivity may be part of the mechanism that leads to hyperactive, overly vigilant responses to threat and maladaptive types of voluntary emotion regulation. These findings support the notion that emotion dysregulation may override cognitive control, emphasizing the importance of emotion regulation in preventing and treating psychopathology (Mueller, 2011).

Depression

The impaired ability to regulate negative affect is central to the maintenance of sad mood in clinical depression (Kovacs, Joormann, & Gotlib, 2008). Several lines of behavioral and neuroimaging research suggest that emotion regulation strategies and executive functions fail in children and adolescents with depression, particularly in the context of negatively valenced information. There are two avenues through which negative stimuli may disrupt cognitive processes for depressed individuals. The first may be that depressed individuals find it difficult to ignore negative cues and stimuli. The other is that negative cues are congruent with their mood and subsequently interfere with their ability to approach information and situations without bias, or think clearly without emotional overload (Lakdawalla, Hankin, & Mermelstein, 2007), which contribute to the development and maintenance of depression. In contrast, the use of cognitive reappraisal has been successfully implemented in the treatment of depression in children (Kovacs et al., 2008).

Behavioral studies measuring response inhibition and working memory show that children and adolescents with depression differ from typically developing youth in their performance on trials that include negatively valenced pictures such as, for example, sad faces (Kyte et al., 2005; Ladouceur et al., 2006; Maalouf et al., 2012). Accuracy in the ability to remember previously viewed happy and sad faces but not fearful or angry faces is also impaired in adolescent depression (Guyer, Choate, Grimm, Pine, & Keenan, 2011). Beyond the laboratory, it has been shown that negative, stressful life events lead to depression via maladaptive forms of cognitive emotion regulation such as self-blame, catastrophizing, and rumination (Stikkelbroek, Bodden, Kleinjan, Reijnders, & van Baar, 2016).

FMRI studies investigating the neural correlates of cognitive emotion regulation in children and adolescents with depression report reduced top-down control and increased limbic activation, similar to findings from adults with major depressive disorder (Beauregard, Paquette, & Lévesque, 2006; Johnstone, van Reekum, Urry, Kalin, & Davidson, 2007). Children with a history of depression have been found to show decreased activity of the left ventrolateral PFC and inferior temporal cortex during reappraisal of sad images compared to children without a history of depression (Belden, Pagliaccio, Murphy, Luby, & Barch, 2015). In a follow-up study of the same sample, Murphy and colleagues (2016) also observed stronger coupling between the amygdala and subgenual ACC during reappraisal of sad images. Connectivity between those regions increased with greater rumination. Other work has shown that current depression in children and adolescents is associated with increased baseline amygdala activity (Belden et al., 2015; Perlman, Huppert, & Luna, 2016) and less reciprocal connectivity between the amygdala and both the insula and medial PFC (Perlman et al., 2012).

Despite a scarcity of fMRI studies examining executive functions in children and adolescents with depression, two studies have reported reduced activation of dorsolateral and ventrolateral PFC regions during working memory and attention tasks (Halari et al., 2009; Vilgis, Chen, Silk, Cunnington, & Vance, 2014). In line with behavioral studies, Colich, Foland-Ross, Eggleston, Singh, and Gotlib (2016) reported atypical behavioral responses when youth were asked to inhibit a response after having been primed with the presence of a sad face. Further, they found attenuated activation in the right dorsolateral PFC in adolescents with major depressive disorder (Colich et al., 2016). This suggests that negative information, or mood-congruent stimuli, interferes with adequate top-down control in adolescents with depression.

While cognitive control and emotion regulation share overlapping neural circuitry, and it is generally assumed that better executive functions correspond to better emotion regulation, and therefore reduced risk of depression, there is relatively little empirical evidence for executive function deficits as a causal factor. It is currently unclear whether executive function deficits in depression represent a problem with the executive system per se or whether deficits arise due to an excessive demand on regulatory resources from other processes (Luciana, 2016). Observations from adult fMRI studies suggest that lateral PFC activation in major depressive disorder is load dependent, such that hyperactivation is observed at low cognitive loads indicating greater demand on a potentially compromised system. In contrast, hypoactivation is observed at high cognitive loads due to a failure to cope with the excessive demand (Walter, Wolf, Spitzer, & Vasic, 2007). This would be in accord with the demand hypothesis. Further, findings that emotion regulation and cognitive control in depression break down specifically in contexts associated with depression congruent affect (i.e. sadness and negative feelings) also suggest that it is excessive demand, rather than regulatory resources per se, that explain cognitive control failures in depression. Negatively valenced material may put greater demand on bottom-up systems such as the amygdala, thereby diminishing resources for top-down systems (Luciana, 2016). This would mean that poor executive function skills are not necessarily a trait characteristic of depression in children and adolescents (Vilgis, Silk, & Vance, 2015), but that depression impairs adequate top-down cognitive control in part due to greater demand on bottom-up systems. Such an account would still allow for a protective role of higher executive function skills against depression in which children and adolescents with greater skills may be better able to counteract bottom-up demands and greater executive function skills are less susceptible to depression symptoms.

Summary and Future Directions

Adolescence marks an important period of development characterized by substantial neurobiological and behavioral changes accompanied by novel social and emotional experiences. Not surprisingly, this period of change and maturation provides a window of vulnerability evidenced by the marked increase in risk for the first onset of a range of mental health disorders including major depressive disorder, anxiety disorders, bipolar disorder, eating disorders, psychosis, and substance abuse (Paus et al., 2008; Luciana & Ewan, Chapter 16, this volume). The prolonged maturation of the PFC and its connections with limbic regions parallels the maturational course of executive functions fundamental to emotion regulation. Maladaptive or less efficient emotion regulation strategies are associated with altered functioning in this network and can be observed in individuals with anxiety and depression. Yet, we are only beginning to understand how executive functions and cognitive emotion regulation strategies manifest across adolescence, taking into account core cognitive and affective processes, their underlying neurobiology and the wider social context in which the individual operates. A better understanding of these factors associated with risk, attenuation, and resilience are fundamental to meeting the needs of youth affected or at risk. Currently, research suggests domain general neurobiological functioning may underlie the cognitive control of executive functions and emotion regulation (Johnson, 2011). It is therefore likely that early interventions and treatment programs targeting cognitive control in general may have beneficial effects for both executive functions and emotion regulation to offset the development of psychopathology.

Future developmental neuroscience research should further explore the associations between traditional executive function tasks, emotion regulation, psychopathology, and social context. Future studies should aim to identify unique and interactive effects of different executive functions and cognitive emotion regulation strategies that current research has not yet well addressed. The underlying cognitive functions associated with psychopathology and emotion dysregulation provide one avenue for targeted prevention and intervention efforts. Indeed, functional improvement associated with these mental health problems can be found through cognitive behavior therapy, and specifically with the use of emotion regulation skills training (Berking et al., 2008; Tolin, 2010). Therefore, emotion regulation is a primary target for interventions designed to help youth who face risk for such problems. However, little is known about how social contexts influence successful engagement of cognitive strategies used to regulate emotions as experienced in naturalistic settings. Indeed, reliance on ecologically valid paradigms in the laboratory, including the neuroimaging environment, would aid in furthering our knowledge of contextual influences. This is an important area for future focus given its potential for translation into interventions to help those vulnerable to developing emotion dysregulation and psychopathology.

References

Aldao, A., Nolen-Hoeksema, S., & Schweizer, S. (2010). Emotion-regulation strategies across psychopathology: A meta-analytic review. *Clinical Psychology Review, 30*, 217–237.

Amstadter, A. (2008). Emotion regulation and anxiety disorders. *Journal of Anxiety Disorders, 22*, 211–221.

Avenevoli, S., Swendsen, J., He, J. P., Burstein, M., & Merikangas, K. R. (2015). Major depression in the National Comorbidity Survey–Adolescent supplement: Prevalence, correlates, and treatment. *Journal of the American Academy of Child and Adolescent Psychiatry, 54*, 37–44.e2.

Banks, S. J., Eddy, K. T., Angstadt, M., Nathan, P. J., & Luan Phan, K. (2007). Amygdala-frontal connectivity during emotion regulation. *Social Cognitive and Affective Neuroscience, 2*, 303–312.

Beauchaine, T. P., & Gatzke-Kopp, L. M. (2012). Instantiating the multiple levels of analysis perspective in a program of study on externalizing behavior. *Development and Psychopathology, 24*, 1003–1018.

Beauchaine, T. P., Gatzke-Kopp, L., & Mead, H. K. (2007). Polyvagal Theory and developmental psychopathology: Emotion dysregulation and conduct problems from preschool to adolescence. *Biological Psychology, 74*, 174–184.

Beauregard, M., Paquette, V., & Lévesque, J. (2006). Dysfunction in the neural circuitry of emotional self-regulation in major depressive disorder. *Neuroreport, 17*, 843–846.

Beesdo, K., Knappe, S., & Pine, D. S. (2011). Anxiety and anxiety disorders in children and adolescents: Developmental issues and implications for DSM-V. *Psychiatric Clinics of North America, 32*, 483–524.

Belden, A. C., Pagliaccio, D., Murphy, E. R., Luby, J. L., & Barch, D. M. (2015). Neural activation during cognitive emotion regulation in previously depressed compared to healthy children: Evidence of specific alterations. *Journal of the American Academy of Child and Adolescent Psychiatry, 54*, 771–781.

Berggren, N., & Derakshan, N. (2013). Attentional control deficits in trait anxiety: Why you see them and why you don't. *Biological Psychology, 92*, 440–446.

Berking, M., Wupperman, P., Reichardt, A., Pejic, T., Dippel, A., & Znoj, H. (2008). Emotion-regulation skills as a treatment target in psychotherapy. *Behaviour Research and Therapy, 46*, 1230–1237.

Best, J., & Miller, P. (2010). A developmental perspective on executive function. *Child Development, 81*, 1641–1660.

Blakemore, S. J., & Choudhury, S. (2006). Development of the adolescent brain: Implications for executive function and social cognition. *Journal of Child Psychology and Psychiatry and Allied Disciplines, 47*, 296–312.

Butler, E. A., Egloff, B., Wilhelm, F. H., Smith, N. C., Erickson, E. A., & Gross, J. J. (2003). The social consequences of expressive suppression. *Emotion, 3*, 48–67.

Casey, B. J. (2015). Beyond simple models of self-control to circuit-based accounts of adolescent behavior. *Annual Review of Psychology, 66*, 295–319.

Clauss, J. A., Benningfield, M. M., Rao, U., & Blackford, J. U. (2016). Altered prefrontal cortex function marks heightened anxiety risk in children. *Journal of the American Academy of Child & Adolescent Psychiatry, 55*, 809–816.

Cohen Kadosh, K., Heathcote, L. C., & Lau, J. Y. F. (2014). Age-related changes in attentional control across adolescence: How does this impact emotion regulation capacities? *Frontiers in Psychology, 5*, 111.

Cole, M. W., Repovš, G., & Anticevic, A. (2014). The frontoparietal control system: A central role in mental health. *The Neuroscientist, 20*, 652–664.

Colich, N. L., Foland-Ross, L. C., Eggleston, C., Singh, M. K., & Gotlib, I. H. (2016). Neural aspects of inhibition following emotional primes in depressed adolescents. *Journal of Clinical Child and Adolescent Psychology, 45*, 21–30.

Derryberry, D., & Reed, M. A. (2002). Anxiety-related attentional biases and their regulation by attentional control. *Journal of Abnormal Psychology, 111*, 225–236.

DeSteno, D., Gross, J. J., & Kubzansky, L. (2013). Affective science and health: The importance of emotion and emotion regulation. *Health Psychology, 32*, 474–486.

Drabant, E. M., McRae, K., Manuck, S. B., Hariri, A. R., & Gross, J. J. (2009). Individual differences in typical reappraisal use predict amygdala and prefrontal responses. *Biological Psychiatry, 65*, 367–373.

Eisenberg, N., Fabes, R. A., Guthrie, I. K., & Reiser, M. (2000). Dispositional emotionality and regulation: Their role in predicting quality of social functioning. *Journal of Personality and Social Psychology, 78*, 136–157.

Etkin, A., Prater, K. E., Hoeft, F., Menon, V., & Schatzberg, A. F. (2010). Failure of anterior cingulate activation and connectivity with the amygdala during implicit regulation of emotional processing in generalized anxiety disorder. *American Journal of Psychiatry, 167*, 545–554.

Evans, G. W., & Fuller-Rowell, T. E. (2013). Childhood poverty, chronic stress, and young adult working memory: The protective role of self-regulatory capacity. *Developmental Science, 16*, 688–696.

Eysenck, M. W., Derakshan, N., Santos, R., & Calvo, M. G. (2007). Anxiety and cognitive performance: Attentional control theory. *Emotion, 7*, 336–353.

Ferri, J., Schmidt, J., Hajcak, G., & Canli, T. (2013). Neural correlates of attentional deployment within unpleasant pictures. *NeuroImage, 70*, 268–277.

Fox, E., Russo, R., Bowles, R., & Dutton, K. (2001). Do threatening stimuli draw or hold visual attention in subclinical anxiety? *Journal of Experimental Child Psychology, 130*, 681–700.

Giedd, J. N., Raznahan, A., Alexander-Bloch, A., Schmitt, E., Gogtay, N., & Rapoport, J. L. (2015). Child Psychiatry Branch of the National Institute of Mental Health longitudinal structural magnetic resonance imaging study of human brain development. *Neuropsychopharmacology, 40*, 43–49.

Gross, J. J. (2014). Emotion regulation: Conceptual and empirical foundations. In J. J. Gross (Ed.), *Handbook of emotion regulation* (2nd ed., pp. 3–20). New York: The Guilford Press.

Gross, J. J., & Jazaieri, H. (2014). Emotion, emotion regulation, and psychopathology: An affective science perspective. *Clinical Psychological Science, 2*, 387–401.

Guyer, A. E., Choate, V. R., Grimm, K. J., Pine, D. S., & Keenan, K. (2011). Emerging depression is associated with face memory deficits in adolescent girls. *Journal of the American Academy of Child and Adolescent Psychiatry, 50*, 180–190.

Guyer, A. E., Lau, J. Y. F., McClure-Tone, E. B., Parrish, J., Shiffrin, N. D., Reynolds, R. C., et al. (2008). Amygdala and ventrolateral prefrontal cortex function during anticipated peer evaluation in pediatric social anxiety. *Archives of General Psychiatry, 65*, 1303–1312.

Guyer, A. E., Silk, J. S., & Nelson, E. E. (2016). The neurobiology of the emotional adolescent: From the inside out. *Neuroscience and Biobehavioral Reviews, 70*, 74–85.

Halari, R., Simic, M., Pariante, C. M., Papadopoulos, A., Cleare, A., Brammer, M., et al. (2009). Reduced activation in lateral prefrontal cortex and anterior cingulate during attention and cognitive control functions in medication-naive adolescents with depression compared to controls 17. *Journal of Child Psychology and Psychiatry, 50*, 307–316.

Hofmann, W., Schmeichel, B. J., & Baddeley, A. D. (2012). Executive functions and self-regulation. *Trends in Cognitive Sciences, 16*, 174–180.

Johnson, M. H. (2011). Interactive specialization: A domain-general framework for human functional brain development? *Developmental Cognitive Neuroscience, 1*, 7–21.

Johnstone, T., van Reekum, C. M., Urry, H. L., Kalin, N. H., & Davidson, R. J. (2007). Failure to regulate: Counterproductive recruitment of top-down prefrontal-subcortical circuitry in major depression. *Journal of Neuroscience, 27*, 8877–8884.

Joormann, J., & Gotlib, I. H. (2008). Updating the contents of working memory in depression: Interference from irrelevant negative material. *Journal of Abnormal Psychology, 117*, 182–192.

Kessler, R. C., Avenevoli, S., & Ries Merikangas, K. (2001). Mood disorders in children and adolescents: An epidemiologic perspective. *Biological Psychiatry, 49*, 1002–1014.

Kohn, N., Eickhoff, S. B., Scheller, M., Laird, A. R., Fox, P. T., & Habel, U. (2014). Neural network of cognitive emotion regulation: An ALE meta-analysis and MACM analysis. *NeuroImage, 87*, 345–355.

Kovacs, M., Joormann, J., & Gotlib, I. H. (2008). Emotion (dys)regulation and links to depressive disorders. *Child Development Perspectives, 2*, 149–155.

Kyte, Z. A., Goodyer, I. M., & Sahakian, B. J. (2005). Selected executive skills in adolescents with recent first episode major depression. *Journal of Child Psychology and Psychiatry and Allied Disciplines, 46*, 995–1005.

Ladouceur, C. D., Dahl, R. E., Williamson, D. E., Birmaher, B., Axelson, D. A., Ryan, N. D., & Casey, B. J. (2006). Processing emotional facial expressions influences performance on a Go/NoGo task in pediatric anxiety and depression. *Journal of Child Psychology and Psychiatry, 47*, 1107–1115.

Lakdawalla, Z., Hankin, B. L., & Mermelstein, R. (2007). Cognitive theories of depression in children and adolescents: A conceptual and quantitative review. *Clinical Child and Family Psychology Review, 10*, 1–24.

Lee, H., Heller, A. S., van Reekum, C. M., Nelson, B., & Davidson, R. J. (2012). Amygdala-prefrontal coupling underlies individual differences in emotion regulation. *NeuroImage, 62*, 1575–1581.

Luciana, M. (2016). Executive function in adolescence: A commentary on regulatory control and depression in adolescents – Findings from neuroimaging and neuropsychological research. *Journal of Clinical Child and Adolescent Psychology, 45*, 84–89.

Luna, B., Marek, S., Larsen, B., Tervo-Clemmens, B., & Chahal, R. (2015). An integrative model of the maturation of cognitive control. *Annual Review of Neuroscience, 38*, 151–170.

Maalouf, F. T., Clark, L., Tavitian, L., Sahakian, B. J., Brent, D., & Phillips, M. L. (2012). Bias to negative emotions: A depression state-dependent marker in adolescent major depressive disorder. *Psychiatry Research, 198*, 28–33.

McRae, K., Ciesielski, B., & Gross, J. J. (2012). Unpacking cognitive reappraisal: Goals, tactics, and outcomes. *Emotion, 12*, 250–255.

Mills, K. L., Goddings, A.-L., Herting, M. M., Meuwese, R., Blakemore, S.-J., Crone, E. A., et al. (2016). Structural brain development between childhood and adulthood: Convergence across four longitudinal samples. *NeuroImage, 141*, 273–281.

Miyake, A., Friedman, N. P., Emerson, M. J., Witzki, A. H., Howerter, A., & Wager, T. D. (2000). The unity and diversity of executive functions and their contributions to complex "frontal lobe" tasks: a latent variable analysis. *Cognitive Psychology, 41*, 49–100.

Monk, C. S., Telzer, E. H., Mogg, K., Bradley, B. P., Mai, X., Louro, H. M. C., et al. (2008). Amygdala and ventrolateral prefrontal cortex activation to masked angry faces in children and adolescents with generalized anxiety disorder. *Archives of General Psychiatry, 65*, 568–576.

Mueller, S. C. (2011). The influence of emotion on cognitive control: Relevance for development and adolescent psychopathology. *Frontiers in Psychology, 2*, 327.

Mueller, S. C., Hardin, M. G., Mogg, K., Benson, V., Bradley, B. P., Reinholdt-Dunne, M. L., et al. (2012). The influence of emotional stimuli on attention orienting and inhibitory control in pediatric anxiety. *Journal of Child Psychology and Psychiatry and Allied Disciplines, 53*, 856–863.

Murphy, E. R., Barch, D. M., Pagliaccio, D., Luby, J. L., & Belden, A. C. (2016). Functional connectivity of the amygdala and subgenual cingulate during cognitive reappraisal of emotions in children with MDD history is associated with rumination. *Developmental Cognitive Neuroscience, 18*, 86–100.

Nelson, B. D., Fitzgerald, D. A., Klumpp, H., Shankman, S. A., & Phan, K. L. (2015). Prefrontal engagement by cognitive reappraisal of negative faces. *Behavioural Brain Research, 279*, 218–225.

Ochsner, K. N., & Gross, J. J. (2008). Cognitive emotion regulation: Insights from social cognitive and affective neuroscience. *Current Directions in Psychological Science, 17*, 153–158.

Paus, T., Keshavan, M., & Giedd, J. N. (2008). Why do many psychiatric disorders emerge during adolescence? *Nature Reviews Neuroscience, 9*, 947–957.

Pe, M. L., Raes, F., & Kuppens, P. (2013). The cognitive building blocks of emotion regulation: Ability to update working memory moderates the efficacy of rumination and reappraisal on emotion. *PLoS One, 8*, e69071.

Perlman, G., Simmons, A. N., Wu, J., Hahn, K. S., Tapert, S. F., Max, J. E., et al. (2012). Amygdala response and functional connectivity during emotion regulation: A study of 14 depressed adolescents. *Journal of Affective Disorders, 139*, 75–84.

Perlman, S. B., Huppert, T. J., & Luna, B. (2016). Functional near-infrared spectroscopy evidence for development of prefrontal engagement in working memory in early through middle childhood. *Cerebral Cortex, 26*, 2790–2799.

Pine, D. S., Guyer, A. E., & Leibenluft, E. (2008). Functional magnetic resonance imaging and pediatric anxiety. *Journal of the American Academy of Child and Adolescent Psychiatry, 47*, 1217–1221.

Pons, F., Harris, P. L., & De Rosnay, M. (2004). Emotion comprehension between 3 and 11 years: Developmental periods and hierarchical organization. *European Journal of Developmental Psychology, 1,* 127–152.

Roy, A. K., Vasa, R. A., Bruck, M., Mogg, K., Bradley, B. P., Sweeney, M., et al. (2008). Attention bias toward threat in pediatric anxiety disorders. *Journal of the American Academy of Child and Adolescent Psychiatry, 47,* 1189–1196.

Silvers, J. A., McRae, K., Gabrieli, J. D., Gross, J. J., Remy, K. A., & Ochsner, K. N. (2012). Age-related differences in emotional reactivity, regulation, and rejection sensitivity in adolescence. *Emotion, 12,* 1235–1247.

Snyder, H. R., Miyake, A., & Hankin, B. L. (2015). Advancing understanding of executive function impairments and psychopathology: Bridging the gap between clinical and cognitive approaches. *Frontiers in Psychology, 6,* 327.

Stikkelbroek, Y., Bodden, D. H., Kleinjan, M., Reijnders, M., & van Baar, A. L. (2016). Adolescent depression and negative life events, the mediating role of cognitive emotion regulation. *PLoS One, 11,* e0161062.

Swartz, J. R., Phan, K. L., Angstadt, M., Klumpp, H., Fitzgerald, K. D., & Monk, C. S. (2014). Altered activation of the rostral anterior cingulate cortex in the context of emotional face distractors in children and adolescents with anxiety disorders. *Depression and Anxiety, 31,* 870–879.

Sylvester, C. M., Corbetta, M., Raichle, M. E., Rodebaugh, T. L., Schlaggar, B. L., Sheline, Y. I., et al. (2012). Functional network dysfunction in anxiety and anxiety disorders. *Trends in Neurosciences, 35,* 527–535.

Thompson, R. A. (1994). Emotion regulation: A theme in search of definition. In N. A. Fox (Ed.), *The development of emotion regulation: Biological and behavioral considerations* (Vol. 59, pp. 25–52). Chicago: University of Chicago Press.

Tolin, D. F. (2010). Is cognitive-behavioral therapy more effective than other therapies? A meta-analytic review. *Clinical Psychology Review, 30,* 710–720.

Vilgis, V., Chen, J., Silk, T. J., Cunnington, R., & Vance, A. (2014). Frontoparietal function in young people with dysthymic disorder (DSM-5: Persistent depressive disorder) during spatial working memory. *Journal of Affective Disorders, 160,* 34–42.

Vilgis, V., Silk, T. J., & Vance, A. (2015). Executive function and attention in children and adolescents with depressive disorders: A systematic review. *European Child & Adolescent Psychiatry, 24,* 365–384.

Visu-Petra, L., Stanciu, O., Benga, O., Miclea, M., & Cheie, L. (2014). Longitudinal and concurrent links between memory span, anxiety symptoms, and subsequent executive functioning in young children. *Frontiers in Psychology, 5,* 443.

Walter, H., Wolf, R. C., Spitzer, M., & Vasic, N. (2007). Increased left prefrontal activation in patients with unipolar depression: An event-related, parametric, performance-controlled fMRI study. *Journal of Affective Disorders, 101,* 175–185.

16

EXECUTIVE FUNCTION AND SUBSTANCE MISUSE

Neurodevelopmental Vulnerabilities and Consequences of Use

Monica Luciana & Emily Ewan

Introduction

The goal of this chapter is to consider whether executive functions and the neural mechanisms that support them are compromised in the context of substance misuse with a particular focus on human adolescence and young adulthood. Executive functions (EFs) comprise a set of regulatory skills, including inhibitory control, flexibility, working memory, and attention that emerge late in infancy and continue to develop into young adulthood (see Cuevas, Rajan, & Bryant, Chapter 1; Chevalier & Clark, Chapter 2; and Crone, Peters, & Steinbeis, Chapter 3, this volume). EFs allow individuals to make adaptive deliberate decisions, to pursue immediate and long-range goals, and to maintain ongoing control over motivational impulses. EF deficits are evident in nearly every form of psychopathology, including substance abuse (Snyder, Miyake, & Hankin, 2015). At present, a Google Scholar search of "executive function and substance abuse" yields nearly 18,000 hits, based on associations observed across the human life span and in animal studies.

Using alcohol as an example to illustrate links between substance misuse and EF, individuals who are prenatally exposed display executive dysfunction as part of a larger syndrome of fetal alcohol effects (Khoury, Milligan, & Girard, 2015). Adolescent and young adult alcohol use in modest amounts (Mahmood, Jacobus, Bava, Scarlett, & Tapert, 2010), in the context of binge drinking (Peeters, Wollebergh, Wiers, & Field, 2014), and in the context of alcohol use disorders (Tapert, Granholm, Leedy, & Brown, 2002) is associated with EF deficits. Adults who consume large quantities of alcohol and who meet formal diagnostic criteria for alcohol dependence may experience disruptions in cognition that most commonly involve impaired explicit memory and executive functions (Bernardin, Maheut-Bosser, & Paille, 2014). Finally, excessive alcohol consumption in men is associated with aging-related cognitive declines (Sabia et al., 2014); outcomes are more variable for women (Lyu & Lee, 2014), with some suggestion of neuroprotective effects. Whether alcohol abuse in older aged individuals is a risk factor for cognitive impairment varies depending on premorbid status (Moussa et al., 2015) and may be mediated by alcohol's impacts on hippocampal volume (Downer, Jiang, Zanjani, & Fardo, 2014). Together, these findings suggest that EFs are compromised through alcohol's effects on the brain and that such effects vary by age and amounts used.

Preclinical studies have outlined potential mechanisms through which such outcomes might occur (Koob & Volkow, 2010). In the context of an emerging addiction, neuroadaptations within limbic, frontal, and striatal circuits lead to a triad of altered brain reward function, increased stress reactivity, and compromised executive function. Theoretically, the negative impacts on executive function involve disruptions in the afferent projections from the prefrontal cortex and insula to the basal ganglia and extended amygdala (Koob & Volkow, 2016), resulting in increasingly greater loss of control over substance intake as it becomes more and more compulsive in nature. This cascade can be viewed along a continuum. Outside of prenatal exposure, a question concerns how early in one's experiences with substances such neuroadaptations can be observed.

Experimentation with substances of abuse often begins in adolescence. According to 2013 statistics maintained by the U.S. Center for Disease Control, 35% of high school students reported using alcohol (the most commonly used substance) within the past month (www.cdc.gov/alcohol/fact-sheets/underage-drinking.htm). Underage drinkers consume more alcohol per drinking session than do adults and tend to do so in the form of binge drinking (defined as consuming more than five drinks in one session (males) or four drinks (females)). Adolescents who start drinking before age 15 are 6 times more likely to develop a substance use disorder (SUD) later in life than those who begin drinking at or after age 21 (SAMHSA, 2014). One potential explanation for this trend is that alcohol compromises the neural circuitry that supports EF, lowering the threshold for developing loss of control over use, a core characteristic of SUDs. This is an elegant model that makes intuitive sense.

A developing brain is a malleable brain given the potential for experience to interfere with the course and timing of neurodevelopmental processes (Kolb et al., 2012). Given their extended developmental trajectory, executive functions are thought to maintain plasticity well into adulthood (Dahlin, Nyberg, Bäckman, & Neely, 2008). Such plasticity confers both benefits and limitations. On the positive side, the extended developmental window allows enriching experiences to enhance regulatory skills via neuroadaptive changes (Dahlin et al., 2008; Selemon, 2013). On the other hand, if an individual engages in activities that are detrimental to optimal brain and behavioral development, those activities, including substance use, could permanently alter the development of regulatory skills and their underlying neural networks due to interference with the intended maturational trajectory (Fontes et al., 2011; Luciana, Collins, Muetzel, & Lim, 2013).

Neuroadaptations Due to Adolescent Substance Use: Interpretive Challenges

Despite this model's intuitive appeal, the human literature is confounded by shortcomings that limit the interpretation of existing data. Many studies are case-control comparisons without any assessment of premorbid function. Longitudinal studies provide a means of examining causal influences, but they are rare. Moreover, sample sizes tend to be small so that epidemiological trends cannot be discerned. Comorbidity is often evident within substance-abusing individuals. Discrete EFs are sometimes reported in isolation, limiting the extent to which specificity of effects can be discerned. Thus, it remains unclear whether alcohol use disrupts EFs due to neuroplastic changes ("neurotoxicity" models as advanced by Koob & Volkow, 2010) or whether individuals with relatively lower levels of EF are vulnerable to disinhibitory behaviors, such that low EF predisposes some people to heavy substance use ("vulnerability" models; cf., McGue, Iacono, Legrand, & Elkins, 2001). Both models may be accurate if high trait levels of disinhibition (e.g., low EF) lead vulnerable individuals to experiment with substances of abuse, to experience lack of behavioral control once use is initiated leading to higher levels of exposure, and then to experience neurotoxic damage to frontostriatal and frontolimbic circuits that support behavioral regulation. Thus, not all individuals are at equivalent risk (Bjork & Pardini, 2015). Premorbidly vulnerable individuals may be particularly likely to experience a downward spiral of executive dysfunction in the context of substance misuse. This cascade is important to understand so that we may ascertain the extent to which substances are neurotoxic when ingested in small amounts and causally implicated in executive dysfunction, to assess how developmental timing interacts with substance-induced neurotoxic effects and to appropriately structure the timing of EF-based intervention strategies in at-risk individuals.

The Neurodevelopment and Expression of Executive Functions

Confirmatory factor analyses suggest that in addition to a common general factor, three-to-four diverse functions underpin many of the tasks currently employed in laboratory-based studies of EF, including behavioral inhibition, working memory/updating, and behavioral flexibility (Miyake et al., 2000; see reviews by Chevalier & Clark, Chapter 2, and Li, Vadaga, Bruce, & Lai, Chapter 4, this volume). The recruitment of these EF functions during times of emotional salience may be particularly challenging during the adolescent period given the many novel motivational contexts in which individuals may find themselves (Luciana & Collins, 2012).

The neurocircuitry that facilitates EFs is extensive, including putative cognitive control circuits spanning superior and dorsolateral prefrontal regions, the anterior cingulate cortex, the parietal lobe, dorsomedial thalamus, and dorsal striatum, but also interconnections between ventral and medial regions of the PFC, ventral striatal and limbic regions that are involved in affective valuation, and brainstem regions that regulate arousal (Arnsten & Rubia, 2012; Holmes, Hollinshead, Roffman, Smoller, & Buckner, 2016). The context-dependent nature of EF recruitment implies that nearly all brain regions are potentially involved in its execution but with the notable involvement of numerous frontal regions.

Executive functions are behaviorally evident during the infant period as an individual learns capacities for self-regulation (see chapter by Cuevas, Rajan, & Bryant, Chapter 1, this volume). EFs show developmental spurts throughout early childhood with rapid accelerations in middle childhood followed by more subtle gains in adolescence and into young adulthood (Luciana, Conklin, Hooper, & Yarger, 2005; see Crone, Peters, & Steinbeis, Chapter 3, this volume). Similarly, neural changes in the structural integrity of the cortical and subcortical circuitry that supports EF are most prominent between early-to-middle childhood with refinements thereafter, perhaps extending well into the third decade of life (Lebel & Beaulieu, 2011; Sowell et al., 2003). Thus, the adolescent to young adulthood period is characterized by continued EF development that may be subject to substance-induced neuroplasticity.

EF Disruption in the Context of Substance Misuse

Various aspects of cognition, including EFs, are relatively impaired in substance users when users are *cross-sectionally* compared to non-using controls. When considered across the life span and across substances, this literature is extensive. In a recent meta-analysis, Roberts, Jones, and Montgomery (2016) compared EF and memory task performance in 1,221 current ecstasy users and 1,242 drug-using controls, finding evidence for overall executive dysfunction in ecstasy users as well as some differential effects across specific subdomains. Ecstasy users, who tended to be polydrug users, showed specific performance deficits in accessing information, switching/behavioral flexibility, and updating, but not in inhibitory control. In contrast, Stephan et al. (2016) meta-analytically examined the effects of alcohol on individual EF subcomponents in 5,140 recently detoxified participants across 77 studies. Findings ranged from a relatively large effect size for inhibition to smaller effects for verbal fluency. Large effect sizes were found for decisional and cognitive impulsivity, while effects for motor impulsivity were more moderate. Other domains that were particularly sensitive to alcohol effects included planning, flexibility, and motivated decision-making.

Focusing specifically on cocaine use, Spronk, van Wel, Ramaekers, and Verkes (2013) found evidence for deficits in shifting and inhibition. In a comprehensive synthesis of the literature across substance types, Fernández-Serrano, Pérez-García, and Verdejo-García (2011) found that numerous domains of EF, including shifting, working memory, and inhibition, were impaired across SUDs. Smith, Mattick, Jamadar, and Iredale (2014) conducted a meta-analysis using fixed-effects models, integrating results from 97 studies that compared groups with heavy substance use with healthy control participants on two measures of response inhibition: the Go/No-Go task and the Stop-Signal task. Inhibitory deficits characterized users of cocaine, MDMA, methamphetamine, tobacco, and alcohol. Similarly, Grant and Chamberlain (2014) reviewed evidence for deficits in impulsive action (the inability to inhibit motor responses) and impulsive choice (preference for immediate smaller rewards to the detriment of long-term outcomes; e.g., delay discounting paradigms) across addictions, finding evidence for deficits in both domains. The authors noted that significant effects appeared to depend on baseline function. In addition, they cautioned that impairments in these domains have frequently been found prior to the development of SUDs. Because findings are largely similar for various substances of abuse and when individuals with different substance use preferences are collapsed into a single group and contrasted with non-users, it can be concluded that deficient EF is consistently observed in the context of active substance misuse.

Brain circuits that underlie EF-based control mechanisms are also distinct in substance-using individuals. A robust finding in adolescent and young adult alcohol users is that the morphology of the frontal lobe is distinct in alcohol users even prior to the onset of frank addiction (see Luciana et al., 2013). At least one study has reported that the morphology of the orbitofrontal region predicts the onset of cannabis use in adolescents (Cheetham et al., 2012). Studies that have focused on white matter pathology are consistent in finding alterations in the superior longitudinal fasciculus, a pathway that interconnects the frontal lobe with other cortical regions, in adolescent substance users (Bava et al., 2009). Functional imaging studies (Courtney, Ghahremani, & Ray, 2013) indicate that the frontostriatal pathway underlying response inhibition is weakened as alcoholism progresses and that fMRI activations during tests of behavioral inhibition predict a worsening course of use over time (Mahmood et al., 2013).

It is common to infer from these group differences that substances of abuse derail EF in a causal manner. Alternatively, it may be that pre-existing EF deficits account for case-control differences observed in cross-sectional studies (see discussion by Snyder et al., 2015) and in longitudinal studies that do not incorporate substance naive baseline samples. Consistent with this hypothesis, Peeters et al. (2012, 2013) found that adolescent drinkers who had poor inhibitory control were more likely to show strong approach tendencies and attentional biases toward alcohol cues than those with better inhibitory skills and that those with poor inhibitory control showed greater alcohol use over a 6-month follow-up period.

It is also challenging to attribute EF deficits solely to the impacts of substances given that users tend to experience other types of life difficulties as well as comorbidities that are independently associated with executive dysfunction. In an analysis of substance users with dual diagnoses, who tend to be highly represented among those with SUDs, Duijkers, Vissers, and Egger (2016) reported that shifting and inhibitory control are the EF domains most often reported to be compromised in dual diagnosis users as compared to healthy controls.

A fundamental thesis of this chapter is that, by definition, the demonstration of disinhibitory or externalizing tendencies as displayed in real-world settings occur along a dimension that encompasses executive dysfunction given that those who display externalizing tendencies are undercontrolled and frequently impulsive. Associations between poor EF and externalizing tendencies can be observed as early as the preschool years (Schoemaker, Mulder, Deković, & Matthys, 2013), in middle childhood (Woltering, Lishak, Hodgson, Granic, & Zelazo, 2016), and in adolescence (Olson, Hooper, Collins, & Luciana, 2007; Young et al., 2009). Together, this literature suggests that disinhibitory forms of psychopathology, including substance misuse, may be dimensionally associated with executive dysfunction.

Disinhibitory Tendencies and the Prediction of Substance Use Initiation

Epidemiological longitudinal studies consistently affirm that disinhibitory or externalizing tendencies in childhood predict a variety of later substance use behaviors in substance naive individuals and in those who are at high risk for substance use disorders (Verdejo-Garcia, Lawrence, & Clark, 2008). For instance, using the Dunedin longitudinal cohort, Caspi, Moffitt, Newman, and Silva (1996) showed that children (primarily males) who were characterized as undercontrolled at ages 3 to 5 were most likely to endorse more alcohol problems at age 21 and were twice as likely to be diagnosed with alcohol dependence. More recently, Groenman et al. (2015) reported similar risks for individuals diagnosed with attention-deficit hyperactivity disorder. McGue et al. (2001) examined over 1,500 participants from the Minnesota Center for Twin and Family Research and found that rates of conduct disorder, oppositional defiant disorder, and other externalizing disorders were significantly increased among early drinkers and that early drinkers scored higher on the oppositional, hyperactive/impulsive, and inattentive teacher rating scales than nondrinkers and also achieved lower grades. Children objectively rated as more inattentive and impulsive at age 11 were more likely to have tried alcohol by age 14. Moreover, hyperactivity and impulsivity, traits that are associated with EF dysfunction, measured at age 14, predicted the later initiation of all types of substance use as well as nicotine dependence and cannabis abuse and dependence (Elkins, King, McGue, & Iacono, 2006). In a cross-sectional survey of over 1,500 nine-year-olds, it was found that low levels of self-reported EF were associated with higher levels of lifetime substance use (Pentz et al., 2015; Riggs, Spruijt-Metz, Chou, & Pentz, 2012), despite substance use being relatively rare in this age group. In this sample, low self-reported EF predicted greater substance use over a 6-month retest interval (Pentz & Riggs, 2013). Nigg et al. (2006) also observed that poor inhibition, as measured by a Stop task, was longitudinally associated with more alcohol- and drug-related problems in a high-risk sample. Thush et al. (2008) and Grenard et al. (2008) found similar associations in a high-externalizing cohort in relation to working memory performance. Together, these studies suggest that risk for alcohol-induced impairments of EFs varies across individuals since those with higher trait levels of disinhibitory and aggressive behaviors are more likely to engage in substance use at an early age (when the brain is still developing), to use heavily (conferring a greater risk for neuroplastic changes; Koob & Volkow, 2010), and to experience fully syndromic patterns of substance use disorder and comorbid psychopathology, each of which is associated with disruptions in cortical architecture and function.

Another line of work has focused on individuals who are at high risk based on a positive family history of substance use problems (Nixon & McClain, 2010). Those considered high risk based on family history demonstrate high levels of sensation-seeking and other externalizing traits (Handley et al., 2011), relatively low scores on EF measures in childhood (Giancola, Martin, Tarter, Pelham, & Moss 1996; Giancola & Tarter, 1999), are more likely to initiate substance use at an early age (Wong et al., 2006), and are more likely to show associations between baseline EF and later substance use (Deckel & Hesselbrock, 1996). Tarter et al. (2003) followed male adolescents with and without positive family histories of SUD and found that disinhibition discriminated the high- and low-risk groups at baseline and interacted with the amount of use at age 16 to predict SUD status 7–9 years after study enrollment. Similarly, Wong et al. (2006) studied a large sample from the Michigan Longitudinal Study who varied in family history of substance use problems. Children of parents with alcohol use disorders were 3 times more likely than those without a family history of alcohol abuse to have initiated alcohol use by age 14.

Brain mechanisms that contribute to these effects have been described. Herting, Fair, and Nagel (2011) reported that adolescents with positive family histories of SUDs have disruptions in cortical and cerebellar white matter microstructure, which, they suggest, contributes to less efficient cortical processing. These authors also found that adolescents with positive family histories showed greater impulsive decision making (Cservenka & Nagel, 2012). Giancola and Parker (2001) studied a high-risk group of males and observed mediation effects such that low laboratory-based EF in the context of difficult temperaments predicted increased aggression and affiliation with delinquent peers at ages 12–14; these, in turn, were related to increased drug use at age 16. Findings were similar in females (Giancola & Mezzich, 2003). Such findings are not limited to behavioral designs. Using a neuroimaging approach, Heitzeg et al. (2014) longitudinally followed a group of 9–12-year-olds and found that brain activation in the left medial frontal gyrus during performance of a go/no-go task predicted problematic alcohol use 4 years later, above and beyond externalizing tendencies. Using a powerful monozygotic twin difference design where adolescent twin pairs differed in their self-reported alcohol use, Malone et al. (2014) demonstrated that twin differences in Iowa Gambling Task performance were attributable to premorbid (e.g., genetic) characteristics rather than to the toxic effects of alcohol use in discordant twin pairs.

Together, these findings are compelling in suggesting that those who are at high risk for substance use initiation in adolescence as a function of family history or because of externalizing traits exhibit EF vulnerabilities *prior* to use onset. While these are not likely to be the only salient predictors of future substance misuse, these vulnerabilities are associated with earlier ages of onset of substance use, with accelerated frequency of use over time, and with use in greater amounts.

Against this backdrop and in the context of age as a significant source of variance in EF, toxic effects of substances may be difficult to detect, especially if substances are ingested in small to moderate amounts.

Growing Evidence for Neurotoxic Impacts of Adolescent Alcohol Use

Relatively few longitudinal studies have been optimized to assess the impacts of premorbid executive dysfunction as well as the effect of substances on the same cognitive measures pre- and post-use. Peeters, Janssen, Monshouwer, Boendermaker, and Pronk (2015) examined 2 EF domains, working memory and inhibitory control, in a high-risk sample of over 500 12–15-year-olds in the Netherlands and found through a survival analysis that relatively weak working memory, as measured by a self-ordered pointing task, predicted both the initiation of the first alcoholic drink and the first binge drinking episode, above and beyond the effect of response inhibition, as measured by the Stroop task, which also predicted initiation of drinking. Impacts on post-drinking executive function were not examined. In contrast, Khurana et al. (2013) studied a community sample that was assessed annually over 4 consecutive years on an EF battery that included working memory tests, delay discounting, and measures of sensation-seeking. Baseline weaknesses in working memory predicted not only concurrent alcohol use but also increased frequency of use over the 4 years of longitudinal testing. Importantly, variations in alcohol use over time were *not* associated with concomitant changes in working memory. The authors suggest that early adolescent alcohol use may be a consequence of pre-existing EF weaknesses. A mechanism for this effect is suggested by another longitudinal study in which it was found that drug-related associations in memory predicted drug use more strongly in students with lower levels of working memory capacity (Grenard et al., 2008), which perhaps extends to other EFs. Consistent with this interpretation, Kim-Spoon et al. (2016) studied 157 early adolescents and found through structural equation modeling that for both behavioral performance and neural activity indicators of inhibitory control, high levels of self-reported behavioral activation predicted earlier onset of substance use among adolescents with low inhibitory control – but not among adolescents with high inhibitory control. Among heavy cannabis users, individual differences in working memory network response had an independent effect on change in weekly cannabis use 6 months later above and beyond baseline levels (Cousijn et al., 2014). Fernie et al. (2013) found similar effects for EF measures of impulsivity in a prospective study.

Several studies have used functional magnetic resonance imaging (fMRI) to assess group differences over time in adolescents who were substance naive at baseline and transitioned (or not) into alcohol use (Norman et al., 2011; Wetherill, Castro, Squeglia, & Tapert, 2013; Wetherill, Squeglia, Yang, & Tapert, 2013). All found that prior to alcohol use initiation, those who went on to initiate use did not activate task-relevant frontoparietal regions important for cognitive control to the same extent as non-initiators.

Brain structural differences are also apparent. Luciana et al. (2013) examined various aspects of brain morphometry and white matter microstructure in a low-risk sample of adolescents who were substance naive at baseline and followed across a 2-year interval during which some transitioned into alcohol use. Alcohol use initiators showed greater-than-expected decreases in cortical thickness in the right middle frontal gyrus from baseline to follow-up as well as blunted development of white matter in the right hemisphere precentral gyrus, lingual gyrus, middle temporal gyrus, and anterior cingulate. Diffusion tensor imaging revealed a relative decrease over time in fractional anisotropy in the left caudate/thalamic region as well as in the right inferior frontal occipital fasciculus. Alcohol initiators did not differ from non-users at the baseline assessment; the groups were largely similar in other premorbid characteristics.

Squeglia et al. (2015) examined gray and white matter volume trajectories in 134 adolescents, of whom 75 transitioned to heavy drinking and 59 remained light drinkers or nondrinkers over roughly 3.5 years. Similar to Luciana et al. (2013), heavy-drinking adolescents showed accelerated gray matter reduction in lateral frontal and temporal regions and attenuated white matter growth of the corpus callosum and pons relative to nondrinkers.

In addition to these recent imaging findings that suggest distinct neurodevelopmental trajectories in alcohol users vs. non-users, some cross-sectional studies and a select number of longitudinal studies of adolescents and young adults have found evidence in experienced users for dose-outcome effects. For instance, relatively greater alcohol consumption has been linked with relatively lower regional brain volumes, with decreases in the microstructural organization of white matter, and with decreased BOLD responses in fMRI paradigms (see Feldstein Ewing, Sakhardande, & Blakemore, 2014, for review). Similarly, a number of recent reports are suggestive of age-of-onset effects on both behavioral and neural outcomes such that an earlier age of substance use onset is associated with relatively worse EF and/or deviant patterns of neural development (Lisdahl, Gilbart, Wright, & Shollenbarger, 2013). These effects of age-of-use-onset could imply that the impacts of substance use are more detrimental to brain development when the substance is introduced relatively early in the adolescent neurodevelopmental sequence. Alternatively, it could be that an earlier age of onset implies a greater lifetime dose that is, cumulatively, detrimental to neural and behavioral function.

Conclusions

The use of alcohol and other substances of abuse often begins during adolescence. Thus, the possibility of substance-induced neurotoxicity and interference with normal patterns of brain and behavioral development is evident. If executive functions are impacted, then there may be a cascade of regulatory failures and deficiencies in cognitive control that render individuals vulnerable to continued misuse of substances, other forms of maladaptive decision-making, or serious persistent psychopathology. Models of addiction propose a cascade of effects as an individual transitions from experimentation with substances of abuse to physical dependence to preoccupation and compulsive use (Koob & Volkow, 2016). This cascade by definition involves compromised executive function.

While there are numerous reports of deficient EF in people who abuse substances relative to those who do not, that literature does not convincingly inform cause–effect analyses given that premorbid status is not typically assessed. As suggested here, premorbid individual differences in executive dyscontrol when measured as early as the preschool years predict an early onset of substance use initiation, escalation of use over time, and the development of substance use disorders. Moreover, individual variations in externalizing traits or tendencies cohere with brain and behavioral deviations that can be similar to what is observed post-substance use. Essentially, it is not clear, when the literature is aggregated, whether EF deficits observed in the context of substance misuse are primarily a cause or a consequence of use. The evidence is compelling that low EF, prior to use initiation, is a significant vulnerability factor in the prediction of which individuals are at greatest risk for later impairments.

Longitudinal studies of large samples are needed to reconcile the question of whether EF is further compromised as addictive behaviors emerge over time. Such studies must include substance-free baseline assessments. Given its epidemiological design, the recently initiated NIH-funded Adolescent-Brain-and Cognitive Development (ABCD) project may ultimately be able to address this question as the study cohort, aged 9–10 at initial enrollment, ages into substance misuse. Of the few longitudinal studies that do exist, evidence for an increasing lack of behavioral control post-use (that is, a *decline* in EF), even during adolescence when cognitive control circuits are still developing, is sparse (cf., Khurana et al., 2013).

However, despite a relative lack of behavioral findings, an increasing number of reports suggest that the brain develops differently in adolescence and into young adulthood in individuals who consume common substances of abuse such as alcohol and marijuana. Moreover, more pronounced deviations are associated in some studies with an earlier age of onset of use or with relatively high frequencies and magnitudes of use, suggesting dose-response effects in support of neurotoxicity models. This is compelling evidence of developmental neuroplasticity of executive control systems in the context of substance misuse. Linking those deviations to expressed behavior has been challenging, perhaps because of the cognitive reserve that characterizes youth and/or perhaps because EF is being measured in the laboratory outside of real-world settings where emotional salience may be more pronounced in biasing behavior.

In conclusion, neurotoxicity models of substance use and EF function have a great deal of support in the preclinical literature and in recent human imaging studies. A goal for future epidemiological studies is to overcome interpretive confounds so that the progression of addiction as it unfolds longitudinally in vulnerable individuals can be better understood in relation to executive dysfunction.

Acknowledgements

This work was supported by grants AA002033 and DA017843 awarded by the National Institutes of Health to M. Luciana.

References

Arnsten, A. F., & Rubia, K. (2012). Neurobiological circuits regulating attention, cognitive control, motivation, and emotion: Disruptions in neurodevelopmental psychiatric disorders. *Journal of the American Academy of Child and Adolescent Psychiatry, 51*(4), 356–367.

Bava, S., Frank, L. R., McQueeny, T., Schweinsburg, B. C., Schweinsburg, A. D., & Tapert, S. F. (2009). Altered white matter microstructure in adolescent substance users. *Psychiatry Research, 173*(3), 228–237.

Bernardin, F., Maheut-Bosser, A., & Paille, F. (2014). Cognitive impairments in alcohol-dependent subjects. *Frontiers in Psychiatry, 5*, 78.

Bjork, J. M., & Pardini, D. A. (2015). Who are those risk-taking adolescents? Individual differences in developmental neuroimaging research. *Developmental Cognitive Neuroscience, 11*(1), 56–64.

Caspi, A., Moffitt, T., Newman, D., & Silva, P. (1996). Behavioral observations at age 3 years predict adult psychiatric disorders. *Archives of General Psychiatry, 53*(11), 1033–1039.

Cheetham, A., Allen, N., Whittle, S., Simmons, J., Yucel, M., & Lubman, D. (2012). Orbitofrontal volumes in early adolescence predict initiation of cannabis use: A 4-year longitudinal and prospective study. *Biological Psychiatry, 71*(8), 684–692.

Courtney, K. E., Ghahremani, D. G., & Ray, L. A. (2013). Frontostriatal functional connectivity during response inhibition in alcohol dependence. *Addiction Biology, 18*(3), 593–604.

Cousijn, J., Wiers, R., Ridderinkhof, R., van den Brink, W., Veltman, D., & Goudriaan, A. (2014). Effect of baseline cannabis use and working memory network function on changes in cannabis use in heavy cannabis users: A prospective fMRI study. *Human Brain Mapping, 35*(5), 2470–2482.

Cservenka, A., & Nagel, B. (2012). Risky decision-making: An fMRI study of youth at high risk for alcoholism. *Alcoholism: Clinical and Experimental Research, 36*(4), 604–615.

Dahlin, E., Nyberg, L., Bäckman, L., & Neely, A. S. (2008). Plasticity of executive functioning in young and older adults: Immediate training gains, transfer, and long-term maintenance. *Psychology and Aging, 23*(4), 720–730.

Deckel, A. W., & Hesselbrock, V. (1996). Behavioral and cognitive measurement predicts scores on the MAST: A 3-year prospective study. *Alcoholism: Clinical and Experimental Research, 20*(7), 1173–1178.

Downer, B., Jiang, Y., Zanjani, F., & Fardo, D. (2014). Effects of alcohol consumption on cognition and regional brain volumes among older adults. *American Journal of Alzheimers Disease and Other Dementias, 30*(4), 364–374.

Duijkers, J. C. L. M., Vissers, C. T. W. M., & Egger, J. I. M. (2016). Unraveling executive functioning in dual diagnosis. *Frontiers in Psychology, 7*, 979.

Elkins, I. J., King, S. M., McGue, M., & Iacono, W. G. (2006). Personality traits and the development of nicotine, alcohol, and illicit drug disorders: Prospective links from adolescence to young adulthood. *Journal of Abnormal Psychology, 115*(1), 26–39.

Feldstein Ewing, S. W., Sakhardande, A., & Blakemore, S. J. (2014). The effect of alcohol consumption on the adolescent brain: A systematic review of MRI and fMRI studies of alcohol-using youth. *NeuroImage: Clinical, 5*, 420–437.

Fernández-Serrano, M. J., Pérez-García, M., & Verdejo-García, A. (2011). What are the specific vs. generalized effects of drugs of abuse on neuropsychological performance? *Neuroscience & Biobehavioral Reviews, 35*(3), 377–406.

Fernie, G., Peeters, M., Gullo, M., Christiansen, P., Cole, J., Sumnall, H., & Field, M. (2013). Multiple behavioral impulsivity tasks predict prospective alcohol involvement in adolescents. *Addiction, 108*(11), 1916–1923.

Fontes, M. A., Bolla, K. I., Cunha, P. J., Almeida, P. P., Jungerman, F., Laranjeira, R. R., et al. (2011). Cannabis use before age 15 and subsequent executive functioning. *The British Journal of Psychiatry, 198*(6), 442–447.

Giancola, P. R., Martin, C. S., Tarter, R. E., Pelham, W., & Moss, H. B. (1996). Executive cognitive functioning and aggressive behavior in preadolescent boys at high risk for substance abuse/dependence. *Journal of Studies on Alcohol, 57*(4), 352–359.

Giancola, P., & Mezzich, A. (2003). Executive functioning, temperament, and drug use involvement in adolescent females with a substance use disorder. *Journal of Child Psychology and Psychiatry, 44*(6), 857–866.

Giancola, P. R., and Parker, A. M. (2001). A six-year prospective study of pathways toward drug use in adolescent boys with and without a family history of a substance use disorder. *Journal of Studies on Alcohol, 62*(2), 166–178.

Giancola, P., & Tarter, R. (1999). Executive cognitive functioning and risk for substance abuse. *Psychological Science, 10*(3), 203–205.

Grant, J. E., & Chamberlain, S. R. (2014). Impulsive action and impulsive choice across substance and behavioral addictions: Cause or consequence? *Addictive Behaviors, 39*(11), 1632–1639.

Grenard, J., Ames, S., Wiers, R., Thush, C., Sussman, S., & Stacy, A. (2008). Working memory capacity moderates the predictive effects of drug-related associations on substance use. *Psychology of Addictive Behaviors, 22*(3), 426–432.

Groenman, A., Oosterlaan, J., Greven, C., Jelle Vuijk, P., Rommelse, N., Franke, B., et al. (2015). Neurocognitive predictors of substance use disorders and nicotine dependence in ADHD probands, their unaffected siblings, and controls: A 4-year prospective follow-up. *Journal of Child Psychology and Psychiatry, 56*(5), 521–529.

Handley, E., Chassin, L., Haller, M., Bountress, K., Dandreaux, D., & Beltran, I. (2011). Do executive and reactive disinhibition mediate the effects of familial substance use disorders on adolescent externalizing outcomes? *Journal of Abnormal Psychology, 120*(3), 528–542.

Heitzeg, M. M., Nigg, J. T., Hardee, J. E., Soules, M., Steinberg, D., Zubieta, J. K., & Zucker, R. A. (2014). Left middle frontal gyrus response to inhibitory errors in children prospectively predicts early problem substance use. *Drug and Alcohol Dependence, 141*(1), 51–57.

Herting, M., Fair, D., & Nagel, B. (2011). Altered fronto-cerebellar connectivity in alcohol-naïve youth with a family history of alcoholism. *NeuroImage, 54*(4), 2582–2589.

Holmes, A. J., Hollinshead, M. O., Roffman, J. L., Smoller, J. W., & Buckner, R. L. (2016). Individual differences in cognitive control circuit anatomy link sensation seeking, impulsivity, and substance use. *The Journal of Neuroscience, 36*(14), 4038–4049.

Khoury, J. E., Milligan, K., & Girard, T. A. (2015). Executive functioning in children and adolescents prenatally exposed to alcohol: A meta-analytic review. *Neuropsychology Review, 25*(2), 149–170.

Khurana, A., Romer, D., Betancourt, L., Brodsky, N., Giannetta, J., & Hurt, H. (2013). Working memory ability predicts trajectories of early alcohol use in adolescents: The mediational role of impulsivity. *Addiction, 108*(3), 506–515.

Kim-Spoon, J., Deater-Deckard, K., Holmes, C. J., Lee, J. I., Chiu, P. H., & King-Casas, B. (2016). Behavioral and neural inhibitory control moderates the effects of reward sensitivity on adolescent substance use. *Neuropsychologia, 91*, 318–326.

Kolb, B., Mychasiuk, R., Muhammad, A., Li, Y., Frost, D. O., & Gibb, R. (2012). Experience and the developing prefrontal cortex. *Proceedings of the National Academy of Science U.S.A., 16*, 109(S2), 17186–17193.

Koob, G. F., & Volkow, N. D. (2010). Neurocircuitry of addiction. *Neuropsychopharmacology, 35*, 217–238.

Koob, G. F., & Volkow, N. D. (2016). Neurobiology of addiction: A neurocircuitry analysis. *The Lancet, 3*(8), 760–773.

Lebel, C., & Beaulieu, C. (2011). Longitudinal development of human brain wiring continues from childhood into adulthood. *The Journal of Neuroscience, 31*(30), 10937–10947.

Lisdahl, K. M., Gilbart, E. R., Wright, N. E., & Shollenbarger, S. (2013). Dare to delay? The impacts of adolescent alcohol and marijuana use onset on cognition, brain structure, and function. *Frontiers in Psychiatry, 4*, 53.

Luciana, M., & Collins, P. F. (2012). Incentive motivation, cognitive control, and the adolescent brain: Is it time for a paradigm shift? *Child Development Perspectives, 6*(4), 392–399.

Luciana, M., Collins, P. F., Muetzel, R. L., & Lim, K. O. (2013). Effects of alcohol use initiation on brain structure in typically developing adolescents. *American Journal of Drug and Alcohol Abuse, 39*(6), 345–355.

Luciana, M., Conklin, H. M., Hooper, C. J., & Yarger, R. S. (2005). The development of nonverbal working memory and executive control processes in adolescents. *Child Development, 76*(3), 697–712.

Lyu, J., & Lee, S. H. (2014). Alcohol consumption and cognitive impairment among Korean older adults: Does gender matter? *International Psychogeriatrics, 26*(2), 335–340.

Mahmood, O., Goldenberg, D., Thayer, R., Migliorini, R., Simmons, A., & Tapert, S. (2013). Adolescents' fMRI activation to a response inhibition task predicts future substance use. *Addictive Behaviors, 38*(1), 1435–1441.

Mahmood, O. M., Jacobus, J., Bava, S., Scarlett, A., & Tapert, S. F. (2010). Learning and memory performances in adolescent users of alcohol and marijuana: Interactive effects. *Journal of Studies on Alcohol and Drugs, 71*(6), 885–894.

Malone, S. M., Luciana, M., Wilson, S., Sparks, J. C., Hunt, R. H., & Thomas, K. M. (2014). Adolescent drinking and motivated decision-making: A co-twin-control investigation with monozygotic twins. *Behavior Genetics, 44*(4), 407–418.

McGue, M., Iacono, W. G., Legrand, L. N., & Elkins, I. (2001). The origins and consequences of age at first drink. I. Associations with substance-use disorders, disinhibitory behavior and psychopathology, and P3 amplitude. *Alcoholism: Clinical and Experimental Research, 25*(8), 1156–1165.

Miyake, A., Friedman, N. P., Emerson, M. J., Witzki, A. H., Howerter, A., & Wager, T. D. (2000). The unity and diversity of executive functions and their contributions to complex "frontal lobe" tasks: A latent variable analysis. *Cognitive Psychology, 41*(1), 49–100.

Moussa, M. N., Simpson, S. L., Mayhugh, R. E., Grata, M. E., Burdette, J. H., Porrino, L. J., & Laurienti, P. J. (2015). Long-term moderate alcohol consumption does not exacerbate age-related cognitive decline in healthy, community-dwelling older adults. *Frontiers in Aging Neuroscience, 6*, 341.

Nigg, J., Wong, M., Martel, M., Jester, J., Puttler, L., Glass, J., et al. (2006). Poor response inhibition as a predictor of problem drinking and illicit drug use in adolescents at risk for alcoholism and other substance use disorders. *Journal of the American Academy of Child and Adolescent Psychiatry, 45*(4), 468–475.

Nixon, K., & McClain, J. A. (2010). Adolescence as a critical window for developing an alcohol use disorder: Current findings in neuroscience. *Current Opinions in Psychiatry, 23*(3), 227–232.

Norman, A. L., Pulido, C., Squeglia, L. M., Spadoni, A. D., Paulus, M. P., & Tapert, S. F. (2011). Neural activation during inhibition predicts initiation of substance use in adolescence. *Drug and Alcohol Dependence, 119*(3), 216–223.

Olson, E. A., Hooper, C. J., Collins, P., & Luciana, M. (2007). Adolescents' performance on delay and probability discounting tasks: Contributions of age, intelligence, executive functioning, and self-reported externalizing behavior. *Personality and Individual Differences, 43*(7), 1886–1897.

Peeters, M., Janssen, T., Monshouwer, K., Boendermaker, W., & Pronk, T. (2015). Weaknesses in executive functioning predict the initiating of adolescents' alcohol use. *Developmental Cognitive Neuroscience, 16,* 139–146.

Peeters, M., Monshouwer, K., Schoot, R. A., Janssen, T., Vollebergh, W. A. M., & Wiers, R. W. (2013). Automatic processes and the drinking behavior in early adolescence: A prospective study. *Alcoholism: Clinical and. Experimental Research, 37*(10), 1737–1744.

Peeters, M., Wiers, R., Monshouwer, K., van de Schoot, R. A. G., Janssen, T., & Vollebergh, W. (2012). Automatic processes in at-risk adolescents: The role of alcohol-approach tendencies and response inhibition in drinking behavior. *Addiction, 107*(11), 1339–1346.

Peeters, M., Wollebergh, W., Wiers, R., & Field, M. (2014). Psychological changes and cognitive impairments in adolescent heavy drinkers. *Alcohol and Alcoholism, 49*(2), 182–186.

Pentz, M. A., & Riggs, N. (2013). Longitudinal relationships of executive cognitive function and parent influence to child substance use and physical activity. *Prevention Science, 14*(3), 229–237.

Pentz, M. A., Shin, H., Riggs, N., Unger, J., Collison, K., & Chou, C. (2015). Parent, peer, and executive function relationships to early adolescent e-cigarette use: A substance use pathway? *Addictive Behaviors, 42,* 73–78.

Riggs, N., Spruijt-Metz, D., Chou, C., & Pentz, M. A. (2012). Relationships between executive cognitive function and lifetime substance use and obesity-related behaviors in fourth grade youth. *Child Neuropsychology, 18*(1), 1–11.

Roberts, C. A., Jones, A., & Montgomery, C. (2016). Meta-analysis of executive functioning in ecstasy/polydrug users. *Psychological Medicine, 46*(8), 1581–1596.

Sabia, S., Elbaz, A., Britton, A., Bell, S., Dugravot, A., Shipley, M., et al. (2014). Alcohol consumption and cognitive decline in early old age. *Neurology, 82*(4), 332–339.

SAMHSA: Substance Abuse and Mental Health Services Administration. (2014). *Results from the 2013 National Survey on Drug Use and Health: Summary of national findings.* NSDUH Series H-48, HHS Publication No. (SMA) 14-4863. Rockville, MD: Substance Abuse and Mental Health Services Administration.

Schoemaker, K., Mulder, H., Deković, M., & Matthys, W. (2013). Executive functions in preschool children with externalizing behavior problems: A meta-analysis. *Journal of Abnormal Child Psychology, 41*(3), 457–471.

Selemon, L. D. (2013). A role for synaptic plasticity in the adolescent development of executive function. *Translational Psychiatry, 3*(3), e238.

Smith, J. L., Mattick, R. P., Jamadar, S. D., & Iredale, J. M. (2014). Deficits in behavioural inhibition in substance abuse and addiction: A meta-analysis. *Drug and Alcohol Dependence, 145*(1), 1–33.

Snyder, H. R., Miyake, A., & Hankin, B. L. (2015). Advancing understanding of executive function impairments and psychopathology: Bridging the gap between clinical and cognitive approaches. *Frontiers in Psychology, 6,* 328.

Sowell, E. R., Peterson, B. S., Thompson, P. M., Welcome, S. E., Henkenius, A. L., & Toga, A. W. (2003). Mapping cortical change across the human life span. *Nature Neuroscience, 6*(3), 309–315.

Spronk, D. V., van Wel, J. H. P., Ramaekers, J. G., & Verkes, R. J. (2013). Characterizing the cognitive effects of cocaine: A comprehensive review. *Neuroscience and Biobehavioral Reviews, 37*(8), 1838–1859.

Squeglia, L. M., Tapert, S. F., Sullivan, E. V., Jacobus, J., Meloy, M. J., Rohlfing, T., & Pfefferbaum, A. (2015). Brain development in heavy-drinking adolescents. *American Journal of Psychiatry, 172*(6), 531–542.

Stephan, R. A., Alhassoon, O. M., Allen, K. E., Wollman, S. C., Hall, M., Thomas, W. J., et al. (2016). Meta-analyses of clinical neuropsychological tests of executive dysfunction and impulsivity in alcohol use disorder. *American Journal of Drug and Alcohol Abuse, 12*, 1–20.

Tapert, S. F., Granholm, E., Leedy, N. G., & Brown, S. A. (2002). Substance use and withdrawal: Neuropsychological functioning over 8 years in youth. *Journal of the International Neuropsychological Society, 8*(7), 873–883.

Tarter, R., Kirisci, L., Mezzich, A., Cornelius, J., Pajer, K., Vanyukov, M., et al. (2003). Neurobehavioral disinhibition in childhood predicts early age at onset of substance use disorder. *American Journal of Psychiatry, 160*(6), 1078–1085.

Thush, C., Wiers, R., Ames, S., Grenard, J., Sussman, S., & Stacy, A. (2008). Interactions between implicit and explicit cognition and working memory capacity in the prediction of alcohol use in at-risk adolescents. *Drug and Alcohol Dependence, 94*(1–3), 116–124.

Verdejo-Garcia, A., Lawrence, A., & Clark, L. (2008). Impulsivity as a vulnerability marker for substance-use disorders: Review of findings from high-risk research, problem gamblers and genetic association studies. *Neuroscience and Biobehavioral Reviews, 32*(4), 777–810.

Wetherill, R. R., Castro, N., Squeglia, L. M., & Tapert, S. F. (2013). Atypical neural activity during inhibitory processing in substance-naïve youth who later experience alcohol-induced blackouts. *Drug and Alcohol Dependence, 128*(3), 243–249.

Wetherill, R. R., Squeglia, L. M., Yang, T. T., & Tapert, S. F. (2013). A longitudinal examination of adolescent response inhibition: Neural differences before and after the initiation of heavy drinking. *Psychopharmacology, 230*(4), 663–671.

Woltering, S., Lishak, V., Hodgson, N., Granic, I., & Zelazo, P. D. (2016). Executive function in children with externalizing and comorbid internalizing behavior problems. *Journal of Child Psychology and Psychiatry, 57*(1), 30–38.

Wong, M., Nigg, J., Zucker, R., Puttler, L., Fitzgerald, H., Jester, J., et al. (2006). Behavioral control and resiliency in the onset of alcohol and illicit drug use: A prospective study from preschool to adolescence. *Child Development, 77*(4), 1016–1033.

Young, S. E., Friedman, N. P., Miyake, A., Willcutt, E. G., Corley, R. P., Haberstick, B. C., & Hewitt, J. K. (2009). Behavioral disinhibition: Liability for externalizing spectrum disorders and its genetic and environmental relation to response inhibition across adolescence. *Journal of Abnormal Psychology, 118*(1), 117–130.

17

TRAJECTORIES AND MODIFIERS OF EXECUTIVE FUNCTION

Normal Aging to Neurodegenerative Disease

G. Peggy McFall, Shraddha Sapkota, Sherilyn Thibeau, & Roger A. Dixon

This chapter addresses our emerging understanding of how executive function (EF) trajectories of normal aging can be related to subsequent transitions of cognitive impairment to two neurodegenerative diseases (NDD). First, we provide our definition of EF in the context of cognitively normal aging. Second, we summarize growing research on risk and protection biomarkers that modify trajectories of EF change from normal to impaired aging in light of a dynamic interactive approach. Third, we examine some specific characteristics of EF in regard to two sets of related conditions: specifically, (1a) Mild Cognitive Impairment (MCI), (1b) Alzheimer's disease (AD); and (2a) Vascular Cognitive Impairment (VCI), (2b) Vascular Dementia (VaD).

Normal Aging

EFs are complex processes by which individuals successfully engage in independent, appropriate, and purposeful behavior (Chevalier & Clark, Chapter 2, this volume; Crone, Peters, & Steinbeis, Chapter 3, this volume; Cuevas, Rajan, & Bryant, Chapter 1, this volume). Along with gradual increases and then declines in EF performance across the life span, age-related changes in the structure of EF are thought to be an important predictor of change associated with exceptional, normal, or compromised cognition across adulthood (de Frias, Dixon, & Strauss, 2009). Using indicators known to define the three functional categories of EF (i.e., shifting, updating, inhibition; Miyake et al., 2000), the structural changes observed across the life span range from EF "differentiation" in children and young adults (Chevalier & Clark, Chapter 2, this volume; Crone et al., Chapter 3, this volume; Cuevas et al., Chapter 1, this volume; Miyake et al., 2000; Wiebe et al., 2010) to EF "dedifferentiation" in older adulthood (Adrover-Roig, Sesé, Barceló, & Palmer, 2012; de Frias, Dixon, & Strauss, 2006, 2009; Li, Vadaga, Bruce, & Lai, Chapter 4, this volume; McFall et al., 2014). Studies exploring the dedifferentiation theory of EF have produced mixed results (see Adrover-Roig et al., 2012; de Frias et al., 2006, 2009; Hull, Martin, Beier, Lane, & Hamilton, 2008; Li et al., Chapter 4, this volume). A cognitive theory of life-span differentiation–dedifferentiation offers a plausible explanation of these mixed results. Specifically, cognitive structure is associated with performance ability and underlying neuronal integrity rather than simply age (Anstey, Hofer, & Luszcz, 2003). Recently, de Frias and colleagues (2006, 2009) confirmed this theory for EF by establishing factorial structure associated with objectively defined cognitive status groups. A large group of older adults who exhibited a one-factor EF latent variable (de Frias et al., 2006) were further stratified by cognitive status (de Frias et al., 2009). Cognitively elite older adults exhibited a three-factor EF latent model, similar to that commonly seen in younger adults. Cognitively normal adults produced good-fitting one- and three-factor EF models. Cognitively impaired adults, on the other hand, were consistent with a one-factor EF model. Individual differences in EF performance and trajectories of change are well established and distinguish normal aging from that of disease states (Harada, Natelson Love, & Triebel, 2013).

The most common age-related EF declines are associated with concept formation, abstraction, mental flexibility, and inhibition, especially after the age of 70 (Harada et al., 2013; Li et al., Chapter 4, this volume). These age-related deficits contribute to decreased abilities in everyday activities that can compromise quality of life and have been linked to prefrontal cortex neurodegeneration (Deak, Freeman, Ungvari, Csiszar, & Sonntag, 2016; Hazlett, Figueroa, & Nielson, 2015; Luszcz, 2011; Turner & Spreng, 2012). EF performance and structural changes may be an important element in the early detection of NDD.

Neurodegenerative Disease

NDD is defined as the age-related and progressive loss in structure or function of nerve cells leading to compromised cognitive or mobility function (Gao, Peng, Deng, & Qing, 2013). The two NDDs selected for this chapter share several common features. Specifically, both AD and VaD, as well as their respective precursor conditions (MCI and VCI), are characterized by "multiple pathogenesis, cellular abnormalities, unexplained phenotypic variability, and fatal outcomes" (Gao et al., 2013, p. 50; Hardy & Gwinn-Hardy, 1998). We have selected AD and VaD because they are the two most prominent and prevalent NDDs, and indeed often appear together (Fotuhi, Hachinski, & Whitehouse, 2009; Iadecola, 2013). Dementia affects more than 46 million people worldwide (Prince et al., 2015). AD is the most common form of dementia, accounting for between 60 and 80% of all dementia cases (Alzheimer's Association, 2016). AD is characterized by the presence of both amyloid and tau neuropathology, accompanied by cognitive decline in multiple domains including memory and EF. VaD is a group of dementias associated with cerebrovascular pathologies and is the second most prevalent form of dementia, accounting for 10% of dementia cases. The definitions of AD and VaD are evolving due to the complex, dynamic, multi-modal, and synergistic interactions of biological, neurological, health, genetic, and environmental factors associated with these NDD. The interesting balance between non-modifiable positive (risk-reducing) and negative (risk-elevating) genetic factors combined with potentially modifiable positive and negative environmental factors throughout the life span can determine where one is situated on the continuum from exceptional cognitive aging to a variety of NDDs (Fotuhi et al., 2009). A dynamic polygon model has been utilized and adapted to (a) demonstrate the active, non-linear interchange between multiple risk factors (e.g., *Apolipoprotein E [APOE]* ε2 or ε4, vascular health, physical activity) and (b) illustrate how such factors are linked to differential trajectories of cognitive decline (see Fotuhi et al., 2009).

Clinical trials for NDD interventions have been frequent (i.e., 413 AD trials between 2002 and 2012) but largely unsuccessful (i.e., 99.6% failure rate; Cummings, Morstorf, & Zhong, 2014). This inability to identify a "cure" for any one NDD has resulted in a research shift toward that of early disease detection and monitoring (Barnes et al., 2010). The goal of early detection is to identify factors that can be targeted through early (pre-diagnosis) intervention in order to prevent or delay disease onset. Perturbations in EF performance (variability, fluctuations, dedifferentiation, decline) appear to be predictive of NDD sometimes as early as other cognitive changes (e.g., memory; Allain, Etcharry-Bouyx, & Verny, 2013; Smits et al., 2015). Historically, EF was thought to be a marker of AD progression – EF was preserved when memory deficits were detected with EF decline indicating more AD pathology (Allain et al., 2013; Hazlett et al., 2015). However, more recently EF decline has been detected several years before AD diagnosis and often in conjunction with memory deficits. This makes EF a promising predictor of MCI, AD, and potentially other NDD. In a recent review, early EF declines were linked to MCI, AD, and VaD (Smits et al., 2015). Finding links between early EF decline and other biomarker signals – both protective and risk conferring – is imperative in understanding the mechanisms underlying NDD and to identify modifiable biomarkers to aid in prevention.

Risk and Protection Biomarkers Influence Executive Function Trajectories

Age-related EF declines and structural changes may be alleviated or exacerbated by protection or risk biomarkers, including health, lifestyle, and genetic factors (Anstey, Cherbuin, & Herath, 2013; Dodge et al., 2014; Harada et al., 2013). Within these clusters are modifiers that are thought to be (a) risk reducing such as *APOE* ε2+ or higher levels of education, (b) risk elevating such as poor vascular health, and (c) risk magnifying such as having both an *APOE* ε4 allele and poor vascular health (e.g., systolic blood pressure, Bender, & Raz, 2012; hypertension, de Frias, Schaie, & Willis, 2014; arterial stiffness, McFall et al., 2014; Raz, Dahle, Rodrigue, Kennedy, & Land, 2011).

It has been hypothesized that maintaining an active lifestyle cognitively, socially, and physically may delay or prevent cognitive decline and dementia (Bielak, 2010; Harada et al., 2013; Small, Dixon, McArdle, & Grimm, 2011). Within the lifestyle domain key modifiers identified in our research have been lifestyle engagement activities. Older adults who were more engaged in lifestyle activities exhibited better EF performance and were better able to maintain cognitive status over time than their less engaged counterparts (de Frias & Dixon, 2014; Thibeau, McFall, Wiebe, Anstey, & Dixon, 2016).

Approximately 60% of the variability in general cognitive ability is thought to be attributable to genetics (Harada et al., 2013). Genetic risk-reducing and risk-elevating effects on EF have been reported. Carriers of the *Insulin degrading enzyme* (*IDE*) G+ genotype (thought to be risk reducing for cognition) exhibited better EF performance at the mean centering age of 75 years than those with risk-elevating homozygote AA genotype (McFall et al., 2014). Lower EF, but not memory, performance has been observed in cognitively intact older adults with an *APOE* ε4+ genotype when compared to those without a ε4 allele (Luck et al., 2015; Raz, Rodrigue, Kennedy, & Land, 2009). There are strong genetic links to NDDs but phenotypes of a given mutation do not predict a single clinical outcome and this variation may be due to epistatic interactions as well as interactions with modifiable environmental markers (Hardy & Gwinn-Hardy, 1998). This highlights the increased importance placed on other risk and protection biomarkers as they modulate EF trajectories from normal aging through impairment and to NDD (Fotuhi et al., 2009).

A Dynamic Interactive Approach to Trajectories of Executive Function Aging

Identification of risk and protection biomarkers associated with EF aging has resulted in productive approaches to determining EF trajectories. Risk indexes using a composite score of risk factors from multiple domains have been found to be predictive of dementia onset (Anstey et al., 2013; Barnes et al., 2010; Kivipelto et al., 2006). Selected predictors of dementia have been found to predict cognitive trajectories as well. In order to develop effective interventions, it is essential to discover and validate the optimal combinations of biomarker risk and protection. An interaction between two or more biomarkers may identify (a) possible mechanisms, (b) potential interventions, and (c) populations most likely to benefit from early detection (Lindenberger et al., 2008; Nagel et al., 2008; Raz et al., 2009, 2011). Due in part to the low penetrance associated with single gene analyses, an interactive approach has been utilized. In this approach genetic risk is assessed in interaction with other known risk biomarkers (Raz & Lustig, 2014). For example, one study found no independent *APOE* effects associated with EF outcomes, but when analyzed in interaction with pulse pressure, *APOE* ε2 was protective against the deleterious effects of vascular health on EF trajectories (McFall, Sapkota, McDermott, & Dixon, 2016). In a study examining the interactive effects of *IDE* × vascular health, the carriers of the *IDE* G+ genotype – protective of EF decrements – showed poorer EF performance and more decline in the presence of poor vascular health (McFall et al., 2014). This suggests that persons with an *IDE* G allele might be targeted for vascular health interventions. Raz and colleagues (2011) reported *COMT* effects on EF performance but only for males. Specifically, male *COMT* Met homozygotes exhibited better working memory performance than male *COMT* Val carriers. Some specific populations may benefit more from lifestyle changes. For example, in a study of gene × lifestyle effects, *IDE* G+ or *BDNF* Met carriers exhibited EF deficits if they had low levels of everyday activity but not when they were more active (Thibeau et al., 2016). Magnified risk for EF decrements have also been observed for gene × gene × age interactions (Nagel et al., 2008). Specifically, *COMT* Val carriers exhibited lower EF performance when they were also *BDNF* Met carriers and this effect was magnified by older age. In a similar approach, cumulative (or additive) genetic risk moderated EF decrements associated with a health-related biomarker such as pulse pressure (McFall et al., 2016) and with another genetic risk factor such as *APOE* (Sapkota, Vergote, Westaway, Jhamandas, & Dixon, 2015). This dynamic interactive approach identifies specific biomarkers that may reveal important mechanistic patterns that will contribute to the development of precision interventions. In addition, specific subgroups are identified that allow researchers to develop and implement precise interventions for those most likely to respond.

The Mild Cognitive Impairment-Alzheimer's Disease Continuum

Mild Cognitive Impairment

MCI is an intermediate stage in the accelerated and expanding cognitive decline and impairment process that typically leads to dementia (Petersen et al., 2014). Recent criteria for MCI classification are: (a) changes in cognitive functioning; (b) impairment in at least one of the following cognitive domains: memory, EF, attention, language, and visuospatial skills; (c) maintenance of activities of daily living; (d) preserved general cognition; and (e) no signs of dementia (Albert et al., 2011). Studies have shown that the fluid nature of the MCI phase can result in (a) maintenance of MCI, (b) transition to dementia, and (c) transitions back to normal cognitive aging (approximately 15%) during follow-up assessments (Koepsell & Monsell, 2012). This may be due to cognitive fluctuations, the impact of underlying medical conditions or medications, misclassification, higher cognitive reserve, and differences in diagnostic criteria. However, MCI patients who revert back to normal cognitive aging remain at a higher risk of converting to dementia in the future. Differences in cognitive trajectories may be due to MCI subtype, of particular interest when examining underlying mechanisms and targeting interventions to specific clinical groups. Therefore, tracking MCI patients longitudinally to identify and understand risk factors along the MCI–AD continuum is necessary to prevent or delay AD onset in the preclinical stages (Albert, Moss, Blacker, Tanzi, & McArdle, 2007; DeCarlo et al., 2016; Dixon et al., 2014; Kirova, Bays, & Lagalwar, 2015; Reinvang, Espeseth, & Westlye, 2013).

Although MCI is often a memory-based phenomena, EF deficits and declines are also part of the story (Albert et al., 2007). Two categories of MCI have been identified based on the affected cognitive domain. Patients are classified with amnestic MCI (aMCI) if they have memory decrements or non-amnestic MCI (naMCI) if the cognitive deficits are in a domain other than memory (e.g., EF, language, visual special abilities; Petersen et al., 2014). These categories are further classified into impairment in one domain (i.e., MCI single-domain) or two or more domains (i.e., MCI multiple-domain). A study examining the four subcategories of MCI showed selective deficits in planning, problem solving, and working memory domains of EF whereas judgment in risk taking and decision making was still intact (Brandt et al., 2009). Notably, those with aMCI multiple-domain deficits exhibited the poorest EF performance of all four groups. Neural circuitry underlying the processes associated with judgment was not affected (i.e., orbitofrontal cortex and its striatal and thalamic connections). However, deficits were seen in neural circuitry underlying problem solving and working memory (i.e., dorsolateral prefrontal cortex). In a longitudinal study of naMCI vs controls, adults showed increased association of EF deficits with white matter radial diffusivity in frontal, cingulate, and entorhinal regions but cortical thinning differences were only observed in the caudal middle frontal region (Grambaite et al., 2011). EF declines may be among the most important predictors of MCI conversion to AD (Gomar et al., 2011).

Genetic risk factors associated with AD are commonly examined to determine if they are also biomarkers of MCI (Brainerd, Reyna, Petersen, Smith, & Taub, 2011; DeCarlo et al., 2016; Dixon et al., 2014). *APOE* ε4+ genotype has been shown to modify EF trajectory in MCI (Albert et al., 2007; Reinvang et al., 2013). Health and lifestyle biomarkers are also found to have independent and interactive (with other biomarker) effects on risk of MCI and later conversion to AD (DeCarlo et al., 2016; Dixon et al., 2014; Reinvang et al., 2013).

Alzheimer's Disease

Although memory deficits and medial temporal lobe pathology are the first deficits observed in AD, recently diagnosed AD patients may also show deficits in EF and other cognitive domains (Allain et al., 2013; Kirova et al., 2015; Marshall et al., 2011; Weintraub, Wicklund, & Salmon, 2012). AD patients exhibit working memory, problem-solving, shifting, and inhibition (suggested to be the most affected) deficits (Allain et al., 2013; Amieva, Phillips, Della Sala, & Henry, 2004). Family history of AD may also significantly predict EF performance. For example, cognitively normal older adults with family history of AD showed poorer performance on the Wisconsin Card Sorting Test than those without family history of AD (Hazlett et al., 2015). As AD progresses, activities of daily living become more compromised and these changes have been linked to poor EF performance rather than memory deficits (Hazlett et al., 2015; Marshall et al., 2011).

Commonly recognized biomarkers of AD are (a) plasma biomarkers such as amyloid-beta levels, (b) cerebral spinal fluid (CSF) markers of amyloid-beta and tau, and (c) imaging biomarkers such as reduced hippocampal volume (Reitz & Mayeux, 2014). Changes in CSF amyloid-beta levels were associated with poor EF performance (Harrington et al., 2013). This corresponds to extracellular amyloid build-up that appears in the basal isocortex. These researchers suggest using EF measures as a screening tool in the detection of AD. Biomarker progression has been reported as most predictive of memory and EF decline in AD (Dodge et al., 2014). For example, 65% of EF decline experienced by AD patients was explained by ventricular volume increases.

Autopsy results show that AD pathology is commonly mixed with VaD pathology (e.g., cerebral infarcts; Schneider, Arvanitakis, Bang, & Bennett, 2007). Although EF performance per se is unlikely to distinguish between AD and VaD (Allain et al., 2013), comparisons are important to explore.

The Vascular Cognitive Impairment: Vascular Dementia Continuum

Vascular Cognitive Impairment

Vascular Cognitive Impairment (VCI) is a heterogeneous group of cognitive disorders that share "evidence of clinical stroke or subclinical vascular brain injury and cognitive impairment affecting at least one cognitive domain" (Gorelick et al., 2011, p. 2677; Moorhouse & Rockwood, 2008). The term VCI has replaced "multi-infarct dementia", and encompasses the range from VCI-no dementia to VaD (Moorhouse & Rockwood, 2008). VCI reflects cognitive impairment due to vascular factors, some of which may be modifiable and responsive to intervention (Iadecola, 2013; Moorhouse & Rockwood, 2008).

Older adults with VCI have a unique cognitive profile compared to those with AD. Specifically, individuals with VCI are largely considered to have better memory performance than those with AD, but worse processing speed and greater EF deficits (Desmond, 2004; Gorelick & Nyenhuis, 2013). In fact, EF deficits are considered to be the cognitive hallmark of VCI (Iadecola, 2013). However, despite the differing cognitive profiles, the diagnostic specificity of VCI is challenging due to the considerable range of cognitive impairments associated with vascular disease. To refine diagnostic criteria and improve characterization of VCI, the Canadian Stroke Network and National Institute for Neurological Disorders and Stroke recommend use of an evaluation battery, which consists of clinical, behavioral, and cognitive measures (Hachinski et al., 2006). Notably, general cognitive screening measures sometimes used for MCI, such as the Mini-Mental State Exam, have proven to be insensitive to the EF deficits experienced in VCI (Bour, Rasquin, Boreas, Limburg, & Verhey, 2010; Ihara, Okamoto, Hase, & Takahashi, 2013). Therefore, test batteries with specific EF tasks have been recommended for screening (Sudo et al., 2013). In addition, neuroradiological evidence of cerebrovascular disease pathology, such as white matter hyperintensities or brain infarcts, should be used for clinical diagnosis of VCI (Gorelick et al., 2011; Hachinski et al., 2006; Sachdev et al., 2014).

Consistent with the dynamic interactive approach to EF aging, the risk factors and predictors of VCI are diverse. They include age, brain atrophy, hypertension, smoking, arterial fibrillation, and white matter lesions and damage (Rincon & Wright, 2013). Recently, vascular lesions in individuals with VCI (i.e., white matter hyperintensities) have been associated with impaired EF performance (Sudo et al., 2013). Additionally, protective factors that reduce ischemic changes in the frontal lobe (i.e., physical activity) have been found to be positively correlated with EF performance in patients with VCI (Ihara et al., 2013). Taken together, recent results provide evidence of interactions between risk and protective factors that may highlight EF as a marker of underlying neurological changes associated with the VCI to VaD transition (Rincon & Wright, 2013).

Vascular Dementia

VaD is regarded as the most acute form of VCI (Gorelick et al., 2011) with multiple levels of classification including subcortical ischemic VaD, strategic infarct dementia, hypoperfusion dementia, hemorrhagic dementia, and dementia caused by specific arteriopathies (O'Brien et al., 2003; Rincon & Wright, 2013). Overall, VaD is caused by interruption of the blood supply to the brain typically due to damaged blood vessels.

Diagnostic criteria for VaD is based on history of clinical stroke or vascular disease and the presence of a cognitive disorder (Gorelick et al., 2011). Unlike

AD, VaD patients exhibit maintenance of memory performance but a higher degree of EF impairment. Although EF deficits in VaD are primarily due to infarctions of the thalamus, caudate, and frontal-subcortical circuit connections, recurrent strokes result in impairments in other cognitive domains (Desmond, 2004). Other vascular changes leading to cognitive impairment include cerebral perfusion, cumulative tissue damage, white matter lesions leading to reduced use of glucose in the frontal lobes, interruptions in brain connectivity, and cerebral atrophy (Iadecola, 2013). Risk factors for VaD, such as hypertension, diabetes, hyperlipidemia, smoking, and hyperhomocystinemia, have been identified to increase dementia risk (Bourdel-Marchasson, Lapre, Laksir, & Puget, 2010; Iadecola, 2013; Sahathevan, Brodtmann, & Donnan, 2012). Ischemic stroke also increases risk, as approximately one-quarter of survivors develop immediate or delayed VCI or VaD (Kalaria, Akinyemi, & Ihara, 2016) and one-third of stroke patients may develop dementia within 3 months of a stroke (O'Brien et al., 2003; Pohjasvaara et al., 1998). Major changes leading to cognitive impairment after stroke include microinfarction, neuronal atrophy, and microvascular changes associated with the blood–brain barrier (Kalaria et al., 2016). Hypertension is one of the leading causes of VaD and often coexists with white matter damage (Buckner, 2004; Söderlund, Nyberg, Adolfsson, Nilsson, & Launer, 2003).

EF activity is largely disrupted in VaD patients because of interruptions to axonal connections between the cerebral cortex and deep gray matter in small vessel disease which mainly affects frontal lobe white matter and certain basal ganglia areas (Series & Esiri, 2012). Decline in vascular functions may affect white matter structures in the frontal-striatal circuits important in EF (Buckner, 2004). Compared with AD patients, VaD patients have preserved long term-memory but show more impairment in EF, and greater ventricular dilation (Gorelick et al., 2011; Smits et al., 2015).

Future Directions in Executive Function Research in Aging and Neurodegenerative Disease

We now note several key avenues for future research in EF in aging and NDD. First, although EF research across the life span has evolved in the last 20 years (also see Allain et al., 2013; Luszcz, 2011; Miyake et al., 2000), attention should continue to focus on improving EF measurements and the use of common measures across different populations (i.e., younger vs. older adults, normal aging older adults vs. AD or VaD patients). Relatively process-pure measures and multiple dimensions in EF should be included to accurately study interindividual differences and intraindividual change in EF in aging. Second, the strengths of using latent EF variables and invariance testing across groups and waves should be emphasized. The latent variable approach can represent the broader EF construct as well as different subdomains (e.g., shifting, inhibition, updating) and has the advantage of reducing measurement error associated with single EF tests. Third, longitudinal follow-ups are essential to identify key EF changes that occur from cognitively normal through impairment (MCI, VCI) and into NDD. Fourth, incorporating a diverse set of NDD-related biomarkers and neuroimaging markers will be a significant factor in determining moderating, modifying, and cumulative risk factors that magnify or attenuate the risk of EF decline. Fifth, a multi-domain lens may also identify pathways and mechanisms representing the synergistic effects of risk factors on EF decline in aging. Developing and applying this dynamic multi-domain approach may lead to early detection of EF decline in the preclinical stages with potential for significant changes through intervention. Sixth, we stress the need for future EF research in other NDD for which EF changes may be among the early pivotal signs of decline in the cognitive impairment process.

Conclusion

In this chapter, we outlined EF structure and function as it relates to cognitively normal and impaired aging. For the latter, we followed two streams of NDD, namely MCI – AD and VCI – VaD. We also explored emerging methodologies that will allow us to improve the precision and timing with which we can detect age-related decrements in EF. Accordingly, we emphasize that EF and aging is a phenomenon of interest along the full continuum from exceptional brain aging to NDD. Utilization of a dynamic interactive approach with the polygon of biomarkers that influence EF changes (normal to impaired) may be pivotal in (a) discovering the mechanisms underlying disease, (b) implementing interventions associated with these mechanisms, and (c) determining populations most likely to benefit from the intervention.

References

Adrover-Roig, D., Sesé, A., Barceló, F., & Palmer, A. (2012). A latent variable approach to executive control in healthy ageing. *Brain and Cognition, 78*, 284–299.

Albert, M., Moss, M. B., Blacker, D., Tanzi, R., & McArdle, J. J. (2007). Longitudinal change in cognitive performance among individuals with mild cognitive impairment. *Neuropsychology, 21*(2), 158–169.

Albert, M. S., DeKosky, S. T., Dickson, D., Dubois, B., Feldman, H. H., Fox, N. C., et al. (2011). The diagnosis of mild cognitive impairment due to Alzheimer's disease: Recommendations from the National Institute on Aging–Alzheimer's Association workgroups on diagnostic guidelines for Alzheimer's disease. *Alzheimer's & Dementia, 7*, 270–279.

Allain, P., Etcharry-Bouyx, F., & Verny, C. (2013). Executive functions in clinical and preclinical Alzheimer's disease. *Revue Neurologique, 169*(10), 695–708.

Alzheimer's Association. (2016). 2016 Alzheimer's disease facts and figures. *Alzheimer's & Dementia, 12*, 459–509.

Amieva, H., Phillips, L. H., Della Sala, S., & Henry, J. D. (2004). Inhibitory functioning in Alzheimer's disease. *Brain, 127*(5), 949–964.

Anstey, K. J., Cherbuin, N., & Herath, P. M. (2013). Development of a new method of assessing global risk of Alzheimer's disease for use in population health approaches to prevention. *Prevention Science, 14*, 411–421.

Anstey, K. J., Hofer, S. M., & Luszcz, M. A. (2003). Cross-sectional and longitudinal patterns of dedifferentiation in late-life cognitive and sensory function: The effects of age, ability, attrition, and occasion of measurement. *Journal of Experimental Psychology: General, 132*(3), 470–487.

Barnes, D. E., Covinsky, K. E., Whitmer, R. A., Kuller, L. H., Lopez, O. L., & Yaffe, K. (2010). Dementia risk indices: A framework for identifying individuals with a high dementia risk. *Alzheimer's & Dementia : The Journal of the Alzheimer's Association, 6*(2), 138–141.

Bender, A. R., & Raz, N. (2012). Age-related differences in memory and executive functions in healthy *APOE* ε4 carriers: The contribution of individual differences in prefrontal volumes and systolic blood pressure. *Neuropsychologia, 50*, 704–714.

Bielak, A. A. M. (2010). How can we not "lose it" if we still don't understand how to "use it"? Unanswered questions about the influence of activity participation on cognitive performance in older age: A mini-review. *Gerontology, 56*, 507–519.

Bour, A., Rasquin, S., Boreas, A., Limburg, M., & Verhey, F. (2010). How predictive is the MMSE for cognitive performance after stroke? *Journal of Neurology, 257*(4), 630–637.

Bourdel-Marchasson, I., Lapre, E., Laksir, H., & Puget, E. (2010). Insulin resistance, diabetes and cognitive function: Consequences for preventative strategies. *Diabetes & Metabolism, 36*(3), 173–181.

Brainerd, C. J., Reyna, V. F., Petersen, R. C., Smith, G. E., & Taub, E. S. (2011). Is the apolipoprotein E genotype a biomarker for mild cognitive impairment? Findings from a nationally representative study. *Neuropsychology, 25*(6), 679–689.

Brandt, J., Aretouli, E., Neijstrom, E., Samek, J., Manning, K., Albert, M. S., & Bandeen-Roche, K. (2009). Selectivity of executive function deficits in mild cognitive impairment. *Neuropsychology, 23*(5), 607–618.

Buckner, R. L. (2004). Memory and executive function in aging and AD: Multiple factors that cause decline and reserve factors that compensate. *Neuron, 44*(1), 195–208.

Cummings, J. L., Morstorf, T., & Zhong, K. (2014). Alzheimer's disease drug-development pipeline: Few candidates, frequent failure. *Alzheimer's Research & Therapy, 6*(4), 37–45.

Deak, F., Freeman, W. M., Ungvari, Z., Csiszar, A., & Sonntag, W. E. (2016). Recent developments in understanding brain aging: Implications for Alzheimer's disease and vascular cognitive impairment. *Journals of Gerontology: Biological Sciences, 71*(1), 13–20.

DeCarlo, C. A., MacDonald, S. W. S., Vergote, D., Jhamandas, J., Westaway, D., & Dixon, R. A. (2016). Vascular health, genetic risk affect mild cognitive impairment status and 4-year stability: Evidence from the Victoria Longitudinal Study. *Journal of Gerontology: Psychological Sciences, 71*(6), 1004–1014.

de Frias, C. M., & Dixon, R. A. (2014). Lifestyle engagement affects cognitive status differences and trajectories on executive functions in older adults. *Archives of Clinical Neuropsychology, 29*, 16–25.

de Frias, C. M., Dixon, R. A., & Strauss, E. (2006). Structure of four executive functioning tests in healthy older adults. *Neuropsychology, 20*, 206–214.

de Frias, C. M., Dixon, R. A., & Strauss, E. (2009). Characterizing executive functioning in older special populations: From cognitively elite to cognitively impaired. *Neuropsychology, 23*, 778–791.

de Frias, C. M., Schaie, K. W., & Willis, S. L. (2014). Hypertension moderates the effect of APOE on 21-year cognitive trajectories. *Psychology and Aging, 29*(2), 431–439.

Desmond, D. W. (2004). The neuropsychology of vascular cognitive impairment: Is there a specific cognitive deficit? *Journal of the Neurological Sciences, 226*(1), 3–7.

Dixon, R. A., DeCarlo, C. A., MacDonald, S. W., Vergote, D., Jhamandas, J., & Westaway, D. (2014). APOE and COMT polymorphisms are complementary biomarkers of status, stability, and transitions in normal aging and early mild cognitive impairment. *Frontiers in Aging Neuroscience, 6*, 236.

Dodge, H. H., Zhu, J., Harvey, D., Saito, N., Silbert, L. C., Kaye, J. A., et al. (2014). Biomarker progressions explain higher variability in stage-specific cognitive decline than baseline values in Alzheimer disease. *Alzheimer's & Dementia, 10*(6), 690–703.

Fotuhi, M., Hachinski, V., & Whitehouse, P. J. (2009). Changing perspectives regarding late-life dementia. *Nature Reviews Neurology, 5*, 649–658.

Gao, A., Peng, Y., Deng, Y., & Qing, H. (2013). Potential therapeutic applications of differentiated induced pluripotent stem cells (iPSCS) in the treatment of neurodegenerative diseases. *Neuroscience, 228*, 47–59.

Gomar, J. J., Bobes-Bascaran, M. T., Conejero-Goldberg, C., Davies, P., Goldberg, T. E., & Alzheimer's Disease Neuroimaging Initiative. (2011). Utility of combinations of biomarkers, cognitive markers, and risk factors to predict conversion from mild cognitive impairment to Alzheimer disease in patients in the Alzheimer's disease neuroimaging initiative. *Archives of General Psychiatry, 68*(9), 961–969.

Gorelick, P. B., & Nyenhuis, D. (2013). Understanding and treating vascular cognitive impairment. *Continuum: Lifelong Learning in Neurology, 19*(2 Dementia), 425–437.

Gorelick, P. B., Scuteri, A., Black, S. E., DeCarli, C., Greenberg, S. M., Iadecola, C., et al. (2011). Vascular contributions to cognitive impairment and dementia: A statement for healthcare professionals from the American Heart Association/American Stroke Association. *Stroke, 42*(9), 2672–2713.

Grambaite, R., Selnes, P., Reinvang, I., Aarsland, D., Hessen, E., Gjerstad, L., & Fladby, T. (2011). Executive dysfunction in mild cognitive impairment is associated with changes in frontal and cingulate white matter tracts. *Journal of Alzheimer's Disease, 27*(2), 453–462.

Hachinski, V., Iadecola, C., Petersen, R. C., Breteler, M. M., Nyenhuis, D. L., Black, S. E., et al. (2006). National Institute of Neurological Disorders and Stroke–Canadian stroke network vascular cognitive impairment harmonization standards. *Stroke, 37*(9), 2220–2241.

Harada, C. N., Natelson Love, M. C., & Triebel, K. L. (2013). Normal cognitive aging. *Clinics in Geriatric Medicine, 29*(4), 737–752.

Hardy, J., & Gwinn-Hardy, K. (1998). Genetic classification of primary neurodegenerative disease. *Science, 282*(5391), 1075–1079.

Harrington, M. G., Chiang, J., Pogoda, J. M., Gomez, M., Thomas, K., Marion, S. D., et al. (2013). Executive function changes before memory in preclinical Alzheimer's pathology: A prospective, cross-sectional, case control study. *PLoS One, 8*, e79378.

Hazlett, K. E., Figueroa, C. M., & Nielson, K. A. (2015). Executive functioning and risk for Alzheimer's disease in the cognitively intact: Family history predicts Wisconsin Card Sorting Test performance. *Neuropsychology, 29*(4), 582–591.

Hull, R., Martin, R. C., Beier, M. E., Lane, D., & Hamilton, A. C. (2008). Executive function in older adults: A structural equation modeling approach. *Neuropsychology, 22*(4), 508–522.

Iadecola, C. (2013). The pathobiology of vascular dementia. *Neuron, 80*(4), 844–866.

Ihara, M., Okamoto, Y., Hase, Y., & Takahashi, R. (2013). Association of physical activity with the visuospatial/executive functions of the Montreal Cognitive Assessment in patients with vascular cognitive impairment. *Journal of Stroke and Cerebrovascular Diseases, 22*(7), e146–e151.

Kalaria, R. N., Akinyemi, R., & Ihara, M. (2016). Stroke injury, cognitive impairment and vascular dementia. *Biochimica et Biophysica Acta (BBA)-Molecular Basis of Disease, 1862*(5), 915–925.

Kirova, A. M., Bays, R. B., & Lagalwar, S. (2015). Working memory and executive function decline across normal aging, mild cognitive impairment, and Alzheimer's disease. *BioMed Research International, 2015*.

Kivipelto, M., Ngandu, T., Laatikainen, T., Winblad, B., Soininen, H., & Tuomilehto, J. (2006). Risk score for the prediction of dementia risk in 20 years among middle aged people: A longitudinal, population-based study. *Lancet Neurology, 5*, 735–741.

Koepsell, T. D., & Monsell, S. E. (2012). Reversion from mild cognitive impairment to normal or near-normal cognition: Risk factors and prognosis. *Neurology, 79*, 1591–1598.

Lindenberger, U., Nagel, I. E., Chicherio, C., Li, S.-C., Heekeren, H. R., & Bäckman, L. (2008). Age-related decline in brain resources modulates genetic effects on cognitive functioning. *Frontiers in Neuroscience, 2*, 2.

Luck, T., Then, F. S., Luppa, M., Schroeter, M. L., Arélin, K., Burkhardt, R., et al. (2015). Association of the apolipoprotein E genotype with memory performance and executive functioning in cognitively intact elderly. *Neuropsychology, 29*(3), 382–387.

Luszcz, M. (2011). Executive function and cognitive aging. In K. W. Schaie & S. L. Willis (Eds.), *The handbook of the psychology of aging* (7th ed., pp. 59–72). San Diego, CA: Academic Press.

Marshall, G. A., Rentz, D. M., Frey, M. T., Locascio, J. J., Johnson, K. A., Sperling, R. A., & Alzheimer's Disease Neuroimaging Initiative. (2011). Executive function and instrumental activities of daily living in mild cognitive impairment and Alzheimer's disease. *Alzheimer's & Dementia, 7*(3), 300–308.

McFall, G. P., Sapkota, S., McDermott, K. L., & Dixon, R. A. (2016). Risk-reducing *Apolipoprotein E* and *Clusterin* genotypes protect against the consequences of poor vascular health on executive function performance and change in nondemented older adults. *Neurobiology of Aging, 42*, 91–100.

McFall, G. P., Wiebe, S. A., Vergote, D., Jhamandas, J., Westaway, D., & Dixon, R. A. (2014). IDE (rs6583817) polymorphism and pulse pressure are independently and interactively associated with level and change in executive function in older adults. *Psychology and Aging, 29*(2), 418–430.

Miyake, A., Friedman, N. P., Emerson, M. J., Witzki, A. H., Howerter, A., & Wager, T. D. (2000). The unity and diversity of executive functions and their contributions to complex "frontal lobe" tasks: A latent variable analysis. *Cognitive Psychology, 41*, 49–100.

Moorhouse, P., & Rockwood, K. (2008). Vascular cognitive impairment: Current concepts and clinical developments. *The Lancet Neurology, 7*(3), 246–255.

Nagel, I. E., Chicherio, C., Li, S.-C., von Oertzen, T., Sander, T., Villringer, A., et al. (2008). Human aging magnifies genetic effects on executive functioning and working memory. *Frontiers in Human Neuroscience, 2*, 1.

O'Brien, J., Erkinjuntti, T., Reisberg, B., Roman, G., Sawada, T., Pantoni, L., et al. (2003). Vascular cognitive impairment. *The Lancet Neurology, 2*(2), 89–98.

Petersen, R. C., Caracciolo, B., Brayne, C., Gauthier, S., Jelic, V., & Fratiglioni, L. (2014). Mild cognitive impairment: A concept in evolution. *Journal of Internal Medicine, 275*(3), 214–228.

Pohjasvaara, T., Erkinjuntti, T., Ylikoski, R., Hietanen, M., Vataja, R., & Kaste, M. (1998). Clinical determinants of poststroke dementia. *Stroke, 29*(1), 75–81.

Prince, M., Wimo, A., Guerchet, M., Ali, G.-C., Wu, Y.-T., Prina, M., & Alzheimer's Disease International. (2015). *World Alzheimer's report 2015: The global impact of dementia – An analysis of prevalence, incidence, cost & trends*. London: Alzheimer's Disease International. Retrieved from www.alz.co.uk/research/world-report-2015.

Raz, N., Dahle, C. L., Rodrigue, K. M., Kennedy, K. M., & Land, S. (2011). Effects of age, genes, and pulse pressure on executive functions in health adults. *Neurobiology of Aging, 32*, 1124–1137.

Raz, N., & Lustig, C. (2014). Genetic variants and cognitive aging: Destiny or a nudge? *Psychology and Aging, 29*(2), 359–362.

Raz, N., Rodrigue, K. M., Kennedy, K. M., & Land, S. (2009). Genetic and vascular modifiers of age-sensitive cognitive skills: Effects of COMT, BDNF, ApoE, and hypertension. *Neuropsychology, 23*(1), 105–116.

Reinvang, I., Espeseth, T., & Westlye, L. T. (2013). APOE-related biomarker profiles in non-pathological aging and early phases of Alzheimer's disease. *Neuroscience and Biobehavioral Reviews, 37*, 1322–1335.

Reitz, C., & Mayeux, R. (2014). Alzheimer disease: Epidemiology, diagnostic criteria, risk factors and biomarkers. *Biochemical Pharmacology, 88*, 640–651.

Rincon, F., & Wright, C. B. (2013). Vascular cognitive impairment. *Current Opinion in Neurology, 26*(1), 29–36.

Sachdev, P., Kalaria, R., O'Brien, J., Skoog, I., Alladi, S., Black, S. E., et al. (2014). Diagnostic criteria for vascular cognitive disorders: A VASCOG statement. *Alzheimer Disease & Associated Disorders, 28*, 206–218.

Sahathevan, R., Brodtmann, A., & Donnan, G. A. (2012). Dementia, stroke, and vascular risk factors: A review. *International Journal of Stroke, 7*(1), 61–73.

Sapkota, S., Vergote, D., Westaway, D., Jhamandas, J., & Dixon, R. A. (2015). Synergistic associations of catechol-O-methyltransferase and brain-derived neurotrophic factor with executive function in aging are selective and modified by apolipoprotein E. *Neurobiology of Aging, 36*, 249–256.

Schneider, J. A., Arvanitakis, Z., Bang, W., & Bennett, D. A. (2007). Mixed brain pathologies account for most dementia cases in community-dwelling older persons. *Neurology, 69*, 2197–2204.

Series, H., & Esiri, M. (2012). Vascular dementia: A pragmatic review. *Advances in Psychiatric Treatment, 18*(5), 372–380.

Small, B. J., Dixon, R. A., McArdle, J. J., & Grimm, K. J. (2011). Do changes in lifestyle engagement moderate cognitive decline in normal aging? Evidence from the Victoria Longitudinal Study. *Neuropsychology, 26*, 144–155.

Smits, L. L., van Harten, A. C., Pijnenburg, Y. A. L., Koedam, E. L. G. E., Bouwman, F. H., Sistermans, N., et al. (2015). Trajectories of cognitive decline in different types of dementia. *Psychological Medicine, 45*(5), 1051–1059.

Söderlund, H., Nyberg, L., Adolfsson, R., Nilsson, L. G., & Launer, L. J. (2003). High prevalence of white matter hyperintensities in normal aging: Relation to blood pressure and cognition. *Cortex, 39*(4), 1093–1105.

Sudo, F. K., Alves, C. O., Alves, G. S., Ericeira-Valente, L., Tiel, C., Moreira, D. M., et al. (2013). White matter hyperintensities, executive function and global cognitive performance in vascular mild cognitive impairment. *Arquivos De Neuro-Psiquiatria, 71*(7), 431–436.

Thibeau, S., McFall, G. P., Wiebe, S. A., Anstey, K. J., & Dixon, R. A. (2016). Genetic factors moderate everyday psychical activity effects on executive functions in aging: Evidence from the Victoria Longitudinal Study. *Neuropsychology, 30*(1), 6–17.

Turner, G. R., & Spreng, R. N. (2012). Executive functions and neurocognitive aging: Dissociable patterns of brain activity. *Neurobiology of Aging, 33*, 826.e1–826.e13.

Weintraub, S., Wicklund, A. H., & Salmon, D. P. (2012). The neuropsychological profile of Alzheimer disease. *Cold Spring Harbor Perspectives in Medicine, 2*(4), a006171.

Wiebe, S. A., Sheffield, T., Nelson, J. M., Clark, C. A. C., Chevalier, N., & Espy, K. A. (2010). The structure of executive function in 3-year-olds. *Journal of Experimental Child Psychology, 108*(3), 436–452.

Wisdom, N. M., Callahan, J. L., & Hawkins, K. A. (2011). The effects of apolipoprotein E on non-impaired cognitive functioning: A meta-analysis. *Neurobiology of Aging, 32*, 63–74.

Index

Page numbers in *italics* denote tables, those in **bold** denote figures.

5-HTTLPR gene 113–15, **115**

active and latent representations 33, 128, 129, 131, 133–5, **134**, 138
active gene–environment correlations 112
activities of daily living, and aging 65, 66, 207
acute physical activity and exercise 189–92
addiction *see* substance misuse
ADHD *see* Attention Deficit Hyperactivity Disorder (ADHD)
adjustment of contention scheduling, and aging 93–4, **94**, 95, 101
adolescence 44–54, **47**, **49**; adversity and stress 147–55; anxiety disorders 234–5, 237–9; cognitive training 48–54, **49**, 202; depression 234–5, 239–41; emotion regulation 233–42; genetic influences *108–9*; neural mechanisms 51–2; physical activity and exercise 189–91, 192–4; substance misuse 247–57; very preterm adolescents 218, 222–3, 225, 226
adversity and stress 21, 147–55, 160, 161, 179
aerobic exercise 153, 188, 191–2, 193, 194, 195, 196
Affordable Care Act, US 154
aging 59–67, 263–72; and bilingualism 177; and cognitive training 66, 200, 206–8; neural mechanisms 91–102, **92**, **94**, **97**, **100**, 207–8; neurodegenerative diseases 264–72; normal 263–4; physical activity and exercise 188, 189, 191–2, 193, 194–5, 196
alcohol misuse 247–57
Alzheimer's disease 193, 264–5, 269
amygdala 236, 237, 238–9, 240, 241, 248
amyloid-beta levels 269
A-not-B tasks: computational models 129–33, **130**; early childhood 30; infancy 12–15, *13*, **16**, 18, 19, *19*, 76–7, 125–6
anoxia 218
ANT *see* Attentional Networks task (ANT)
anterior cingulate cortex 250; aging 94, 98–100, **100**; bilingualism 177; childhood 33, 36, 82, 83; emotion regulation 236, 237, 238, 240; infancy 21; substance misuse 255
anterior insula 81, 83, 84
antisaccade tasks *13*, 107
anxiety disorders 234–5, 237–9
Apolipoprotein E (APOE) genotypes 266, 267, 268
apoptosis 218
Attention Deficit Hyperactivity Disorder (ADHD) 36, 37, 48–9, 54; cognitive training 200–1, 202–3; genetic

influences 114; neural mechanisms 85–6; and substance misuse 252
attentional control 12, 219, **220**; anxiety disorders 237–8; childhood 29, 31, 35; dividing attention and aging 60, 64–5, 66; infancy 11, 17, 21, 29; training 37, 50; very preterm children 221–3, 224
Attentional Networks task (ANT) 174–5, 176, 178
attentional shifting strategies 233, 235, 237
autism spectrum disorders 36, 85–6, 116
AX-CPT tasks 97–8, **97**
axonal growth 18

Baddeley, A. D. 61, 83
basal ganglia 176, 222, 223, 248, 271
Bell, M. A. 76–7, 82
Bialystok, E. 175, 179
bilingualism 172–82
binge drinking 247–8, 254
Blair, C. 149, 150, 167–8
blood-oxygen-level dependent (BOLD) response 96, 222–3, 256
body mass index 116
BOLD *see* blood-oxygen-level dependent (BOLD) response
brain derived neurotrophic factors (BDNF) 188–9, 267
brain imaging *see* neuroimaging studies
brain injury, very preterm children 217, 218–19, 222–3, 224
brainstem 250
Buschkuehl, M. 207
Buss, A. T. 33, 84

Canadian Stroke Network 270
candidate gene studies 112–15, **115**
cannabis use 251, 253, 255
CANTAB visuospatial working memory task 225
cardiovascular hypothesis 194
caregiving 21–2, 36–7, 149–50, 154, 160–8
Carlson, S. 30, 174, 179
Caspi, A. 114, 252
catecholamine 113, 189
catechol-o-methyltransferase (COMT) gene 21, 112–13, 267
caudal superior frontal sulcus 84
caudate nucleus 84, 222, 255, 271
cell adhesion molecule 2 (CADMA2) 116
Center for Disease Control, US 248
Central Benefit Model 196
cerebellum 60, 222, 223

cerebral blood flow 189, 218–19, 270–1
cerebral oxygenation 66, 188, 189, 190–1
childhood *see* early and middle childhood
chronic physical activity and exercise 192–5
Clark, C. A. C. 225
cocaine use 251
cognitive behavior therapy 241–2
cognitive flexibility 12, 77, 124, 219, 220, **220**; adolescence 45–6, 51; and aging 60, 62–3, 65, 94–5, 100–1, 191–2, 193, 194, 269; childhood 30, 31, 32–3, 35, 78–80, **79**, 126, 161, 202; computational models 125, 126, **127**, 133–9, **134**; and emotion regulation 234, 235, 237; genetic influences *109*, *110*, 111, 113, 114–15, **115**, 116; language switching 173, 176, 181–2; neural mechanisms 51, 78–80, **79**; neurodegenerative diseases 269; and parental absence 161; and physical activity 190, 191–2, 193, 194; and substance misuse 250, 251, 252; very preterm children 224–6; *see also* task switching
cognitive impairment, very preterm children 217–27
cognitive learning, adolescence 46, **47**, 51
cognitive reappraisal strategies 233, 235, 237–8, 239
cognitive training 17, 200–8; adolescents 48–54, **49**, 202; children 37, 166, 201–4; compensation vs. magnification accounts 54; older adults 66, 200, 206–8; young adults 204–6
Cohen, J. D. 1
Cohorts for Heart and Aging Research in Genomic Epidemiology (CHARGE) 116
Colcombe, S. 188, 194
Colorado Longitudinal Twin Sample 111
complex span tasks 61–2
computational models 14, 23, 124–40; connectionist 126–8, 129–31, **130**, 132, 133–5, **134**, 138; dynamic field 128–9, **130**, 131–3, **134**, 135–8, 139
computer-based training 37, 153, 202–3
COMT *see* catechol-o-methyltransferase (COMT) gene
conflict monitoring theory 98–9
conflict processing 99, 177
connectionist models 126–8, 129–31, **130**, 132, 133–5, **134**, 138
context processing deficit theory 95

Context Processing Theory of Cognitive Control 59
control coordination 35–6; *see also* dual mechanisms of control theory
coping strategies 152
corpus callosum 218, 255
cortisol 21, 151
creative insight 46
cue encoding 101
cue processing 34–5
Cuevas, K. 154, 165–6

Day-Night task 82
DCCS *see* Dimensional Change Card Sort (DCCS)
de Frias, C. M. 60, 264
dedifferentiation, age-related 60, 263–4
delayed gratification 163
delayed-memory search tasks: computational models 129–33, **130**; early childhood 30; infancy 12–15, *13*, **16**, 18, 19, *19*, 76–7, 125–6
dementia 264–72
dendritic growth 18, 76
depression 162, 234–5, 239–41
developmental disorders 36, 37, 48–9, 54; cognitive training 200–1, 202–3; genetic influences 114, 116; neural mechanisms 85–6; and substance misuse 252
developmental psychopathology 233–42
DFT *see* dynamic field theory (DFT)
Diamond, A. 14–15, 18, 76, 77, 166
differential susceptibility theory 151–2
differentiation of executive function with age 32–3
diffusion tensor imaging (DTI) 18–19, 91, 255
Digit Span task 66, *108*, *110*
Dimensional Change Card Sort (DCCS) 30, 31, 78–80, **79**, 126, **127**; computational models 125, 133–8, **134**
discrimination tasks 194
disinhibitory tendencies, and substance misuse 249, 252–4, 256
divergent thinking 46
divided attention tests, and aging 60, 64–5, 66
DNA methylation 22
dopamine 20–1, 52–3, 112, 113, 189
dorsolateral prefrontal cortex 75, 250; adolescence 51; adversity and stress 151; aging 62, 97–8, **97**, 268; childhood 78, 80, 82, 83, 85; cognitive training 52,

53; emotion regulation 236, 237, 238, 240; infancy 12, 18, 76; neurodegenerative diseases 268
Dowsett, S. M. 201
drug misuse *see* substance misuse
DTI *see* diffusion tensor imaging (DTI)
dual mechanisms of control theory 95, 96; *see also* control coordination
dual-tasking: and aging 60, 64–5, 66; physical activity and exercise 190, 191
Dunedin longitudinal cohort 252
Durston, S. 33, 81–2
dynamic field theory (DFT) 128–9, **130**, 131–3, **134**, 135–8, 139

early adulthood *see* young adults
early and middle childhood 29–37, 126; adversity and stress 21, 147–55, 160, 161, 179; anxiety disorders 234, 238, 239; bilingualism 172–82; cognitive training 37, 166, 201–4; depression 234, 239–40, 241; emotion regulation 234, 235, 238, 239–40, 241; genetic influences *108–9*, 152, 165–6; interventions 37, 150, 153, 154, 166–7, 201–4; neural mechanisms 75–86, **79**; parental influences 36–7, 149–50, 154, 160–8; physical activity and exercise 189–91, 192–4; very preterm children 217–27
Early Head Start interventions 166
economic circumstances 21, 147–55, 160, 161, 179
ecstasy use 250
electroencephalography (EEG) 18–20, *19*, 76–7, 82, 91, 208; *see also* event-related potential (ERP)
emotion regulation 233–42
emotion regulation skills training 241–2
energizing processes, and aging 93–4, **94**, 97–8, **97**
Engel de Abreu, P. M. J. 175, 179
Engle, R. W. 12
environmental factors 21–2, 36–7, 147–55, 160–8; *see also* gene–environment interaction
epigenetic changes 22
ERP *see* event-related potential (ERP)
error-related negativity (ERN) 98–9; *see also* event-related potential (ERP)
Espinet, S. D. 80
event-related potential (ERP): aging 95–6, 97, 98–100, **100**, 101; childhood 80

everyday life functioning, and aging 65, 66, 207
Executive Control System framework 219–27, **220**
executive function training *see* cognitive training
exercise and physical activity 188–96
experiential canalization model 150
externalizing tendencies, and substance misuse 249, 252–4, 256
Ezekiel, F. 80

family chaos 162–3
family history: of Alzheimer's Disease 269; of substance misuse 253–4
family influences *see* parental influences
far transfer effects 37, 48, 49, **49**, 51, 53, 66, 201, 204, 205, 206
feedback negativity (FN) 98, 99–100, **100**
fetal alcohol effects 247
flanker tasks 45; aging 98, 99; bilingualism 173, 174, 175, 178, 179, 181; physical activity and exercise 194–5
flexibility, cognitive *see* cognitive flexibility
fluid intelligence 153, 202, 204, 205, 206
fMRI *see* functional magnetic resonance imaging (fMRI)
fNIRS *see* functional near-infrared spectroscopy (fNIRS)
fractional anisotropy 60, 255
Friedman, N. P. 45, 107, 111
frontal lobe 1, 29, 75, 126, 147, 160, 250; adolescence 51, 54; adversity and stress 150, 151; aging 59–60, 61, 62, 63, 64, 66, 91, 92, **92**, 95, 97–8, **97**, 99, 101, 264, 268, 270, 271; anxiety disorders 238, 239; bilingualism 176, 177; childhood 33, 78–86; cognitive training 52, 53; compensatory brain activity 61, 64; computational models 135, 136, 137–8, 139; depression 240; emotion regulation 236, 237, 238, 239, 240, 241; genetic influences 112, 113; infancy 12, 18–20, *19*, 21, 75–7; neurodegenerative diseases 268, 270, 271; physical activity and exercise 188, 190–1; substance misuse 248, 251, 254, 255; very preterm children 221, 222, 223, 225
frontal-subcortical circuit 271
frontolimbic circuits 239, 249
fronto-parietal network 75, 78; adolescence 51; adversity and stress 151;
childhood 33, 80, 81, 83, 84, 85; infancy 20, 22, 77, 85; substance misuse 255
frontostriatal circuits 135, 223, 249, 251, 271
functional magnetic resonance imaging (fMRI) 76, 80, 81, 82, 83, 91, 96, 97, 98, 139, 177, 208, 222–3, 238, 239–41, 251, 255–6
functional near-infrared spectroscopy (fNIRS) 20, 66, 76, 78–9, **79**, 81, 84, 126, 137–8

gambling tasks 98, 99–100, 254
Garon, N. 17
Gathercole, S. E. 50
Gazzaley, A. 96, 97
gene–environment interaction 112, 114–15, **115**, 117–18, 265
genetic influences 20–1, 106–18, *108–9*, *110*, **115**, 152, 165–6; neurodegenerative diseases 265–7, 268
genome-wide association studies (GWAS) 115–16, 117
Gift Delay task 174, 176
goal identification 34–5
goal setting 219, 220, **220**, 226
goal-oriented parental behaviors 163, 164, 167
Goldman-Rakic, P. S. 76, 77
Go/NoGo task: aging 63–4; and bilingualism 176–7; childhood 81–2; genetic influences *108*, *110*; substance misuse 251, 254; training 205
gray matter: and aging 62, 92–3, 271; bilingualism 177; and socioeconomic status 160; and substance misuse 255–6; very preterm children 219, 222, 223, 224
gross motor skills training 195
GWAS *see* genome-wide association studies (GWAS)

haemorrhage, intraventricular 218–19, 224
Hasher, L. 59
Hawthorne effect 203
Hayling Sentence Completion task 63
Head Start interventions 150, 166
heart rate 20
hemodynamic response 76, 137–8, 139, 195
heritability estimates 106–12, *108–9*, *110*
hippocampus 236, 248, 269
Hiraki, K. 78–9, **79**
Hitch, G. J. 61

Holmboe, K. 21
home environment quality 21–2, 36–7, 149–50, 154, 160–8
Home Observation Measurement of Environment (HOME) 161
home-based interventions 154, 166
hormonal changes, puberty 44
Huizinga, M. 45
Hulme, C. 203, 204
hyperactivity *see* Attention Deficit Hyperactivity Disorder (ADHD)
hypofrontality hypothesis 190–1
hypoxia 218
hypoxic-ischemic injury 218–19

iADL *see* instrumental activities of daily living (iADL)
if–then logical processes, and aging 100–1
impulsivity: and substance misuse 252–3, 255; very preterm children 222
infancy 11–23, *13*, **16**, *19*, 29, 125–6; interventions 154; neural mechanisms 18–20, *19*, 75–7
inferior frontal cortex 81, 223
inferior frontal gyrus 82, 83, 84
inferior frontal junction 78, 80
inferior frontal occipital fasciculus 255
inferior parietal cortex 80, 82, 223
inferior temporal cortex 240
inflammatory reactions 218–19
information processing 219–20, **220**; very preterm children 223
inhibitory control 77, 124; adolescence 45, 46, 51; and aging 60, 63–4, 66, 92–4, **92**, **94**, 95–6, 191–2, 193, 269; and bilingualism 173, 174, 175, 176, 178; childhood 30–1, 32–3, 81–3, 161, 163, 193, 202; computational models 129–33, **130**, 138–9; and depression 239; and emotion regulation 234, 235, 237, 239; genetic influences *108*, *110*; infancy 12–15, *13*, 17, 18, *19*, 20, 21, 125–6; neural mechanisms 81–3, 92–3, **92**, **94**, 95–6; neurodegenerative diseases 269; and over-stimulation 163; and parental absence 161; and physical activity 191–2, 193; and substance misuse 250, 251, 252, 254, 255; very preterm children 222–3
inhibitory control training 201, 202, 204, 205, 206
inhibitory deficit hypothesis 95
instrumental activities of daily living (iADL), and aging 65, 66, 207

insula 81, 83, 84, 236, 240, 248
insulin degrading enzyme (IDE) G+ genotype 266, 267
insulin-like growth factor (IGF-1) 189
integrate-and-fire models 129, 139
intellectual disabilities 37, 54
interactive specialization theory 51–2, 53
International Society on Infant Studies 22
interventions: adversity and stress 150, 153, 154; emotion regulation 241–2; genetically sensitive populations 118; neurodegenerative diseases 265; parenting 166–7; physical activity and exercise 193–5, 196; *see also* cognitive training
intraventricular haemorrhage 218–19, 224
Iowa Gambling Task 254
ischaemia 218–19
ischemic stroke 271

Johnson, M. H. 51, 85, 234, 241

Karbach, J. 50, 206–7
Kendler, K. S. 117–18
Kleibeuker, S. W. 46, **47**
Klingberg, T. 48–9, 202
Koob, G. F. 249, 256
Kramer, A. F. 63, 188, 194
Kray, J. 62
Krott, A. 181

language ability 33–4; *see also* bilingualism
language switching 173, 176, 181–2
latent and active representations 33, 128, 129, 131, 133–5, **134**, 138
lateral frontal cortex 126, 136, 255
lateral prefrontal cortex 52, 75, 250; adolescence 51; adversity and stress 151; aging 62, 92, **92**, 97–8, **97**, 99, 268; childhood 78, 79, 80, 81, 82–3, 84, 85; cognitive training 52, 53; emotion regulation 236, 237, 238, 240; infancy 12, 18, 76; neurodegenerative diseases 268
Lee, K. 32–3
life-span theories of development 200
lifestyle biomarkers, neurodegenerative diseases 266, 268
limbic regions 239, 240, 241, 248, 249, 250
Lindenberger, U. 62
lingual gyrus 255
Liu-Ambrose, T. 194–5, 196
Livesey, D. J. 201

Luk, G. 177
Luszcz, M. 67

McGue, M. 249, 252
MCI *see* Mild Cognitive Impairment (MCI)
magnetic resonance imaging (MRI): functional 76, 80, 81, 82, 83, 91, 96, 97, 98, 139–40, 177, 208, 222–3, 238, 239–41, 251, 255–6; structural 76, 91, 92
Marcovitch, S. 33
Maternal, Infant, and Early Childhood Home Visiting (MIECHV) program, US 154
maternal depression 162
mathematics 50, 51
medial frontal cortex 20, 82, 98, 99, 223
medial frontal gyrus 254
medial frontal negativities (MFN) 98, 99
medial prefrontal cortex 76, 112, 236, 240, 250
Melby-Lervag, M. 203, 204
Meltzoff, A. N. 174, 179
mental health 233–42
mental representations 33–4
metacognition development 36
Michigan Longitudinal Study 253
midbrain 80
middle frontal gyrus 255
middle temporal gyrus 255
Mild Cognitive Impairment (MCI) 264, 265, 267–8
Miller, E. K. 1
Miller, S. E. 33
mindfulness 153
Mini-Mental State Exam 270
Minnesota Center for Twin and Family Research 252
mixing costs 101
Miyake, A. 1, 45, 64, 77, 93, 220
mobility in aging 65–6, 196
monitoring processing, and aging 93–4, **94**, 95, 98–100, **100**
Morton, J. B. 33, 80, 133–5, **134**
MRI *see* magnetic resonance imaging (MRI)
Mulder, H. 224–5, 226
Munakata, Y. 33, 129–31, **130**, 133–5, **134**
muscular strength 193
myelination 18, 76, 111

N boxes/pots tasks *13*
National Institute for Neurological Disorders and Stroke 270

National Institute of Child Health and Human Development (NICHD) 162
n-back tasks *109*, *110*, 113, 204
near transfer effects 48, **49**, 201, 204, 205, 206, 207
neural mechanisms: adolescence 51–2; aging 91–102, **92**, **94**, **97**, **100**, 207–8; anxiety disorders 238–9; bilingualism 176–7; childhood 75–86, **79**; and cognitive training 207–8; depression 239–41; emotion regulation 236–7, 238–41; infancy 18–20, *19*, 75–7; neurodegenerative diseases 268; substance misuse 247–57
neurodegenerative diseases 264–72
neurodevelopmental impairment, very preterm children 217–27
neuroimaging studies 51, 52, 75, 208; aging 59–60, 63, 64, 66, 91–2, 96, 97–8, **97**; bilingualism 177; childhood 78–9, **79**, 80, 81, 82, 83, 84, 126, 160; cognitive training 208; and computational models 135, 137–8, 139–40; emotion regulation 236–7, 238, 239–41; infancy 18–19, *19*, 76–7, 126; physical activity and exercise 194–5; substance misuse 251, 254, 255–6; very preterm children 222–3
neuromuscular fitness 193
neurotoxicity model of substance misuse 249, 254–6
neurotransmitters 20–1, 52–3, 112, 113, 189
neurotrophic factors 188–9
NICHD Early Child Care Research Network 162
Nigg, J. 253
NIRS *see* functional near-infrared spectroscopy (fNIRS)
Noble, K. G. 150
norepinephrine 112, 113, 189
Norman, D. A. 93

object retrieval tasks *13*
occipital cortex 20, 52, 223
oddball task 194
older adults *see* aging
orbitofrontal cortex 112, 223, 236, 237, 251, 268
over-stimulation 163

parallel distributed processing (PDP) models 126–8, 129–31, **130**, 132, 133–5, **134**, 138

parental absence 161
parental education 148, 149, 150
parental influences 21–2, 36–7, 149–50, 154, 160–8
parental scaffolding 149–50, 163, 164, 167
parental well-being 162
parietal lobe 29, 75, 78, 126, 223, 250; adolescence 51, 54; adversity and stress 151; aging 63, 101; childhood 33, 80, 81, 82, 83, 84–5; computational models 136, 137, 139; emotion regulation 236; infancy 19, 20, 22, 77, 85; substance misuse 255
Parker, A. M. 253–4
passive gene–environment correlations 112
Paxton, J. L. 97–8, **97**
Pennington, B. F. 12
periventricular leukomalacia 218
perseverative errors: early childhood 30, 78–9, **79**; infancy 14, 15, 18, 76, 126
Peters, S. 46, **47**
phenylketonuria 21
physical activity and exercise 188–96
Play and Learning Strategies (PALS) 166
Pliatsikas, C. 177
polygenic scores 117
pons 255
posterior parietal cortex 75, 78, 85
poverty 21, 147–55, 160, 161, 179
precentral gyrus 83, 255
prefrontal cortex 1, 29, 75, 147, 160, 250; adolescence 51, 54; adversity and stress 150, 151; aging 59–60, 61, 62, 63, 64, 66, 91, 92, **92**, 97–8, **97**, 99, 264, 268; anxiety disorders 238, 239; bilingualism 176, 177; childhood 33, 78–86; cognitive training 52, 53; compensatory brain activity 61, 64; depression 240; emotion regulation 236, 237, 238, 239, 240, 241; genetic influences 112, 113; infancy 12, 18, 21, 75–7; neurodegenerative diseases 268; physical activity and exercise 188, 190–1; substance misuse 248, 251; very preterm children 221, 222
Prefrontal Cortex Function Theory of Cognitive Aging 59
premotor cortex 66, 83, 84
preschool interventions 150, 153, 154
pre-supplementary motor region 80, 83
preterm children 217–27
proactive control 35–6
problem solving 164, 224, 268, 269

Processing Resource Deficit Model 59
processing speed 220; adolescence **47**; and aging 60, 61; childhood 31, 33–4; genetic influences 116; infancy 17; neurodegenerative diseases 270; very preterm children 223
prohibition tasks *13*
prospective memory 203–4
protection biomarkers, neurodegenerative diseases 265–7, 268, 269
psychological refractory period (PRP) paradigm 64–5
psychopathology: developmental 233–42; substance misuse 247–57
puberty 44

Random Number Generation task 195
Rapport, M. D. 203
Raver, C. C. 150
reactive control 35–6
reading span task 61
reinforcement learning 98, 99–100
relationship-based parental behaviors 163–4, 165, 167
resilience of executive functioning 117
resistance exercise 191–2, 194–5
risk biomarkers, neurodegenerative diseases 265–7, 268, 269
Roalf, D. R. 46, **47**
Roberts, R. J., Jr. 12
Rovee-Collier, C. 22, 23

scaffolding, parental 149–50, 163, 164, 167
Scaffolding Theory of Aging Cognition (STAC) 61
school performance 50, 51
school-based interventions 166
self-control 37, 163, 222
self-monitoring 222
self-regulation 21, 150, 152, 153, 162–3
serotonin 112, 113
serotonin transporter (5-HTTLPR) gene 113–15, **115**
SES *see* socioeconomic status (SES)
Shallice, T. 93
shifting *see* task switching
Simon task 173, 174–5, 178, 181
single parents 161
Smith, L. B. 14
socioeconomic status (SES) 21, 147–55, 160, 161, 179
Spencer, J. P. **134**, 136, 137
STAC *see* Scaffolding Theory of Aging Cognition (STAC)

Stocco, A. 176
stop-signal tasks 45, 63, 107, *108*, 175, 205, 251
strength training 191–2, 194–5
stress and adversity 21, 147–55, 160, 161, 179
stress hormones 21, 151
striatum 250; adolescence 52–3, 54; aging 60, 268; bilingualism 176; childhood 80; neurodegenerative diseases 268; substance misuse 248, 249; very preterm children 223, 225
stroke 270–1
Stroop tasks 45, 82–3; aging 63, 65, 93–5, 97, 98, 99; bilingualism 173, 178, 180; genetic influences 107, *108*, *110*; and physical activity 190, 191–2, 196; substance misuse 254; training 205
structural equation modeling 23, 138–9, 204
structural magnetic resonance imaging (MRI) 76, 91, 92
Stuss, D. T. 93, **94**
substance misuse 247–57
superior frontal sulcus 80, 84
superior longitudinal fasciculus 251
superior parietal cortex 63, 80
superior prefrontal cortex 250
supervisory system framework 93–102, **94**, **97**, **100**
switching *see* task switching
synaptic pruning 76, 111
synaptogenesis 18, 76

tai chi interventions 195
task switching 12, 77, 124, 219, 220, **220**; adolescence 45–6, 51; and aging 60, 62–3, 65, 94–5, 100–1, 191–2, 193, 194, 269; childhood 30, 31, 32–3, 35, 78–80, **79**, 126, 161, 202; computational models 125, 126, **127**, 133–9, **134**; and emotion regulation 234, 235, 237; genetic influences *109*, *110*, 111, 113, 114–15, **115**, 116; language switching 173, 176, 181–2; neural mechanisms 51, 78–80, **79**; neurodegenerative diseases 269; and parental absence 161; and physical activity 190, 191–2, 193, 194; and substance misuse 250, 251, 252; very preterm children 224–6; *see also* cognitive flexibility
task switching training 202, 204, 205, 208

task-cueing paradigm 62
task-oriented parental behaviors 163, 164, 167
temperament 21, 152
temporal lobe 19, 81, 126; computational models 136, 137; depression 240; neurodegenerative diseases 269; substance misuse 255; very preterm children 222–3
temporal processing, and aging 92–3, **92**
thalamus 60, 80, 222, 250, 255, 268, 271
Thelen, E. **130**, 131–2
Tower of London (ToL) task 45–6
Trail Making tests 65, *110*, 113, 192
training *see* cognitive training
transfer effects 37, 48–9, **49**, 51, 53, 66, 201, 204–8
Travers, S. 101
twin studies 106–12, *108–9*, *110*, 117, 254

Vascular Cognitive Impairment (VCI) 264, 269–70
Vascular Dementia 264, 265, 270–1
vascular endothelial growth factor (VEGF) 189
vascular health 266, 270–1
VCI *see* Vascular Cognitive Impairment (VCI)
ventrolateral prefrontal cortex 75; adolescence 51; adversity and stress 151; childhood 78, 82, 85; emotion regulation 236, 238, 240
verbal fluency 31, 65, 113, 250
verbal working memory 161, 206, 225
Verburgh, L. 189–90, 193–4
Verhaeghen, P. 60, 61, 64, 206, 207
very preterm children 217–27
visuospatial working memory: neural mechanisms 83–5; very preterm children 224–5
Voelcker-Rehage, C. 195
Volkow, N. D. 249, 256
vulnerability model of substance misuse 249, 252–4, 256
Vygotsky, L. S. 163

Wasylyshyn, C. 62–3
white matter: and aging 60, 62, 268, 270, 271; and socioeconomic status 151; and substance misuse 251, 253, 255–6; very preterm children 218, 219, 221, 222, 223, 224, 225

Wisconsin Card Sorting Task (WCST) 45–6, 65, 75, 78, 92–3, **92**, *109*, 113, 269
Wolfe, C. D. 82
Wood, D. 164
Woodward, L. J. 225
working memory 77, 124; adolescence 45, 46, **47**, 51; and aging 61–2, 66, 92–3, **92**, 195, 268, 269; childhood 31–2, 35, 37, 83–5, 161; computational models 129–33, **130**, 138–9; and depression 239; and emotion regulation 234, 235, 237, 239; genetic influences *108–9*, *110*, 111, 113, 114; infancy 12–15, *13*, 17, 18, 19, *19*, 20, 21, 125–6; neural mechanisms 83–5, 92–3, **92**; neurodegenerative diseases 268, 269; and parental absence 161; and physical activity 190, 192, 195; and substance misuse 250, 251, 254–5; very preterm children 224–5
working memory training: adolescents 48–54, **49**, 202; children 202, 204; older adults 66, 206–7, 208; young adults 204–5

Yes-No task 82
young adults 45, 46, 60; adversity and stress 151, 152; anxiety disorders 238; bilingualism 174, 177, 178; cognitive training 204–6; neural mechanisms 83, 96, 97–8, **97**, 99, 101; physical activity and exercise 189–91, 192–3, 194; substance misuse 247–57

Zacks, R. T. 59
Zelazo, P. D. 80, 137
Zhou, B. 181
Zinke, K. 207

PGMO 07/18/2018